RESIDENT DUTY HOURS

ENHANCING SLEEP, SUPERVISION, AND SAFETY

Cheryl Ulmer, Dianne Miller Wolman, Michael M. E. Johns, *Editors*

Committee on Optimizing Graduate Medical Trainee (Resident) Hours and Work Schedules to Improve Patient Safety

INSTITUTE OF MEDICINE
OF THE NATIONAL ACADEMIES

THE NATIONAL ACADEMIES PRESS
Washington, D.C.
www.nap.edu

THE NATIONAL ACADEMIES PRESS 500 Fifth Street, N.W. Washington, DC 20001

NOTICE: The project that is the subject of this report was approved by the Governing Board of the National Research Council, whose members are drawn from the councils of the National Academy of Sciences, the National Academy of Engineering, and the Institute of Medicine. The members of the committee responsible for the report were chosen for their special competences and with regard for appropriate balance.

This study was supported by Contract No. HHSP233200700003T between the National Academy of Sciences and the Agency for Healthcare Research and Quality. Any opinions, findings, conclusions, or recommendations expressed in this publication are those of the author(s) and do not necessarily reflect the view of the organizations or agencies that provided support for this project.

Library of Congress Cataloging-in-Publication Data

Resident duty hours : enhancing sleep, supervision, and safety / Committee on Optimizing Graduate Medical Trainee (Resident) Hours and Work Schedules to Improve Patient Safety ; editors, Cheryl Ulmer, Dianne Miller Wolman, Michael M.E. Johns.
p. ; cm.
Includes bibliographical references and index.
ISBN 978-0-309-12776-9 (hardcover)
1. Residents (Medicine)—United States. 2. Fatigue. 3. Sleep deprivation. 4. Hours of labor. 5. Hospitals—United States—Safety measures. 6. Medical errors. I. Ulmer, Cheryl. II. Wolman, Dianne Miller. III. Johns, Michael M. E. IV. Institute of Medicine (U.S.). Committee on Optimizing Graduate Medical Trainee (Resident) Hours and Work Schedules to Improve Patient Safety.
[DNLM: 1. Internship and Residency—standards—United States—Guideline. 2. Education, Medical, Graduate--standards—United States—Guideline. 3. Medical Errors—prevention & control—United States—Guideline. 4. Patient Care—standards—United States—Guideline. 5. Sleep Deprivation—prevention & control—United States—Guideline. 6. Work Schedule Tolerance—United States—Guideline. W 20 R4335 2009]
RA972.R465 2009
610.71'55—dc22

2009003372

Additional copies of this report are available from the National Academies Press, 500 Fifth Street, N.W., Lockbox 285, Washington, DC 20055; (800) 624-6242 or (202) 334-3313 (in the Washington metropolitan area); Internet, http://www.nap.edu.

For more information about the Institute of Medicine, visit the IOM home page at: **www.iom.edu.**

Printed in the United States of America

The serpent has been a symbol of long life, healing, and knowledge among almost all cultures and religions since the beginning of recorded history. The serpent adopted as a logotype by the Institute of Medicine is a relief carving from ancient Greece, now held by the Staatliche Museen in Berlin.

Photo Credit: Front cover reprinted with permission from Emory University Photo/Video, 2008.

Suggested citation: IOM (Institute of Medicine). 2009. *Resident Duty Hours: Enhancing Sleep, Supervision, and Safety.* Washington, DC: The National Academies Press.

"Knowing is not enough; we must apply.
Willing is not enough; we must do."
—Goethe

INSTITUTE OF MEDICINE
OF THE NATIONAL ACADEMIES

Advising the Nation. Improving Health.

THE NATIONAL ACADEMIES

Advisers to the Nation on Science, Engineering, and Medicine

The **National Academy of Sciences** is a private, nonprofit, self-perpetuating society of distinguished scholars engaged in scientific and engineering research, dedicated to the furtherance of science and technology and to their use for the general welfare. Upon the authority of the charter granted to it by the Congress in 1863, the Academy has a mandate that requires it to advise the federal government on scientific and technical matters. Dr. Ralph J. Cicerone is president of the National Academy of Sciences.

The **National Academy of Engineering** was established in 1964, under the charter of the National Academy of Sciences, as a parallel organization of outstanding engineers. It is autonomous in its administration and in the selection of its members, sharing with the National Academy of Sciences the responsibility for advising the federal government. The National Academy of Engineering also sponsors engineering programs aimed at meeting national needs, encourages education and research, and recognizes the superior achievements of engineers. Dr. Charles M. Vest is president of the National Academy of Engineering.

The **Institute of Medicine** was established in 1970 by the National Academy of Sciences to secure the services of eminent members of appropriate professions in the examination of policy matters pertaining to the health of the public. The Institute acts under the responsibility given to the National Academy of Sciences by its congressional charter to be an adviser to the federal government and, upon its own initiative, to identify issues of medical care, research, and education. Dr. Harvey V. Fineberg is president of the Institute of Medicine.

The **National Research Council** was organized by the National Academy of Sciences in 1916 to associate the broad community of science and technology with the Academy's purposes of furthering knowledge and advising the federal government. Functioning in accordance with general policies determined by the Academy, the Council has become the principal operating agency of both the National Academy of Sciences and the National Academy of Engineering in providing services to the government, the public, and the scientific and engineering communities. The Council is administered jointly by both Academies and the Institute of Medicine. Dr. Ralph J. Cicerone and Dr. Charles M. Vest are chair and vice chair, respectively, of the National Research Council.

www.national-academies.org

COMMITTEE ON OPTIMIZING GRADUATE MEDICAL TRAINEE (RESIDENT) HOURS AND WORK SCHEDULES TO IMPROVE PATIENT SAFETY

MICHAEL M. E. JOHNS (*Chair*), Chancellor, Emory University, Atlanta, GA
JAMES BAGIAN, Chief Patient Safety Officer, Director, VA National Center for Patient Safety, Department of Veterans Affairs, Ann Arbor, MI
JAYANTA BHATTACHARYA, Assistant Professor, Center for Primary Care and Outcomes Research, Stanford University, CA
MAUREEN BISOGNANO, Executive Vice President and Chief Operating Officer (COO), Institute for Healthcare Improvement, Cambridge, MA
PASCALE CARAYON, Procter & Gamble Bascom Professor in Total Quality, Department of Industrial and Systems Engineering, and Director, Center for Quality and Productivity Improvement, University of Wisconsin–Madison
JORDAN J. COHEN, Professor, Medicine and Public Health, George Washington University, Washington, DC
DAVID F. DINGES, Professor and Chief, Division of Sleep and Chronobiology, Department of Psychiatry, University of Pennsylvania School of Medicine, Philadelphia
JAVIER A. GONZALEZ DEL REY, Professor of Pediatrics and Director, Pediatric Residency Programs, Cincinnati Children's Hospital Medical Center, Ohio
PETER J. KOLESAR, Professor Emeritus and Research Director, Deming Center for Quality, Productivity and Competitiveness, Columbia University, New York, NY
BRIAN W. LINDBERG, Executive Director, Consumer Coalition for Quality Health Care, Washington, DC
KENNETH M. LUDMERER, Professor of Medicine and Professor of History, Washington University, St. Louis, MO
DANIEL MUNOZ, Fellow, Division of Cardiology, Johns Hopkins University School of Medicine, Baltimore, MD
CHRISTOPHER S. PARSHURAM, Director, Center for Safety Research, Assistant Professor, Department of Critical Care Medicine, Hospital for Sick Children, and Departments of Paediatrics, Health Policy Management and Evaluation, University of Toronto, ON, Canada
ANN E. ROGERS, Associate Professor, University of Pennsylvania School of Nursing, Philadelphia
DENISE M. ROUSSEAU, H. J. Heinz II Professor of Organizational Behavior and Public Policy and Director, Project of Evidence-Based Organizational Practices, Carnegie Mellon University, Pittsburgh, PA
EDUARDO SALAS, Pegasus Professor and University Trustee Chair, Department of Psychology and Institute for Simulation and Training, University of Central Florida, Orlando
BRUCE SIEGEL, Director, Center for Health Care Quality, George Washington University School of Public Health and Health Services, Washington, DC

IOM Study Staff

Cheryl Ulmer, Study Co-director
Dianne Miller Wolman, Study Co-director
Michelle Bruno, Research Associate
Cassandra Cacace, Senior Program Assistant
Roger Herdman, Director, Board on Health Care Services

Reviewers

This report has been reviewed in draft form by individuals chosen for their diverse perspectives and technical expertise, in accordance with procedures approved by the National Research Council's Report Review Committee. The purpose of this independent review is to provide candid and critical comments that will assist the institution in making its published report as sound as possible and to ensure that the report meets institutional standards for objectivity, evidence, and responsiveness to the study charge. The review comments and draft manuscript remain confidential to protect the integrity of the deliberative process. We wish to thank the following individuals for their review of this report:

BARBARA LEE BASS, Methodist Institute for Technology, Innovation & Education (MITIE)™, Department of Surgery, The Methodist Hospital, Houston, TX, and Weill Medical College of Cornell University, New York
LINDA EMANUEL, Buehler Center on Aging, Health and Society, Feinberg School of Medicine, Northwestern University, Chicago, IL
KATHLYN E. FLETCHER, Primary Care Division, Clement J. Zablocki VA Medical Center, Milwaukee, WI
DORRIE K. FONTAINE, University of Virginia School of Nursing, Charlottesville, VA
ROBERT L. HELMREICH, Human Factors Research Project, The University of Texas at Austin, Austin, TX
STEVEN K. HOWARD, VA Palo Alto Health Care System Anesthesia Service and Stanford University School of Medicine Anesthesia Service, Palo Alto, CA

SANDEEP JAUHAR, Heart Failure Program, Long Island Jewish Medical Center, New York, NY
ERIC B. LARSON, Group Health Center for Health Studies, Seattle, WA
ARTHUR A. LEVIN, Center for Medical Consumers, New York, NY
STEVEN LIPSTEIN, BJC HealthCare, St. Louis, MO
ALAN R. NELSON, American College of Physicians, Fairfax, VA
TERRANCE D. PEABODY, Section of Orthopaedic Surgery and Rehabilitation Medicine, The University of Chicago Medical Center, Chicago, IL
DEBORAH E. POWELL, University of Minnesota Medical School, Minneapolis, MN
RANGARAJ RAMANUJAN, Owen Graduate School of Management, Vanderbilt University, Nashville, TN
ROGER R. ROSA, National Institute for Occupational Safety and Health, Washington, DC
DAVID P. STEVENS, Quality Literature Program, Dartmouth Institute for Health Policy and Clinical Practice, Lebanon, NH
KEVIN VOLPP, Center for Health Equity Research and Promotion (CHERP), Philadelphia VA Medical Center, and Center on Health Incentives, Leonard Davis Institute for Health Economics, University of Pennsylvania School of Medicine and the Wharton School, Philadelphia, PA
JAMES K. WALSH, Sleep Medicine and Research Center, St. Luke's Hospital, Chesterfield, MO

Although the reviewers listed above have provided many constructive comments and suggestions, they were not asked to endorse the conclusions or recommendations nor did they see the final draft of the report before its release. The review of this report was overseen by Donald Steinwachs, Health Services Research and Development Center, Department of Health Policy and Management, Bloomberg School of Public Health, Johns Hopkins University, Baltimore, MD, and Adel A. F. Mahmoud, Woodrow Wilson School and Department of Molecular Biology, Princeton University, Princeton, NJ. Appointed by the National Research Council and the Institute of Medicine, they were responsible for making certain that an independent examination of this report was carried out in accordance with institutional procedures and that all review comments were carefully considered. Responsibility for the final content of this report rests entirely with the authoring committee and the institution.

Foreword

Most physicians can recall moments of total exhaustion during their residency, when they had been working steadily on patients around the clock, and other moments of total exhilaration, such as when they realized a critically ill patient would pull through. The intense residency learning period that follows medical school is an integral part of a physician's professional development and essential preparation for clinical practice. Physicians may have very strong feelings about how well our own training experience prepared us and ways in which it could have been improved. We may have memories of mistakes we made during training and wonder whether they could have been prevented had we consulted the attending earlier, received more information during the handover, remembered a critical test, or correctly calculated the dose of medication. Today, with deeper appreciation of risks to patients, we may wonder how the work environment of residents can be redesigned to enhance patient safety and whether this can be done while preserving or, even better, while enhancing the learning to be a doctor that is at the heart of any residency training program.

The Institute of Medicine (IOM) appointed the Committee on Optimizing Graduate Medical Trainee (Resident) Hours and Work Schedules to Improve Patient Safety, at the request of Congress and the Agency for Healthcare Research and Quality, to weigh these questions. Specifically, the committee examined whether residents' duty hours and schedules could be improved to reduce sleep deprivation, performance degradation, and the risk of error, while ensuring that residents have sufficient time to receive the necessary training and experience. The IOM has a history of reports on medical education, training, and the healthcare workforce, as well as a long-standing concern for patient safety. The current committee builds on the Quality Chasm series of reports, beginning with *To Err Is Human* in 1999 and *Crossing the Quality Chasm* in 2001, that produced many evidence-based recommendations to inform medical education, safety, and work systems redesign.

This study stirred considerable interest, concern, and debate among physician educators, residents, and patient interest groups. The first set of common national duty hour standards for all types of residencies was implemented just 5 years ago, in 2003. Although limited data directly assess the impact of these regulations, the committee was able to utilize a robust body of evidence on sleep, fatigue, and human performance. Importantly, the committee considered various aspects of residency beyond duty hours, such as the educational process and work environment, in search of ways to improve the learning experience for residents and maximize the value to patients of their hours on duty.

I am grateful to the committee and to the staff who supported its work for their conscientious deliberation and concrete guidance. I hope this report stimulates a spirited discussion and prompts needed improvements in residency training.

Harvey V. Fineberg, M.D., Ph.D.
President, Institute of Medicine
November 2008

Preface

Graduate medical education (GME), also known as residency training, has evolved significantly over the last century since first initiated in its modern form at the Johns Hopkins Hospital. The processes of accreditation of training programs by the Accreditation Council for Graduate Medical Education (ACGME) and of certification of graduates by specialty certifying boards are also progressively evolving. In 2003, the ACGME promulgated national guidelines regarding resident duty (work) hours that, for the first time across all specialties, limited the number of hours per week that a resident could work to the same common limits. Since then there has been much interest in the extent and effects of implementation of the 2003 guidelines, as well as continuing concerns about resident fatigue and its relationship to patient safety.

This committee was asked to synthesize evidence on the relationship of medical resident duty hours and schedules to healthcare safety and to develop strategies for implementing optimal resident work schedules. The committee understood that proposed strategies must take into account the learning and experience that residents must achieve during their training, with recommendations structured to optimize both the quality of care and the educational objectives.

The committee includes experts with experience in medical care and medical education as well as a variety of disciplines such as organization change, patient safety, and human factors engineering. Through scheduled workshops and written submissions, the committee was privileged to hear from a wide array of knowledgeable and interested individuals and organizations who helped broaden our perspective on the issues.

The result of our study and deliberations is a series of recommendations concerning adjustments to residency duty hours and schedules, resident supervision, education, and training program oversight and management. The report also includes suggested strategies, practices, interventions, and tools that we believe can be helpful in achieving improved outcomes on the critical metrics of patient safety and effective learning.

Patient safety continues to be a serious problem in the United States. Many factors affect safety; fatigue is one. Redesigning resident duty hours and other aspects of GME could contribute to improved safety. There is no question that the evidence base is still nascent and much more research must be done. The committee reviewed the scientific literature on sleep and human performance as well as evidence that continues to emerge concerning the benefits to patient safety, resident learning, and overall resident work life of well-structured limits to resident duty hours. The evidence was sufficient to recommend action now. Providing safe patient care during residency is a matter not just of hours at work, but also of the amount of effective supervision, sleep obtained, and a balanced workload. Recommended changes to these elements of GME are all interrelated and should be considered together.

The committee well understands that implementing more circumscribed limits on resident duty hours carries real costs and significant challenges. Resident work restrictions can create new costs in terms of personnel and systems required to compensate for fewer hours worked per resident. There can also be added risks to patient safety from related issues, including increased "handoffs" among providers and breaks in continuity of care. New administrative costs can be incurred from changes to scheduling, management, and reporting requirements. Society should weigh the real and hoped-for benefits of further reform against these costs.

Our responsibility is to understand and act upon the best evidence available in achieving the goals of patient safety and the best possible preparation of health professionals. Many of the committee's recommendations are synergistic, working together to promote safer conditions. Some of the suggested changes could be implemented quickly and at relatively low cost; others will require an investment that should be supported by all funders of GME.

It is clear that the issue of resident work hours is but one of a constellation of related issues that go to the heart of how our healthcare systems and our training systems should be organized, implemented, and evaluated. Changes in resident duty hours, schedules, and related strategies require adjustments throughout caregiving and educational programming and processes. We believe that the lens provided by this report can therefore be very useful in sharpening the focus on the kinds of process improvements, new systems, and new thinking and modeling that can lead to reducing the quality chasm.

I wish to thank the committee members, all of whom contributed to the formulation of this report, and especially our staff for their dedicated and tireless efforts on behalf of safer conditions and quality care for our educational and patient care systems.

Michael M. E. Johns, M.D., *Chair*
Committee on Optimizing Graduate Medical
Trainee (Resident) Hours and Work Schedules
to Improve Patient Safety

Acknowledgments

The committee and staff are grateful to many individuals and organizations who contributed to this study. The individuals below testified before the committee during public workshops or open meetings:

Vineet Arora, Pritzker School of Medicine, University of Chicago
H. Hunt Batjer, American Board of Neurological Surgery
Lisa Bellini, University of Pennsylvania School of Medicine
Barbara K. Chang, Department of Veterans Affairs
Carolyn Clancy, Agency for Healthcare Research and Quality
Felicia Cohn, University of California–Irvine School of Medicine
Charles Czeisler, Brigham and Women's Hospital, Boston
Charles Daschbach, St. Joseph's Hospital and Medical Center, Phoenix
Christian de Virgilio, University of California–Los Angeles/Harbor-UCLA Medical Center
Pamela L. Dyne, University of California–Los Angeles/Olive View
Michael Ehlert, American Medical Students Association
Ethan Fried, St. Luke's–Roosevelt Hospital Center
Paul Friedmann, Accreditation Council for Graduate Medical Education
Jimmy Hara, Kaiser Permanente Los Angeles Center for Medical Education
Gerald B. Healy, American College of Surgeons
Elizabeth M. Imholz, Consumers Union of the United States, Inc.
Christopher Landrigan, Brigham and Women's Hospital, Boston
Miechal Lefkowitz, Center for Medicare and Medicaid Services

L. Toni Lewis, Committee of Interns and Residents/SEIU
Richard J. Liekweg, University of California–San Diego Medical Center
Peter Lurie, Public Citizen
Thomas Nasca, Accreditation Council for Graduate Medical Education
Mark Noah, Cedars-Sinai Medical Center
Lawrence M. Opas, Los Angeles County and University of Southern California Medical Center
Ingrid Philibert, Accreditation Council for Graduate Medical Education
Sunny Ramchandani, Residents and Fellows Section, American Medical Association
Bernard F. Ribeiro, The Royal College of Surgeons of England
Mark R. Rosekind, Alertness Solutions
Paul Schyve, Joint Commission
Arpana Vidyarthi, University of California–San Francisco School of Medicine
Kevin Volpp, University of Pennsylvania Wharton School of Business
Steven Weinberger, American College of Physicians
Debra Weinstein, Association of American Medical Colleges
Kevin Weiss, American Board of Medical Specialties
Tom Whalen, American College of Surgeons

We also thank the following people who provided essential information and generously shared their expertise and time to help the committee: Carlo DiMarco, American Osteopathic Association; Karen Fisher, David Longnecker, Erica Steinmetz, and Sunny Yoder, Association of American Medical Colleges; Daniel Goodenberger, University of Nevada School of Medicine; Jennifer Jolly, Gerald Maguire, and Christopher Wall, University of California–Irvine School of Medicine; Alexander Khalessi, University of Southern California; Lee Learman, University of California–San Francisco; Adam Levine and Paul Rockey, American Medical Association; Graham McMahon, Brigham and Women's Hospital; David Meltzer, University of Chicago; David Nashel, Veterans Affairs Medical Center, Washington, DC; Susan Okie, New England Journal of Medicine; Parveen Parmar, University of California–Los Angeles/Olive View Hospital; Terrance Peabody and Vincent Pellegrini, American Orthopaedic Association; Ingrid Philibert, Accreditation Council for Graduate Medical Education; Deborah Powell, New Zealand Resident Doctors Association; Meilan and Michael Rutter, Cincinnati Children's Hospital Medical Center; Sandy Shea, Committee on Interns and Residents–Service Employees International Union; Robert Wachter, University of California–San Francisco; Veronica Wilbur, IPRO; and Arezou Yaghoubian, University of California–Los Angeles Medical Center.

Consultants Teryl Nuckols and Jose Escarce, both of the David Geffen School of Medicine at University of California–Los Angeles and RAND, were generous in assisting the committee in analyzing the potential costs of its recommendations, and Daniel Polsky, University of Pennsylvania, reviewed their work for the committee.

Funding for this study was provided by the Agency for Healthcare Research and Quality (AHRQ). The committee appreciates its financial support for the project as well as the substantive support from AHRQ staff, Eileen Hogan and James Battles.

Many within the Institute of Medicine (IOM) were helpful during the study process and of assistance to the study staff. The project staff benefited from the assistance of two IOM interns, Adam Schickedanz, University of California–San Francisco School of Medicine, and Melissa Crocker, Children's National Medical Center, and the contributions of Ann Page, Michele Orza, and Samantha Chao at the initiation of the project. The staff would especially like to thank Clyde Behney, Patrick Burke, Linda Kilroy, Bill McLeod, Abbey Meltzer, Judith Salerno, Bronwyn Schrecker-Jamrock, Christine Stencel, Janet Stoll, Lauren Tobias, Jackie Turner, and Harvey Fineberg.

Contents

Boxes, Figures, and Tables

Summary

Chapter 1

Chapter 2

Abstract

The Institute of Medicine's Committee on Optimizing Graduate Medical Trainee (Resident) Hours and Work Schedules to Improve Patient Safety evaluated the literature concerning (1) the impact of current residents' duty hours on patient safety and (2) the relationship of hours of work and sleep to performance. The principal aim of residency training in the United States is to prepare young doctors for the safe, independent practice of medicine once they are on their own. While they are in training, residents are often required to be on duty for long hours. Many medical educators believe that extensive duty hours during training are essential to provide residents with the rich educational experience necessary to achieve professional competence in the complexities of diagnosis and treatment of patients. In 2003 the Accreditation Council for Graduate Medical Education (ACGME) adopted common program requirements to restrict resident workweeks to an average of 80 hours over 4 weeks and the longest consecutive period of work to 30 hours, as well as other limits.

Based on its review of the scientific evidence, the committee recognized that it should focus on increasing opportunities for sleep during resident training to prevent acute and chronic sleep deprivation and to minimize fatigue-related errors, rather than on simply reducing total duty hours. It recommends a protected sleep period of 5 hours during any work shift beyond 16 hours duration. This on-duty sleep period should be counted toward the weekly maximum of 80 hours averaged over 4 weeks. The ACGME and residency programs should also

- increase the opportunity for sleep each day by having defined periods off between shifts,

- increase the number and regularity of days free from work for "catch-up sleep" and recovery to minimize cumulative sleep loss,
- limit any additional paid healthcare work (moonlighting) that residents undertake, and
- provide safe transportation to any resident too fatigued to drive home safely.

The committee sees benefits in continuing ACGME monitoring of recommended duty hours because of the relationship of duty hours to education. However stronger enforcement and whistle-blower processes need to be adopted by ACGME and supported by outside oversight to promote adherence and to protect residents who report pressure to violate rules.

Duty hour limits should be accompanied by specialty-specific workload reductions and additional funding to avoid unintended consequences on patient safety and residents' safety and education. The committee recommends improvements in the content of residents' work, a patient workload appropriate to learning and observation of duty hours, and better supervision with more frequent consultations between residents and their supervisory attending physicians. Greater supervision, especially of first-year residents, could intercept errors before they harm patients (e.g., having to spend more days in a hospital because a resident did not order a diagnostic test). Such "near-miss" experiences then become opportunities for learning. In addition, residents should be trained in systems for quality improvement and error reporting.

A handover, the transfer of patient information and responsibility for patient care from one healthcare provider or team of caregivers to another, is identified as a time when lack of clear communication can contribute to error, but it can also be a time for learning and the interception of errors. The committee recognizes that it is vital for residents to learn how to perform handovers most effectively because handover frequency increased after the 2003 duty hour limits and may increase further with new duty hour parameters. Shift changeovers should be scheduled so that there is adequate overlap time to conduct effective handovers.

To meet the committee's recommendations, additional financial and human resources will have to be obtained and existing ones applied differently. Some resident work could be transferred to other clinicians, additional residents, and support staff. According to an economic estimate of select scenarios commissioned by the committee, the annual national costs of personnel substitution could be around $1.7 billion. This represents approximately 0.4 percent of Medicare outlays. Additional funds for graduate medical education (GME) would be needed to support recommendations not contained in the economic model. The committee strongly urges Congress and all potential GME and research funding sources to support the recommendations and the evaluations necessary to monitor and assess their full effect. More knowledge of the details of implementation will help avoid unintended consequences and refine duty hour and educational requirements over time.

Educating resident physicians is an exceedingly critical function of the health system to ensure safe, high-quality health care to patients in the future. A fundamental requirement of resident education is in-depth, firsthand experience caring for actual patients. Ensuring the safety and well-being of patients who participate in the education of residents is of utmost importance. One must look beyond hours of work alone as a risk factor and put in place practices (e.g., time for sufficient sleep, enhanced supervision, appropriate workload, unambiguous handovers) to minimize other contributors to errors so that the patient care environment can be made safer.

Summary

STUDY SCOPE AND OVERVIEW

The Institute of Medicine (IOM), at the request of Congress, and under a contract with the Agency for Healthcare Research and Quality (AHRQ), formed a consensus committee to "1) synthesize current evidence on medical resident schedules and healthcare safety, and 2) develop strategies to enable optimization of work schedules to improve safety in the healthcare work environment. The strategies recommended will take into account the learning and experience that residents must achieve during their training. The recommendations will be structured to optimize both the quality of care and the educational objectives." (See Appendix A.) AHRQ expressed interest in total resident duty hours and how they were scheduled, and included both in the selection of the committee name: Committee on Optimizing Graduate Medical Trainee (Resident) Hours and Work Schedules to Improve Patient Safety. Given the charge outlined in its statement of task, the committee additionally focused on limited aspects of graduate medical education and the resident work environment related to hours, schedules, and patient safety.

The committee first reviewed graduate medical training in the United States and the views of various stakeholders toward the current Accreditation Council for Graduate Medical Education (ACGME) duty hour limits (Chapter 1), data on resident adherence to the limits and ACGME monitoring practices (Chapter 2), and resident duty hour limits in other countries (Appendix C). The committee then reviewed evidence on sleep, fatigue, work, and performance, relative to errors and safety, and came to the following conclusion: There is considerable scientific evidence that 30 hours of continuous time awake, as is permitted and common in current resident work schedules, can result in fatigue. There is also extensive research that shows that fatigue is an unsafe condition that contributes to reduced well-being for residents and increased errors and accidents (Chapters 5, 6, and 7). A detailed examination of the scientific literature on fatigue and hours of work identified prevention of sleep deprivation as a fundamental way to optimize resident work schedules and prevent or minimize fatigue, while ensuring the learning and experience that residents must achieve during their training (Chapter 7). Studies find that fatigued residents can make more errors and have more accidents, but there are simply too few data to reliably estimate the extent to which errors in performance by fatigued residents affect patients and cause them harm (Chapter 6). Evidence also suggests additional ways to improve learning and safety in the healthcare work environment, through adjustment of resident workload, increased supervision (Chapters 3 and 4), and other systems changes to enhance patient safety (Chapter 8). Additional resources will be required to achieve the committee's recommended adjustments to resident duty hours (Chapter 9).

Human beings deprived of sleep exhibit decreased cognitive performance and alertness and increased likelihood of making errors. Although some people are more vulnerable than others to sleep loss, everyone is adversely affected by lack of adequate sleep. The committee evaluated the current ACGME duty hours from the perspective of how well they prevent acute and chronic sleep deprivation. It has concluded that greater attention should be focused on increasing the opportunities for sleep during resident training to prevent fatigue-related errors, rather than on simply reducing total duty hours. The recommended fatigue prevention and mitigation approach preserves options to address individual training program needs to have residents available for patient care at night and to allow for continuity of patient care on admitting days through extended duty periods. The evidence concerning resident safety and the risk of causing errors when fatigued argues for strong and prompt action.

The committee has also concluded that solely regulating resident duty hours and increasing adherence to them would be insufficient to improve conditions for resident and patient safety. The committee firmly believes that a number of additional interrelated changes are needed: more direct supervision of junior residents, adjustment of residents' workload, providing sufficient time for residents to reflect on their clinical experiences, and improved patient transfers. These necessary accompaniments to duty hour reform are worth implementing even under existing duty hour limits. A stronger culture of safety in hospitals and enhanced teamwork in patient care can also contribute to safety. The committee noted that 8 years after the publication of the 2000 IOM report *To Err Is Human* (IOM, 2000), patient safety in hospitals remains a very serious problem that goes well beyond the subset of hospitals that train residents. Adequate and reliable national data necessary to identify the scope of the problem and track progress are not available.

BACKGROUND

The principal aim of residency training in the United States is to prepare young doctors for the safe, independent practice of medicine once they are on their own. An important part of graduate medical training is that it exposes residents to the demands of real-life practice, including the long work hours of physicians (50 to 60 hours a week on average, with a certain percentage working more than 80 hours a week). In 2003 the ACGME adopted a set of duty hour regulations limiting resident workweeks to an average of 80 hours over 4 weeks, among other limits (ACGME, 2003). The 80-hour average was established as a maximum workweek, not a required workweek. Many medical educators believe that these extensive duty hours during training are essential to provide residents with the rich educational experience necessary to achieve professional competence in the complexities of diagnosis and treatment of patients.

Residents play a significant role in the health system. They are frequently the frontline physician-level staff on duty 24/7 in teaching hospitals. Residency continues to consist largely of an apprenticeship approach to learning through service to hospital inpatients and outpatients under the guidance of their attending physician. It is during rotations on inpatient services that residents are more likely to have 80-hour workweeks.

The academic health centers in which most residents train are known for their cutting-edge, quality care for many conditions. During training, residents care for a large number of patients. All current and potential consumers of health care benefit from their services as well as from the graduate medical system that trains future physicians. Doctors in training, while paid an annual salary, cost the institution less than other fully trained clinicians (e.g., nurse practitioners, physician assistants, attending physicians) who could perform some or all of residents' services, when their salary is calculated on an hourly basis because they work such long hours.

As the committee examined alternative resident duty hours and schedules, it was aware of the tension between the educational objectives of medical residency and the economic incentives of training institutions. Both society at large and the training institutions benefit from residents' service at relatively low cost. An institution's decision about when to assign residents to perform service tasks and when to use other healthcare professionals depends on both the costs and the availability of a workforce with appropriate skills.

THE HISTORY OF RESIDENT DUTY HOUR REGULATION

The work of the committee follows previous modifications in residents' duty hours. Before 2003, the on-duty hours of first- and second-year residents frequently exceeded a mean of 80 hours per week (e.g., neurosurgery residents reported averaging 110 hours per week) (Baldwin et al., 2003).

The genesis of widespread public concern about resident duty hours was the death of 18-year-old Libby Zion in the emergency room of a New York City hospital in 1984. Her family charged that her death was due to inadequate care provided by overworked and undersupervised medical residents. A grand jury did not charge any of the residents but concluded that the long duty hours of residents are counterproductive to both patient care and resident learning. The Bell Commission was formed to review these issues and recommended that resident duty hours in New York be limited and supervision increased (Bell, 2003). Since then the focus of attention has been on regulating duty hours, yet the Bell Commission's greatest concern was actually with the supervision of residents by more experienced physicians.

After the Bell Commission, resident unions, some residents' organizations, and public interest groups advocated for national duty hour limits common to all specialties. Public Citizen petitioned the U.S. Occupational Safety and Health Administration to regulate resident hours as a worker safety issue, and Congress introduced legislation that would have the U.S. Department of Health and Human Services regulate resident hours and impose fines for institutional violations. These legislative proposals would have provided incremental funding to help institutions adjust to the limits. These proposals were not adopted. In 2003 the ACGME promulgated nationwide requirements common to all specialties limiting the workweek to an 80-hour average. Although for a sound educational rationale some programs can obtain an exemption for up to 88 hours per week, relatively few programs (primarily neurosurgery programs) have received this exemption (ACGME, 2003).

IMPLEMENTATION OF 2003 DUTY HOUR RULES

Residency programs changed in a variety of ways to accommodate the 2003 ACGME rules. Some residency programs redesigned their schedules or shifted tasks from interns to more senior residents or faculty; others hired substitutes for some of the residents' workload (e.g., support staff, nurse practitioners, physician assistants, hospitalists, moonlighting residents and fellows). Still others reconfigured their programs to eliminate nighttime coverage by residents, restricted which services would be part of resident training programs (e.g., retreating from some affiliations), or even considered no longer having training programs. The committee has reviewed reports and heard testimony on particular programs' adaptations, but no one has conducted a national data-driven assessment across specialties of how adoption of the 2003 ACGME duty hour requirements has changed residency programs.

In assessing the influence of the 2003 duty hour limits to date, the following are five key questions:

1. ***Have resident duty hours actually been reduced?*** Yes, it appears so from a single national study and from individual program reports. The best available national data across multiple specialties from the first year of implementation (2003-2004) show that the workweek of interns, who typically had the longest duty hours, was reduced from an average of 70.7 to 66.6 hours per week. However, 43 percent of interns reported having violated the 80-hour rule when averaged over 4 weeks (Landrigan et al., 2006). No more recent, reliable, national data are available to determine average hours worked by training year or specialty or the reasons for violations when they occur. Reports from individual programs, ACGME surveys and accreditation visits, and annual reviews of compliance in the State of New York also indicate that violations persist, particularly of the 30-hour extended duty rule and the required opportunities for rest and recovery from fatigue. Reasons given by residents for violating the duty hour limits include workload pressures, individual patient circumstances, or the desire of residents to stay in order to participate in the continuing care of their patients.
2. ***Have patient outcomes improved?*** A few large-scale nationwide studies show slight improvements in mortality for some medical, but not surgical, patients in teaching-intensive hospitals and no worsening of mortality in teaching hospitals after the introduction of the 2003 limits (Shetty and Bhattacharya, 2007; Volpp et al., 2007a,b). One cannot attribute these improvements to duty hour reduction per se because numerous quality improvement initiatives were introduced in teaching hospitals over the same period; however, these studies show no evidence of harm as measured by mortality rates. Individual site-specific and specialty-specific studies focus on their success in restructuring programs to maintain previous levels of patient outcomes; these studies tend to be too small to detect statistically significant changes in mortality or do not control for external trends in quality improvement.

3. ***Is resident fatigue from long duty hours among the most significant risks to patient safety?*** Residents report that fatigue decreases the quality of care they deliver and contributes to error, as does high workload. Patient safety is affected by many factors, and the research data available did not make it possible for the committee to assess the current level of all risks to patients or the degree to which fatigued residents contribute to patient harm. Only one randomized controlled trial compared shifts of up to 16 hours and scheduled work of 60-63 hours per week to a schedule with extended duty periods up to 30 hours and scheduled work weeks averaging 77-81 hours. This study reported no statistically significant difference in patient safety as measured by preventable adverse events (Landrigan et al., 2004). However, in the more traditional schedule with longer duty hours, residents made more serious medical errors (Landrigan et al., 2004) and had a higher rate of attentional failure (Lockley et al., 2004). The committee believes there is enough evidence from studies of residents and additional scientific literature on human performance and the need for sleep to recommend changes to resident training and duty hours aimed at promoting safer working conditions for residents and patients by reducing resident fatigue.
4. ***Have educational outcomes been affected?*** Residency training takes 3 to 7 years, depending on the specialty being pursued; the first cohort of 3-year residents trained entirely under 2003 ACGME limits finished in June 2006. Data on board certification pass rates for this cohort are just beginning to emerge. Thus, it is impossible at this time to determine if there has been a consistent trend across specialties.
5. ***Has resident quality of life improved?*** In general, the perception of residents and faculty, as reported in the literature and testimony before the committee, is that resident quality of life and work-life balance have improved with the advent of the 2003 duty hour limits. Eighty hours a week is still a demanding schedule, and a number of single-institution and specialty-specific studies show that residents report high rates of stress, depression, and burnout. However, studies also suggest that factors beyond duty hours, such as work intensity, contribute to the resident's emotional state.

THE NEXT ERA OF REFORM FOR BETTER EDUCATION AND PATIENT SAFETY

ACGME and its constituent stakeholders adopted the 80-hour workweek in 2003 as a national standard for all graduate medical training in the

United States. Countries under the European Work Time Directive currently have fewer weekly hours for their training programs; the European goal is 48 hours per week by 2009. Elsewhere, New Zealand has a 72-hour limit, and Manitoba, Canada, an 89-hour limit. Foreign nations have had trouble implementing their significantly reduced duty hour targets, and some of their efforts appear to have had unintended consequences, such as exacerbating workforce shortages and reducing the amount of time for residents to learn and for surgeons to gain operative experience (see Appendix C). The committee concludes from these international experiences that no single model from another country is directly and completely applicable to the U.S. system of care.

The past 5 years since the ACGME duty hour rules were implemented have been a period of change and adjustment for training programs in the United States. Many programs have replaced scheduling and staffing models adopted in the initial year, and they continue to refine them in their efforts to improve educational value, quality of patient care, and service coverage. Research studies tend to report institution-specific adaptations, and there are few national data or rigorous analyses of different scheduling models across institutions or specialties. However, based on the collective field experiences of programs, the committee concluded that some degree of flexibility in duty hour scheduling would have to be retained.

COMMITTEE FINDINGS AND RECOMMENDATIONS

The evidence and rationale behind each recommendation can be found in the chapter cited prior to the recommendation.

Preamble to Recommendations

To promote conditions for safe medical care, improve the education of doctors in training, and increase the safety of residents and the general public, the committee offers the following recommendations, which should be implemented with all deliberate speed. While some recommendations should be implemented immediately, changes to duty hours, adjustments in workload, and the funding needed for these changes might require an integrated phase-in. The recommendations will require additional resources—both financial and human. Without the necessary restructuring in resource allocation, attempts to implement the recommendations will fail to have the desired benefits and could even reduce patient safety. The committee believes that the Accreditation Council for Graduate Medical Education and the other organizations charged to implement aspects of the recommendations should begin their work with urgency, and that action on all recommendations should be taken within 24 months.

Preventing and Mitigating Fatigue

A robust evidence base linking fatigue with decreased performance in both research laboratory and clinical settings has convinced the committee to focus on how to prevent fatigue when possible and how to mitigate fatigue when residents must be on duty by allowing for sleep during extended duty periods and adequate time for recovery sleep while off duty. Reducing total duty hours from an 80-hour average is one way that might be expected to allow more sleep, but evidence suggests it is an indirect and inefficient approach given the moderate correlation that exists between resident duty hours and sleep time. Prolonged wakefulness in excess of 16 hours at work, reduced or disturbed periods of sleep, more consecutive days or nights of work, shift variability, and the volume of work all increase fatigue and thus can contribute to errors. Meeting daily and weekly sleep needs helps prevent fatigue and diminished performance and contributes to an enhanced ability to learn and remember.

Residency programs should increase the opportunity for sleep each day, utilize strategic naps and longer sleep periods at work, increase the number and frequency of days free from work for "catch-up sleep" and recovery, and minimize cumulative sleep loss in a week based on rest and recovery factors. Published research from the sleep literature supports the specific actions contained in the committee's adjustments to duty hours, including limiting the amount of time a resident is continuously working each day to no more than 16 hours unless a 5-hour protected period for sleep is provided. This in-house sleep period during extended duty of 30 hours should count against total duty hours as sleep during night shifts or overnight call periods does now. Table S-1 compares the elements in the committee's recommendations to current ACGME rules. (See also Chapter 7.)

The recommendations permit flexibility in several ways under the new duty hour parameters set out below. Although the scientific evidence base establishes that human performance begins to deteriorate after 16 hours of wakefulness, the committee does not believe that limiting all shifts to a maximum of 16 hours would address the educational needs of all specialties. So extended duty periods of up to 30 hours (the current limit) are allowed with the inclusion of a sleep period to address acute sleep deprivation. Additionally, there is the possibility of nonroutine exemptions from individual limits for the safety of unstable patients and exceptional learning experiences with the expectation that residents will be closely supervised when these learning experiences extend beyond hour limits, and ACGME already sponsors research projects to test innovations for scheduling alternatives. Further, the committee has retained the maximum of an 80-hour-a-week average, rather than reduce it, to continue to allow each specialty and program site to have what they determine are sufficient

TABLE S-1 Comparison of IOM Committee Adjustments to Current ACGME Duty Hour Limits

	2003 ACGME Duty Hour Limits	IOM Recommendation
Maximum hours of work per week	80 hours, averaged over 4 weeks	No change
Maximum shift length	30 hours (admitting patients up to 24 hours then 6 additional hours for transitional and educational activities)	• 30 hours (admitting patients for up to 16 hours, plus 5-hour protected sleep period between 10 p.m. and 8 a.m. with the remaining hours for transition and educational activities) • 16 hours with no protected sleep period
Maximum in-hospital on-call frequency	Every third night, on average	Every third night, no averaging
Minimum time off between scheduled shifts	10 hours after shift length	• 10 hours after day shift • 12 hours after night shift • 14 hours after any extended duty period of 30 hours and not return until 6 a.m. of next day
Maximum frequency of in-hospital night shifts	Not addressed	4 night maximum; 48 hours off after 3 or 4 nights of consecutive duty
Mandatory time off duty	• 4 days off per month • 1 day (24 hours) off per week, averaged over 4 weeks	• 5 days off per month • 1 day (24 hours) off per week, no averaging • One 48-hour period off per month
Moonlighting	Internal moonlighting is counted against 80-hour weekly limit	• Internal and external moonlighting is counted against 80-hour weekly limit • All other duty hour limits apply to moonlighting in combination with scheduled work
Limit on hours for exceptions	88 hours for select programs with a sound educational rationale	No change
Emergency room limits	12-hour shift limit, at least an equivalent period of time off between shifts; 60-hour workweek with additional 12 hours for education	No change

hours to achieve their learning goals. The committee does not believe that all specialties and rotations will require this lengthy workweek. Any Residency Review Committee that sets educational standards for its specialty in conjunction with ACGME may choose to create more restrictive duty hour limits if it considers changes to be necessary for its particular circumstances (e.g., severity of patient cases, constancy of high-intensity work). For example, this has been done in emergency medicine, which limits shift length to 12 hours, totaling 60 hours per week, plus 12 hours for education; the committee does not recommend any change in the hours for emergency medicine. (See Chapter 7.)

Residency programs will have to continue to redesign schedules and handover practices to promote patient safety. They may need to use night floats or other backup mechanisms, such as onsite attending-level supervision, when residents are required to have a scheduled sleep period. The committee understands the challenges of changing individual and institutional behaviors and the importance of changing professional attitudes to promote personal responsibility for one's own safety and that of others by obtaining necessary sleep. With implementation of the new duty hour adjustments, monitoring is necessary to identify and address unintended scheduling consequences that provide fewer educational experiences for residents (e.g., excessive nighttime work, expanded cross-coverage). (See Chapter 7.)

Recommendation: ACGME should adopt and enforce requirements for residency training that adhere to the following principles: duty hour limits and schedules should promote the prevention of sleep loss and fatigue; additional measures should mitigate fatigue when it is unavoidable (e.g., during night work and extended duty periods); and schedules should provide for predictable, protected, and sufficient uninterrupted recovery sleep to relieve acute and chronic sleep loss, promote resident well-being, and balance learning requirements. Programs should design resident schedules using the following parameters:

- **Duty hours must not exceed 80 per week, averaged over 4 weeks.**
- **Scheduled continuous duty periods must not exceed 16 hours unless a 5-hour uninterrupted continuous sleep period is provided between 10 p.m. and 8 a.m. This period must be free from all work and call, and used by the resident for sleep in a safe and sleep-conducive environment. The 5-hour period for sleep must count toward total weekly duty hour limits. Following the protected sleep period, a resident may continue the extended duty period up to a total of 30 hours, including any previous work time and the sleep period.**

- Residents should not admit new patients after 16 hours during an extended duty period.
- Extended duty periods (e.g., 30 hours that include a protected 5-hour sleep period) must not be more frequent than every third night with *no* averaging.
- After completing duty periods, residents must be allowed a continuous off-duty interval of
  - o A minimum of 10 hours following a daytime duty period that is not part of an extended duty period,
  - o A minimum of 12 hours following a night float or night shift work that is not part of an extended duty period, and
  - o A minimum of 14 hours following an extended duty period, and residents should not return to service earlier than 6 a.m. the next day.
- Night-float or night-shift duty must not exceed four consecutive nights and must be followed by a minimum of 48 continuous hours off duty after three or four consecutive nights.
- At least one 24-hour off-duty period must be provided per 7-day period without averaging; one additional (consecutive) 24-hour period off duty must be provided to ensure at least one continuous 48-hour period off duty per month.
- In exceptional circumstances requiring the resident's physical presence to ensure patient safety or to engage in a critical learning opportunity, program faculty may permit, but not require, residents to remain on duty beyond the scheduled time; programs must record for ACGME review the nature of each exception allowed. These exceptions are not to become routine practice. Residency Review Committees should determine at the time of program re-accreditation whether the documented exceptions to scheduled duty hours warrant citation.
- The ACGME should develop criteria for granting individual programs waivers from one or more of the above scheduling parameters; such criteria should be formulated *only* to accommodate rare, well-documented circumstances in which patient safety and/or educational requirements of specific programs outweigh the advantages of full compliance with the committee's recommendations and cannot be addressed by means other than the requested waiver(s); programs that are granted waivers (if any) and the nature of those waivers should be posted on the public access portion of the ACGME website. Included in the application for waiver should be a long-term plan that articulates how the program will work to avoid a permanent need for the requested waiver. All waivers should be monitored

and reviewed on an annual basis to determine suitability for renewal.

- **Programs should provide annual formal education for residents and staff on the adverse effects of sleep loss and fatigue and on the importance of and means for their prevention and mitigation.**
- **Sponsoring institutions and programs should ensure that their practices promote and ensure that residents take the required sleep during extended duty periods.**

Given the committee's intent to reduce fatigue and improve learning during residency, it believes that moonlighting by residents, which can interfere with already limited opportunities for sleep, must be addressed. Moonlighting outside of residency training would cut into the strategically designed periods for rest and sleep and could reduce residents' readiness for their primary duties. Limits on resident duty hours designed to protect patients and residents should extend to any additional paid healthcare work that residents undertake. This requirement, built into the residency contract, would emphasize that residents ultimately have a responsibility to exhibit professional commitment and to avoid additional obligations that increase their fatigue level and interfere with their capacity to learn and to provide safe patient care. (See Chapter 7.)

Recommendation: The ACGME should immediately amend its current requirements on moonlighting by

- **Requiring that any internal and external moonlighting for patient care adhere to the duty hour limits listed above (e.g., 80 hours and all other limits), even if the program has received an exception to schedule longer hours; and**
- **Requiring that sponsoring institutions, if they choose to permit moonlighting, include provisions in resident contracts that (1) a resident must request prospective, written permission from the program director for moonlighting; and (2) resident performance will be monitored to ensure that there is no adverse effect of moonlighting on resident performance.**

Improving Adherence to Current Duty Hours

ACGME is currently responsible for assessing adherence to duty hours rules along with the educational aspects of graduate medical training as part of its *announced* onsite accreditation review and via surveys of residents. In 2006-2007, ACGME reported that 8.8 percent of programs were substantially noncompliant with some aspect of duty hour limits. This is likely an underestimation of noncompliance—probably due to the current disincentive for residents to report violations because it puts their training program at risk of disaccreditation. The committee concludes that ACGME monitoring of duty hours needs to be strengthened by adding *unannounced* visits and increasing their frequency to deter violations. Additionally, the incentives need to be realigned, perhaps through fines for continued violations and improved protections for residents who report pressure to violate limits.

The committee sees benefits in continuing ACGME monitoring because of the value of maintaining the integration of duty hours with educational program monitoring and the need to expedite a stronger process. Rather than establish a new entity, ACGME could move more quickly to enhance its enforcement and whistle-blower processes, since it already has several years of experience and has the infrastructure in place. The committee noted that the experiences of other countries and other industries with government regulation does not ensure full adherence to duty hour limits. To further address concerns raised to the committee about ACGME as the sole duty hour monitoring agency and to tie duty hours to patient safety reviews, it considered the pros and cons of involving other organizations in monitoring responsibilities. The committee recommends a complementary oversight role for both the Centers for Medicare and Medicaid Services (CMS) and the Joint Commission. CMS could conduct or contract for periodic evaluations of adherence to resident duty hours, the effectiveness of ACGME monitoring practices, and the acceptability of program rationales for exceptions to duty hour limits. Similarly, the Joint Commission could integrate duty hour oversight by monitoring the contribution of fatigue to patient safety events in the tracer cases that it reviews during hospital accreditation site visits. (See Chapter 2.)

Recommendation: ACGME and residency programs should ensure adherence to the current limits now, and to any new limits when implemented, by strengthening their current monitoring practices. To provide additional support, the Centers for Medicare and Medicaid Services and the Joint Commission should take an active oversight role:

- **ACGME should maintain responsibility for duty hour monitoring and should enhance its procedures by including unannounced visits for monitoring duty hours and regular collection of sufficient data to understand when and why limits are violated.**
- **Sponsoring institutions should provide for confidential, protected reporting of duty hour violations by residents through their com-**

pliance office or by an entity above the program level that does not have direct responsibility over the residency programs.

- **ACGME should strengthen its complaint procedures to provide more confidentiality and protection to persons reporting violations of duty hours, as well as other violations of residency rules.**
- **The Centers for Medicare and Medicaid Services should assess the reliability of ACGME procedures and data and should sponsor periodic independent reviews of ACGME's duty hour monitoring to determine the characteristics of and reasons for violations.**
- **The Joint Commission should seek to ensure that duty hour monitoring is linked to broader activities to improve patient safety in hospitals, including the use of ACGME's adherence data as part of the Joint Commission's hospital surveys and accreditation actions.**

Improving the Safety of Residents and the Public

The degree of fatigue experienced by residents places them at risk for workplace and driving injuries. At work, physical injuries commonly occur while caring for patients, such as accidental needlesticks or exposure to blood-borne pathogens. Driving home after an extended duty period or a night shift can be hazardous to both residents and the public because residents are more likely to be involved in a crash at those times. The committee recognizes that steps to reduce fatigue such as the 5-hour protected sleep period may not be put in place immediately, making it particularly important to provide safe transportation options now to and from work for residents working extended duty periods. Education should also be provided for residents to understand the risks they pose to themselves and others if they choose to drive. The committee recognizes that there may be alternative solutions (e.g., providing space to allow residents to sleep before driving home after long shifts), but there should be monitoring and evaluation to ensure usage of alternatives and reduction in opportunities for unsafe driving. (See Chapter 5.)

Recommendation: The committee recommends that sponsoring institutions immediately begin to provide safe transportation options (e.g., taxi or public transportation vouchers) for any resident who for any reason is too fatigued to drive home safely.

Optimizing Resident Education for Resident Learning and Patient Safety

One of the unintended consequences of the 2003 duty hour limits has been work compression (i.e., residents have to care for the same number of patients in less time), which is basically an increased workload. Economic pressures continue to tilt the balance between learning and service in many residency programs too far toward service delivery and away from education. To improve the quality of care delivered to current and future patients and to meet long-term educational objectives, the committee recommends improvements in the content of residents' work, a patient workload and intensity appropriate to learning, and more frequent consultations between residents and their supervisors. The committee believes that better-educated residents will contribute to increased safety for future patients. Educational research demonstrates that a manageable workload contributes to effective learning because of human limits on cognitive capacity, the necessity for well-timed periods of reflection, and the need for sleep in order to consolidate learning.

There are more than 26 types of residency specialties (e.g., surgery, pediatrics, anesthesiology, emergency medicine), and each has a different mix of patient characteristics, flow of work, and types of interventions. Residency Review Committees (RRCs) are in a better position than this committee to determine proficiency requirements for the individual specialties and to set appropriate caseload limits that support learning for each year of residency. The committee notes that the ACGME's internal medicine RRC is the only discipline thus far to set caseload caps for its residents. Other RRCs should gather and analyze the data needed to establish guidelines for caseload, as a start toward making the number of patients that residents care for more transparent and reducing unjustified variability within a specialty across the country while permitting necessary adjustments for individual program circumstances.

Reducing resident duty hours and workload within those hours should not mean that residency training must be lengthened, although some disciplines may choose to do so. Having better ways to identify and assess mastery of a specialty (e.g., use of simulators) and maximizing the learning content of each resident's clinical experiences, rather than relying on "time in service" as a proxy for determining true competence, would be a major advance in medical education. Also, the committee emphasizes that the reduction of work with little or no educational value (e.g., making follow-up appointments) opens time for education, caring for additional patients, and compliance with duty hours. (See Chapter 3.)

Recommendation: To ensure that residency programs fulfill their core educational mission, ACGME should require that institutions sponsoring residency programs appropriately adjust resident workload by

- **Providing support services and redesigning healthcare delivery systems to minimize the current level of residents' work that is of limited or no educational value, is extraneous to their graduate**

medical education program's educational goals and objectives, and can be done well by others; and
- **Providing residents with adequate time to conduct thorough evaluations of patients and for reflective learning based on their clinical experiences.**

ACGME should require each Residency Review Committee to define and then require appropriate limits on the caseload (e.g., patient census, number of admissions, number of surgical cases to assist per day, cross-coverage) that can be assigned to a resident at a given time, taking into consideration the severity and complexity of patient illness and the level of residents' competency.

In the Libby Zion case, the grand jury said, "A hospital is . . . a place where the learning process should continue under strict supervision. Thus, medical decisions, whether in an emergency room or on a hospital floor should not be made by inexperienced interns and junior residents without in-person consultations with more senior physicians . . ." (Bell, 2003). Better supervision not only provides educational benefits, but also increases the likelihood of intercepting potential errors, better patient outcomes, less test ordering, more resident comfort with performing procedures, fewer delays in diagnosis and test ordering, more widespread use of care guidelines, and potentially lower costs.

Although reimbursement policies require residents to consult with their supervising attending physicians on their assessment of a patient and the proposed treatment plan, residents too often lack adequate communication with them except in the operating room where they are more likely to be directly supervised. Protocols should be developed and implemented to have the supervisor reach out and periodically check with the resident on duty, thus increasing the willingness of residents, especially first-year residents, to contact their supervisors. (See Chapter 4.)

Recommendation: To increase patient safety and enhance education for residents, the ACGME should ensure that programs provide adequate, direct, onsite supervision for residents. The ACGME should require

- **The Residency Review Committees, in conjunction with teaching institutions and program directors, to establish measurable standards of supervision for each level of doctor in training, as appropriate to their specialty; and**
- **First-year residents not to be on duty without having immediate access to a residency program-approved supervisory physician in-house.**

Deploying Learning Systems for Handovers and Error Detection, Correction, and Reporting

A handover is the transfer of patients' information and responsibility for their care from one healthcare provider or team of caregivers to another. Handovers are considered critical moments in the continuity of patient care and have been identified as a significant source of hospital errors, often related to poor communication. Learning how to conduct better handovers and intercept errors before they reach patients would enhance the performance of all staff, not only residents. Yet because handover frequency increased with the reduction of duty hours in 2003 (and likely with protected sleep periods as well), the committee concludes that it will be vital for residents to learn how to perform them most effectively. Residents will need to be trained in practicing structured handover procedures, with their attending physicians helping them learn to anticipate the key information that needs to be passed from one shift to another. It will be important to schedule shift changes so that there is an adequate overlap of time to conduct effective handovers. (See Chapter 8.)

Recommendation: Teaching hospitals should design, implement, and institutionalize structured handover processes to ensure continuity of care and patient safety.

- **Programs should train residents and teams in how to hand over their patients using effective communications.**
- **Programs should schedule an overlap in time when teams transition on and off duty to allow for handovers.**
- **The process should include a system that quickly provides staff and patients with the name of the resident currently responsible in addition to the name of the attending physician.**

Residents also need to be taught error detection, correction, reporting, and monitoring in order to participate fully in the hospital's quality improvement efforts. Although residents admit to making errors, the reason for the error is often not traceable to individual negligence, fatigue, or lack of knowledge, but rather to shortcomings in the system in which the resident works (e.g., unsafe medication labeling, excessive workload leading to rushing). Residents (and others) are also reluctant to report errors if the environment is punitive. Residency programs could become leaders by helping their institution develop a culture of safety and integrating residents into its quality improvement efforts. (See Chapter 8.)

Recommendation: Graduate medical education-sponsoring institutions should fully involve residents in their safety reporting, learning, and quality improvement systems, and this should become an important part of the residents' educational experience.

Obtaining Additional Resources for Implementation

Sponsoring institutions incurred substantial costs when adapting to the 2003 ACGME duty hour rules; some major teaching hospitals report an additional $1 million to $7 million each in annual costs. No specific national funds were allocated for implementation, but many hospitals were able to offset the costs through enhanced revenues or reduced expenditures elsewhere. To meet the committee's recommended duty hour changes, additional financial and human resources would have to be obtained and existing ones applied differently. Some resident work could be transferred to other clinicians and support staff, but programs in some areas might be constrained by shortages of nurses, physician assistants, and nurse practitioners or by lack of funds to hire additional personnel. The committee estimated that annual national costs of personnel to substitute for the reduced resident work could be approximately $1.7 billion, according to an economic model of selected scenarios. This range represents approximately 0.4 percent of the Medicare budget (CBO, 2008). While some institutions would be able to find some or all of the necessary financial and human resources, other institutions would need outside assistance to help implement the recommendations.

To avoid having residents bear the burden of implementing the duty hour recommendations by increasing their workload again, and increasing the risk to patient safety, additional funds for graduate medical education (GME) are needed from all existing as well as new sources. The committee strongly urges Congress and all potential GME funding sources to consider various mechanisms to support the recommended changes. For example, some possible considerations include increasing the pool of federally supported residency positions (perhaps through changes to the current Medicare cap on positions), enhancing Medicare's direct medical education payments, and greater support for residency training through private insurers. (See Chapter 9.)

Recommendation: All financial stakeholders in graduate medical education, such as the Centers for Medicare and Medicaid Services, Department of Veterans Affairs, Department of Defense, Health Resources and Services Administration, states and local governments, private insurers, and sponsoring institutions, should financially support the changes necessitated by the committee's recommendations to promote

patient safety and resident safety and education, with special attention to safety net hospitals.

- **An independent convening body should bring together all the major funders of graduate medical education to examine current financing methodologies and develop a coordinated approach to generate needed resources.**

Closing the Gap in Knowledge

Gaps in the available evidence base hampered the committee's work. Given concerns that the medical community has expressed about the 2003 changes in duty hours, the committee was disappointed with the lack of any comprehensive attempt to document changes in residency programs and their impact, if any, on educational outcomes and patient safety. The committee believes that its recommendations can be implemented now without years of additional research because the adjustments for duty hours are rooted in a solid evidence base. Going forward, there should be a plan to evaluate key indicators and a process to document future changes by specialty. Monitoring is important for early detection of any unintended consequences that might indicate a need to fine tune the recommendations over time. Prospective studies that have attempted to evaluate the effects of duty hours on patient safety generally have had sample sizes that lacked sufficient power to determine whether significant changes in errors (especially preventable adverse events), mortality, or other measures of patient harm occurred. Prospective studies of the implementation of the committee's recommendations should be planned, conducted, and funded; consideration of any future adjustments to duty hours would then have a more comprehensive database as a foundation for recommendations. (See Chapter 9.)

Recommendation: To gather the data necessary to monitor implementation of these recommendations and to prepare for future adjustments as needed to achieve the desired objectives, ACGME should convene a meeting of stakeholders and potential funders to set priorities for research and evaluation projects. The Centers for Medicare and Medicaid Services, Agency for Healthcare Research and Quality, National Institutes of Health, Department of Defense, Department of Veterans Affairs, and other funders should support this work as a high priority.

CONCLUSION

Educating resident physicians is an exceedingly important function of the health system; it is essential for ensuring safe, high-quality health care to patients in the future. A fundamental requirement of resident education is in-depth, firsthand experience caring for actual patients. Ensuring the safety and well-being of patients who participate in the education of residents is of the utmost importance. During acquisition of the competencies required for independent practice, residents are going to make errors but they should not result in harm to patients. One must look beyond hours of work alone as a risk factor during training and put in place practices (e.g., time for sufficient sleep, enhanced supervision, appropriate workload, unambiguous handovers) that will minimize other contributors to error (fatigue, insufficient knowledge to arrive at a diagnosis, excessive workload that leads to rushing, failure to communicate key clinical data). Fortunately, these factors can be addressed, and in doing so, the patient care environment can be made safer. The committee recognizes that full implementation of all its recommendations will take some time to be phased in.

The aim in adjusting duty hours and recommending other improvements is to develop training institutions that provide the best health care in safe environments for patients and the optimal learning environment for residents. The issues surrounding residency education and duty hours should be revisited in a few years to assess the changes put in place and their impact. The committee hopes that by spurring more rigorous monitoring and evaluation there will be further identification of best practices that result in improved patient and resident safety. Duty hour requirements should evolve to incorporate new scientific evidence as well as changing circumstances in the U.S. healthcare system.

REFERENCES

ACGME (Accreditation Council for Graduate Medical Education). 2003. *Common program requirements for duty hours*. Chicago, IL: ACGME.

Baldwin, D. C. J., S. R. Daugherty, R. Tsai, and M. J. J. Scotti. 2003. A national survey of residents' self-reported work hours: Thinking beyond specialty. *Academic Medicine* 78(11):1154-1163.

Bell, B. M. 2003. Reconsideration of the New York State laws rationalizing the supervision and the working conditions of residents. *Einstein Journal of Biological Medicine* 20(1):36-40.

CBO (Congressional Budget Office). 2008. The budget and economic outlook: An update. Washington, DC: Congressional Budget Office. Supplemental data of mandatory outlays. http://www.cbo.gov/budget/factsheets/2008b/medicare.pdf (accessed September 23, 2008).

IOM (Institute of Medicine). 2000. *To err is human: Building a safer health system*. Washington, DC: National Academy Press.

Landrigan, C. P., J. M. Rothschild, J. W. Cronin, R. Kaushal, E. Burdick, J. T. Katz, C. M. Lilly, P. H. Stone, S. W. Lockley, D. W. Bates, and C. A. Czeisler. 2004. Effect of reducing interns' work hours on serious medical errors in intensive care units. *New England Journal of Medicine* 351(18):1838-1848.

Landrigan, C. P., L. K. Barger, B. E. Cade, N. T. Ayas, and C. A. Czeisler. 2006. Interns' compliance with Accreditation Council for Graduate Medical Education work-hour limits. *JAMA* 296(9):1063-1070.

Lockley, S. W., J. W. Cronin, E. E. Evans, B. E. Cade, C. J. Lee, C. P. Landrigan, J. M. Rothschild, J. T. Katz, C. M. Lilly, P. H. Stone, D. Aeschbach, C. A. Czeisler, and Harvard Work Hours-Health and Safety Group. 2004. Effect of reducing interns' weekly work hours on sleep and attentional failures. *New England Journal of Medicine* 351(18):1829-1837.

Shetty, K. D., and J. Bhattacharya. 2007. Changes in hospital mortality associated with residency work-hour regulations. *Annals of Internal Medicine* 147(2):73-80.

Volpp, K. G., A. K. Rosen, P. R. Rosenbaum, P. S. Romano, O. Even-Shoshan, A. Canamucio, L. Bellini, T. Behringer, and J. H. Silber. 2007a. Mortality among patients in VA hospitals in the first 2 years following ACGME resident duty hour reform. *JAMA* 298(9):984-992.

Volpp, K. G., A. K. Rosen, P. R. Rosenbaum, P. S. Romano, O. Even-Shoshan, Y. Wang, L. Bellini, T. Behringer, and J. H. Silber. 2007b. Mortality among hospitalized Medicare beneficiaries in the first 2 years following ACGME resident duty hour reform. *JAMA* 298(9):975-983.

1

Background and Overview

After graduation from medical school, residency training provides an in-depth experience in learning the science and art of medicine for a specific discipline (e.g., general surgery, internal medicine, pediatrics). This chapter gives a brief history and overview of graduate medical training in the United States, with a focus on resident duty hours. Key issues related to patient safety, resident safety, and resident learning are introduced, including adherence to current duty hour limits, the relationship of duty hours to patient and resident safety, and the importance of training for achieving the competence to practice medicine independently at the end of residency. These issues are reviewed in greater detail in subsequent chapters. Based on evidence in the later chapters, the committee recommends adjustments to current duty hours and other steps to improve the education of future physicians, which should enhance the safety of residents and their current and future patients.

The primary mission of graduate medical education (GME) is to train the next generation of physicians. To achieve this, graduates of medical school spend 3 to 7 additional years in residency training, becoming skilled doctors prepared for independent practice in different disciplines. This evolution takes place in environments both exhilarating and exhausting, being a preparation for similar challenges and rewards later in their career (Ludmerer, 1999). Residency training is essential for professional development both educationally and practically, since all state medical boards require at least 1 year of graduate medical training before a doctor is eligible for an unrestricted license to practice medicine (AMA, 2008; FSMB, 2007).

Early in the 20th century, graduate medical trainees, also known as residents or house staff, actually resided at the hospital where they provided "on-call" medical service as part of their training. Until the recent duty hour reforms, resident work schedules frequently totaled 90 hours or more per week, made up of 36-hour shifts separated by 12 hours or less of rest. Proposals to reduce resident duty hours have been met with concern within the medical community that physician training could be compromised. Given that residents provide a source of inexpensive labor for medical institutions, questions have also been raised about whether duty hours are inflated to meet the service needs of hospitals without a commensurate increase in educational value for residents. The overall structure of funding for GME in the United States has promoted hospital-based training for residents. Teaching hospitals and the public have come to depend on residents to deliver services around the clock, and providing substitutes for their time is expensive. Attending physicians (physicians who supervise the care provided by residents) also benefit from having residents onsite to facilitate patient care rather than always being present themselves.

In 1984, the length of resident duty hours came under public scrutiny when Libby Zion died after being seen in an emergency room in New York City. A subsequent grand jury investigation highlighted the risks to patient care posed by inadequately supervised and fatigued residents. Following the recommendations of the Bell Commission's review of emergency care, New York State limited resident duty hours in 1989 to 80 hours a week (averaged over 4 weeks), affecting medical facilities statewide where approximately 15 percent of all residents in the country trained (IPRO, 2007). The duty hour limits set in New York later became the basis for national reform. In 2003, the Accreditation Council for Graduate Medical Education (ACGME) mandated an 80-hour weekly average for all residents along with implementing other minimum requirements for time off from the hospital; these are discussed more fully in Chapter 2 (ACGME, 2007b).

This chapter outlines the scope of issues studied by the Institute of Medicine (IOM) Committee on Optimizing Graduate Medical Trainee (Resident) Hours and Work Schedules to Improve Patient Safety. The committee's name highlights the group's task of looking at residents' total hours of work and the distribution of those hours (schedule) over a period of time. This chapter provides background on the general nature of residency training, areas of residency specialization, and number of residents and training programs in the country. To provide some context, the total workweek hours of residents are compared with the total duty hours of physicians in practice. Finally, the chapter highlights concerns with respect to duty hours that surfaced in the scientific and academic medicine literature and in testimony to the committee.

CHARGE TO COMMITTEE

Congress, through the Subcommittee on Oversight and Investigations of the House Committee on Energy and Commerce, requested that the Department of Health and Human Services (HHS) sponsor a study by the IOM to examine the relationship between resident duty hours and patient safety. The subcommittee had been investigating preventable medical errors and asked if the duty hours of physicians and residents are among the most serious threats to patient safety (Dingell et al., 2007).

The IOM, under a contract with the Agency for Healthcare Research and Quality (AHRQ), HHS, agreed to form a consensus committee and conduct a study of residents to

> 1) synthesize current evidence on medical resident schedules and healthcare safety, and
>
> 2) develop strategies to enable optimization of work schedules to improve safety in the healthcare work environment, ... [and] Consider also evidence on the safety of the residents, the education and training experience of the residents, the quality of the interactions from both the resident and patient perspective, and other aspects of safety and quality of care such as care hand-offs and transitions.
>
> (See Appendix A for the complete Statement of Task.)

Residency continues to consist of an apprenticeship approach to learning through service to hospital inpatients and their clinic patients under the guidance of attending faculty physicians. Residents play a significant role in the healthcare system. They are frequently the primary frontline physician-level staff on duty around the clock in teaching hospitals. Residents are exposed to disease pathology and learn effective management of both common and unusual illnesses and injuries. The resident's dual roles of learner and provider of care mean that changes in duty hours have implications for educational quality and access to care, as well as both short-term and long-term patient safety. Adjusting hours without other compensatory system changes can endanger these desirable outcomes.

Because residents provide their valuable skills for a relatively modest taxable stipend (e.g., for 2007-2008 the mean stipend for a first-year resident was $44,747), reductions in duty hours also have economic implications (AAMC, 2007a). A reduction in residents' duty hours creates a demand for other clinicians (e.g., more attending physicians, more nurse practitioners, more residents, more support staff) to fill the resulting gaps in medical care, and there are already projected shortages of some clinicians (AAMC, 2008b; ANSR, 2008; HHS, 2006). The committee therefore felt that its evaluation of the effects of resident duty hour changes should be viewed in the broader context of the possible educational, safety, and resource consequences.

In meeting its charge, the committee carefully examined a broad range of pertinent evidence.[1] Since the 2003 ACGME duty hour limits for residents have been in effect for only 5 years, there are limited data concerning the impact of these changes. The committee represented diverse points of view and vigorously debated the issues and relevance of studies. The existing body of evidence was sufficient for the committee to find common ground and conclude that steps should be taken to strengthen resident education and establish an environment that would better protect residents from fatigue and patients from the potential for resident errors. The committee sought to prevent acute and chronic sleep deprivation in residents when possible and, when it is not, to reduce the risks to patients and residents resulting from residents' loss of sleep and fatigue.

GRADUATE MEDICAL TRAINING

Teaching hospitals provide a setting for the clinical education of residents as well as other health profession students (e.g., nursing, pharmacy) and have different degrees of organizational integration with medical schools (e.g., shared ownership or affiliations but separate funding and governance). Residents primarily take care of inpatients and outpatients at these hospitals as well as spend some time in community-based settings, such as community health centers and physicians' offices, depending on their areas of study. These academic institutions often combine teaching, patient care, and medical research. There are approximately 1,100 hospitals participating in medical education, but three-fourths of all residency training takes place in about 275 hospitals and health systems in the United States. Highly specialized services (e.g., transplant services, interventional cardiology, neonatal intensive care units [ICUs], burn care units, regional trauma centers, AIDS services) are more concentrated into this smaller set of teaching institutions than other teaching and non-teaching hospitals. The median number of residents at each of the 275 institutions in 2005 was 173,

---

[1]Bibliographic searches were conducted of the primary biomedical bibliographic databases, MEDLINE, EMBASE, CINHAL, and PsychInfo. The searches included articles from January 1980 to January 2008. The terms used for these searches, many in combination with each other, included resident(s), residency, internship, fatigue, sleep, sleep disorders, burnout, mood, depression, work schedule(s), work hours, 80-hour workweek, adverse events, medical errors, job satisfaction, handoffs, handovers, transitions, mortality, patient outcomes, patient safety, quality of care, medical education, graduate medical education, workload, and performance. Publications dated after January 2008 were added to the evidence base of this report as they became available or were brought to the attention of IOM staff.

while other teaching hospitals have a median of 21 residents (AAMC, 2004, 2008a; Council of Teaching Hospitals and Health Systems, 2008).

A resident is any physician enrolled in a GME program (ACGME, 2007a). First-year residents were, until recently, more often called interns, and they are also referred to as PGY-1 (postgraduate year 1) residents in many programs and research articles. Residents in the later years of practice are termed PGY-2, PGY-3, and so on to distinguish the year of postgraduate training. Since residents are in training, they must function under the qualified supervision of faculty and more senior residents. As residents demonstrate knowledge and skills appropriate to progressive levels of their postgraduate training, they are given increasing responsibility for the care of patients, larger patient loads, and greater authority to make final patient management decisions (ACGME, 2007a). A patient may not always be able to distinguish resident trainees from other physicians because those in residency have an M.D. or a D.O. degree.

By tradition and as necessitated by the nature of their responsibilities, the length and rigor of formal education and training for physicians is among the most challenging of any job or profession in the United States (U.S. Department of Labor, 2007). Residency is a period of intensive supervised learning in a real-world environment where critical skills and competencies are developed, including the needed professionalism that is the hallmark of a caring, competent, and dedicated physician. Residency training has periods during which prolonged duty hours are perceived as necessary to achieve the educational goals—this is more the case for some specialties than others. Trainees spend years preparing for the opportunity to train as resident physicians—through 4 years of a premedical curriculum in college and 4 more years of challenging medical school study and testing. Only about half of the applicants to medical school are accepted (AAMC, 2008d), but more than 90 percent of the students accepted will graduate (AAMC, 2007b). It is at this point that most graduates will enter a residency in their chosen area of specialization.

Two main tracks exist for preparing doctors in the United States: graduates of allopathic schools receive an M.D. (doctor of medicine) degree and graduates of osteopathic schools receive a D.O. (doctor of osteopathic medicine) degree. Graduates of both types of schools pursue graduate medical training before being licensed to practice independently. In 2003, 99 percent of U.S. allopathic and 43 percent of osteopathic school graduates undertook allopathic residencies (Shannon, 2007). The focus of this report is on the allopathic residencies accredited by the ACGME.

For academic year 2008-2009, 15,242 U.S. medical school seniors were matched to one of the 22,240 available first-year U.S. residency positions available across the United States (about 68 percent of the available first-year positions) based on applicant preferences and how the training program ranked them (AAMC, 2008e; National Residency Match Program, 2008). An insufficient number of U.S. medical school graduates are available to fill all residency positions. In 2006-2007, 66.5 percent of all graduate medical trainees graduated from U.S. medical schools, 26.9 percent were international medical graduates (non-Canadian), 6.3 percent graduated from schools of osteopathy, and 0.3 percent were from Canadian medical schools (Brotherton and Etzel, 2007). Efforts are under way to increase the number of U.S. medical school graduates (AAMC, 2008d).

Total Number of Residents and U.S. Training Programs

Nearly 105,000 graduate medical trainees were at various stages of their residency training in the 2007-2008 academic year. Residents work in public and private, teaching and community hospitals across the country, affiliated with more than 8,500 distinct accredited residency programs. ACGME reviews and evaluates each residency program on average every 3 to 4 years with site visits and resident interviews to examine the content of training and to ensure compliance with educational and duty hour requirements (ACGME, 2007a, 2008). Although residency programs are regulated by this private sector organization, they have grown and are maintained with substantial federal and state funding support, particularly with money from the Medicare program, the Department of Veterans Affairs, the Health Resources and Services Administration, the Department of Defense, and various state and local programs.

In 2006-2007, of the total number of residents there were 89,269 in medical specialty programs, the first stage of graduate medical training (Brotherton and Etzel, 2007). "Specialty" medical training programs are available in the fields listed in Table 1-1, and many of these specialties are combined in 19 additional residency designations, such as internal medicine and pediatrics, internal medicine and psychiatry, and internal medicine and emergency medicine. After having completed a specialty residency, 15,610 residents in 2006-2007 trained in advanced subspecialty programs, also called fellowships, and these resident trainees are sometimes referred to as "fellows." The 26 types of specialty training programs listed in Table 1-1 offer subspecialty fellowships in about 100 areas, ranging from vascular surgery to geriatrics to pediatric endocrinology. ACGME duty hour rules apply to both specialty and subspecialty residents.

Resident Learning Environment

At the best of times, residency training provides daily intellectual stimulation and gratification in solving complex problems and making a difference in patients' lives. Delivering a baby, teaching skills to medical students,

TABLE 1-1 U.S. Resident Training Programs by Specialty and Resident Physicians on Duty[a]

Specialty	Number of Resident Physicians	Number of Accredited Programs
Internal medicine	22,099	386
Family medicine	9,456	464
Pediatrics	7,964	201
Surgery (general)	7,651	251
Anesthesiology	4,970	131
Obstetrics and gynecology	4,739	250
Psychiatry	4,613	181
Emergency medicine	4,379	140
Radiology, diagnostic	4,368	188
Orthopedic surgery	3,187	152
Pathology	2,310	150
Neurology	1,507	122
Otolaryngology	1,292	104
Ophthalmology	1,225	117
Physical medicine and rehabilitation	1,167	79
Dermatology	1,069	112
Urology	992	118
Neurological surgery	881	97
Plastic surgery	609	89
Radiation oncology	556	79
Preventive medicine	285	74
Surgery (thoracic)	282	85
Allergy and immunology	274	71
Nuclear medicine	143	61
Medical genetics	77	47
Surgery (colon and rectal)	71	45

[a]As of December 1, 2006.
SOURCE: Brotherton and Etzel, *JAMA*, 2007 298:1081-1096. Copyright © 2007, American Medical Association.

removing a gall bladder, performing a lumbar puncture that confirms a case of treatable meningitis, diagnosing an unusual ailment, or running a code blue to resuscitate a patient in cardiac arrest before the attending arrives—all can provide a sense of accomplishment. Surgical residents gain confidence as they learn surgical principles and perfect their technique. Residents learn how to talk with patients and families both when the news is good and when it is not. There are also quiet moments—holding the hand of a dying patient or waiting by the bedside to see if a patient is responding to treatment. Each specialty will have a different complement of illnesses and injuries, tests and procedures, but they have in common the need to learn how to communicate with patients and other members of their medical teams.

Residents train as a team of doctors, with supervision provided by residents further along in training and by attending physicians (senior physicians, most often faculty of medical schools, who have completed residency training and are ultimately responsible for patient care decisions). Attendings provide the daily instruction during teaching rounds by guiding residents through consideration of possible diagnoses and management plans, and in the case of surgery, they spend much of their time in direct instruction and observation of procedures. Attendings have other "teachable moments" during the day when residents present summaries about newly admitted patients. Residents have other sources of learning, including the medical literature, lectures on important concepts, grand rounds, educational conferences, and interactive skills-training exercises. Residents are also part of a larger team or system of care that includes nursing, administrative, and other staff who are integral to the hospital's care mission (e.g., pharmacy, laboratory, transport, social work, nutrition, administration) and provide other valuable sources of information for resident learning.

Residents rotate through a variety of teams or services, usually spending 4 weeks per rotation. It is important for the reader of this report to recognize that each rotation as well as each specialty has different duty hour demands. An ambulatory care rotation might have five 8-hour shifts totaling a 40-hour workweek. On an emergency department (ED) rotation, there might be five 12-hour shifts (day or night) in the ED plus another 12 hours for instructional activities, totaling 72 hours. An ICU rotation might have some duty periods that are 10 hours long alternating with 30-hour duty periods, totaling 80 hours or more over the course of a week. Each year of training has a different set of rotations, and residents have greater choice of rotations as they progress through training, allowing them to concentrate on areas that they plan to pursue at the end of training or supplementing their knowledge in areas where improvement is needed.

DUTY HOUR DEMANDS IN THE MEDICAL PROFESSION

Graduate medical training programs have a tradition of requiring long hours. In 1998-1999, residents in surgical specialties were still regularly clocking more than 100 hours per week in their PGY-1 and PGY-2 training years. First-year residents in 8 of 12 specialties surveyed at that time had an average workweek of more than 80 hours, and the average across all 12 types of programs for these interns was 83 hours (Baldwin et al., 2003). Medical educators expect that ample duty hours provide residents with the needed time to think through diagnoses, manage patient plans, and gain a rich educational experience as they follow patients closely during the critical first day of admission, learn from exposure to a great variety of patient cases, increase technical skills as they participate in many procedures, and develop communication skills. Long hours during training have also become a rite of passage to the profession, testing residents' stamina, resilience, and dedication and reinforcing the idea that medical practice requires a certain mental toughness.

In 2003, ACGME set an 80-hour workweek average as the limit for all graduate medical specialties, although a few programs received exemptions to have an 88-hour week average. The best available nationwide data on how many hours residents work under the new ACGME limits come from the initial year of implementation and apply only to first-year residents. On average, PGY-1 residents (interns) in more than 13 types of programs reported a mean of 66.6 hours weekly in 2003-2004 (Landrigan et al., 2006).[2] This was a decline of 4.1 hours from 2002-2003 when interns reported 70.7 hours on average (Baldwin et al., 2003). Interns, those with the least experience, tend to work longer hours than residents in the same field but in later years of specialty training.

An important part of graduate medical training is that it exposes residents to the demands of real-life practice, apart from the training environment, including the necessity to attend to ill patients at all hours even when away from the training environment. This is often inconvenient and involves night and weekend work. According to the U.S. Bureau of Labor Statistics, many physicians work long and unpredictable hours around the clock once they finish their graduate medical training—longer hours than most other workers in the United States. The Bureau of Labor Statistics estimates that more than one-third of fully licensed physicians work 60 hours or more per week (U.S. Department of Labor, 2007). Physicians in office-based practice report an average workweek of 60 hours for surgical specialties and 50 hours for primary care (e.g., family practitioners, internists), but the workweek is as long for some internal medicine specialties as for surgeons (e.g., mean of 60 hours per week for cardiologists and nephrologists) (Weiss, 2006). Certain types of specialties are more likely to report working 80 hours per week or more. Of the 23 specialties examined in 2005, the following reported 15 percent or more of their office-based practice members working more than 80 hours: urologists (15 percent), obstetricians-gynecologists (OB/GYNs; 16 percent), pulmonologists (16 percent), hematologists-oncologists-immunologists (17 percent), infectious disease specialists (17 percent), general surgeons (19 percent), cardiologists (20 percent), neurosurgeons (23 percent), and thoracic surgeons (33 percent) (Weiss, 2005). Similarly, residency programs have different work

---

[2]This national sample of interns was self-selected.

requirements depending on the specialty but now have common duty hour limits.

SCOPE AND ORGANIZATION OF REPORT

Oral presentations before the committee and additional submitted testimony from stakeholders and experts raised a broad array of issues and perspectives for consideration relative to resident duty hours and patient safety. It became obvious early in the study process that appropriately balancing these issues would present a challenge, as would reaching a consensus within the committee. All of the testimony provided a useful guide for the committee's research, indicating areas for in-depth study and potential sources of data and evidence. The major topic areas raised included the following:

- Current duty hours and adherence to them
- Educational needs
- Resident safety and well-being
- Patient safety
- Economic implications

The committee heard from diverse speakers: patient advocates; an ethicist; residents in training; resident and medical school student representatives; residency program directors of several specialties; administrators in charge of all graduate medical training programs at their institution; hospital executives and financial officers from institutions with residents; scientists who study sleep, fatigue, and human performance; physician specialty societies; the president of the Royal College of Surgeons; representatives of national organizations involved in GME, including matching medical school graduates to residencies, and accreditation of programs; major funders of GME; and the Joint Commission. The presentations of the speakers that appeared before the committee are available on the project website, www.iom.edu/residenthours (see Appendix F for the public agenda for committee meetings).

Current Duty Hours and Adherence to Them

Currently, the ACGME sets standards for GME including duty hour limits, and it monitors how well residency programs adhere to those rules and the educational standards set for each specialty. Questions surfaced about whether there was a scientific basis for the 2003 ACGME rules now in place and how the length of the workweek and the number of consecutive duty hours compared with the experiences of other countries and other industries. Some speakers were in favor of changing certain elements of the existing duty hour limits, especially the extended duty period of 30 hours and averaging provisions (CIR/SEIU, 2007; Landrigan, 2007; Public Citizen, 2007), while others thought any further change in hours premature (AAMC, 2008c; ABNS, 2008; ACGME, 2007a, 2008; ACS, 2008). Moreover, there was uncertainty about the actual number of hours that residents currently work per week, how often the limits are violated, and the reasons for those violations (e.g., patient care needs, excessive workload) and whether residents were staying of their own volition versus program or institutional pressures (ACGME, 2008; AMA Resident/Fellow Section, 2007; American Medical Student Association, 2007; Arora, 2007; CIR/SEIU, 2007; Resident Panel, 2008; Vidyarthi, 2007).

The committee was asked by some presenters to determine whether the current ACGME procedures for assessing adherence to duty hours were sufficiently rigorous (CIR/SEIU, 2007; Public Citizen, 2007). The ACGME's position was that hours should not be viewed in isolation but as one of many pieces of information integral to assessing the quality of an educational program (including supervision and institutional support) and the quality of patient care delivered (AAMC, 2008c; ACGME, 2007a, 2008). Some organizations argued that ACGME's data collection methods were insufficient to adequately enforce work hour limits, resulting in identification of too few violations of duty hours (CIR/SEIU, 2007; Public Citizen, 2007). Additionally, speakers pointed to disincentives to accurate duty hour reporting among residents who feared that their training program could lose accreditation or that they would personally face retribution (AMA Resident/Fellow Section, 2007; American Medical Student Association, 2007).

Chapter 2 examines past and present duty hour limits in the United States and what is known about the monitoring and compliance issues outlined above. Further, it briefly examines the duty hour limits set for other safety-sensitive industries such as aviation and trucking and the efforts at regulation in those work environments. Appendix C draws lessons from the experiences of other countries that have mandated significantly reduced resident duty hours (e.g., by 2009, Europe will reduce duty hours to 48 hours per week). Lessons drawn from the experiences of other countries relate to the period needed for phase-in of requirements, workforce implications, scheduling adaptations, and the need to modify educational programs to adapt to reduced duty hours.

Educational Needs

The quality of the educational experience of residents today is a "determinant of patient safety and health care quality for decades to come" (AAMC, 2008c). Educators asked the committee to consider the long-term goal of ensuring patient safety by producing competently trained physicians versus the potential risks to patients that they treat during residency training. The sparsity of data on educational outcomes since duty hour reform is partly due to the fact that the first cohorts of residents fully trained under the 2003 limits are now completing their residencies (ABMS, 2008; AMA Resident/Fellow Section, 2007; American Orthopaedic Association, 2008). In considering the initial data, it is important to recognize that adaptations to the limits may have evolved over time and conclusions drawn from early implementation studies may not paint an accurate picture of the current situation. Duty hours were not uniformly implemented at the same rate or in the same manner across all programs; some programs have modified their schedules several times over the past few years before moving to full compliance.

Reducing hours means having fewer residents available for duty at any single time. The committee heard testimony that this has led to increased resident workload as sufficient substitutes for residents were not always added (AAMC, 2008c; Arora, 2007; Bellini, 2008; Vidyarthi, 2007). Although the intensity of work for residents may have increased during the hours they are on service, there are also questions about the educational value of some of the work they currently perform (e.g., blood drawing, routine scheduling of appointments) (Bellini, 2008). Consequently, the committee heard that when residents make mistakes, heavy workload and inadequate supervision are factors as well as fatigue (Bellini, 2008; Consumers Union of the United States, 2008).

The 2003 ACGME duty hour limits resulted in some residents having less opportunity to observe patients' care from beginning to end and to observe changes in the course of their illness and recovery, all of which are considered by many educators to be essential for quality patient care and effective education (AAMC, 2008c). Therefore, the challenge to the committee was, on the one hand, to suggest ways to minimize any risks of extended duty hours for patients and residents, while, on the other hand, suggesting ways to maximize the presumed educational and patient safety benefits of uninterrupted continuity of resident involvement (especially early in the course of illness or immediately after surgery).

Chapter 3 examines the work and learning environment of residents, types of strategies used to try to conform to the 2003 rules, and how the content of resident work and the patient caseload have consequences for adhering to duty hour limits. Chapter 4 draws upon the educational research literature on how people learn, basic tenets underlying GME (assumption of responsibility, time for reflection, continuity of care), and the importance of supervision for resident training. It also considers the necessity for new models of instruction and assessment for GME within reduced duty hours.

Resident Safety and Well-Being

The overall sense from testimony to the committee from a wide spectrum of sources was that reduced duty hours had improved resident quality of life (AAMC, 2008c; ABMS, 2008; ACGME, 2008). Residents themselves said that ACGME 2003 duty hours allowed them to spend more time with family, catch up on personal chores (e.g., do laundry, pay bills), and participate in more leisure activities (e.g., exercise, social events with friends). Some of the committee members heard residents' testimony that the time off afforded under current limits was not necessarily spent sleeping because time off is still relatively limited (Resident Panel, 2008). Concerns remain about personal safety issues for residents due to fatigue, including driving incidents and needlestick injuries (AMA Resident/Fellow Section, 2007; CIR/SEIU, 2007; Landrigan, 2007; Public Citizen, 2007). Chapter 5 examines what is known about risks for residents, associated with working long hours and having limited sleep, in terms of their physical and mental health, personal relationships, and professional interactions.

Patient Safety

The committee appreciates that a complex set of issues is associated with considering the short- and long-term safety implications of making any adjustments to resident duty hours (Cohn, 2008). Carolyn Clancy, director of AHRQ and the sponsor of the IOM study, crystallized the challenge before the committee from her perspective (Clancy, 2007):

> At some point . . . we have to acknowledge the fact that a human being can work only so long without sleep deprivation becoming a factor. Research shows that we do not do well in transitions of care . . . but limiting these transitions by having duty hours that are not compatible with human physiology is not the answer.

The public perception of an appropriate number of hours for doctors to work often differs from the current reality of residency training (Public Citizen, 2007). One speaker quoted findings from a National Sleep Foundation phone poll of 1,010 Americans in 2001 who were asked about the likelihood of their actions if they learned that "the doctor that is about to perform their surgery has been on duty for 24 consecutive hours." Sixty-five percent indicated they would very likely feel anxious about their safety, and 45 percent indicated they would very likely ask for another doctor (National Sleep Foundation, 2002; Public Citizen, 2007). Thus, the benefits for residents of long continuous duty hours thought desirable by educators are not always clear to the public and those being treated. Another study at three institutions found that nearly one-quarter of internal medicine inpatients surveyed were concerned about resident fatigue and about discontinuity of care due to patient handovers; these patients tended to be the same set with worries about their care. Patients reported their perception of how many hours residents work in a week (60 hours) and how long they should work (51 hours). In actuality, residents at the three institutions studied worked 67 to 69 hours per week (Fletcher et al., 2007).

All who spoke to the issue agreed that shorter duty hours have resulted in more handovers of care, which have been associated with increased risks to patient safety particularly due to poor communication of essential information (AAMC, 2008c; ACGME, 2007a). This risk may be due as much to the fact that handover techniques are not standardized or optimized for their intended purpose, rather than the fact that handoffs are inherently a systemic hazard. Handovers can also be viewed as a time to reassess patient care and catch previous errors. As is the case for fully trained physicians in practice, residents clearly cannot be on duty 24/7. Handovers of patient care to competent colleagues at appropriate intervals are essential, and strengthening these transfers of patients from one clinician to another is necessary.

Several researchers presented information on schedule changes, increases in the number of handovers, error rates, and mortality data for the committee to consider (Arora, 2007; Czeisler, 2007; de Virgilio, 2008; Landrigan, 2007; Vidyarthi, 2007; Volpp, 2007). Various observers questioned whether there was sufficient evidence to link resident duty hours to direct harm for patients and whether enough is known about the offset in the risks of more handovers versus less fatigued residents (AAMC, 2008c; AMA Resident/Fellow Section, 2007; Volpp, 2007). Furthermore, a sleep scientist indicated that experience in other industries suggests that limiting duty hours alone is not sufficient to ensure safety. Multiple factors related to time worked, recovery sleep, frequency of shift changes, night work, and the length of continuous work all contribute to the level of human performance. One type of schedule is unlikely to fit the needs of all residency program operations (Rosekind, 2008). Whether advocating for shorter shifts or not, speakers encouraged the committee to examine information from sleep science when considering any adjustments to duty hours (AAMC, 2008c; ACGME, 2007a; Bellini, 2008; CIR/SEIU, 2007; Czeisler, 2007).

Chapter 6 examines what is known about errors in hospital care attributable to fatigued residents and the effects of the 2003 duty hours on patient outcomes. A set of interrelated studies on resident hours of work and sleep are examined in depth to determine what lessons might be learned about resident error and patient safety. Chapter 7 covers what is known about preventing acute or chronic sleep loss and its effects on making errors and what the implications would be for the redesign of resident duty hours and schedules, and the chapter includes the committee's recommendations for adjusting duty hour limits. Chapter 8 discusses strategies to incorporate residents into an enhanced culture of safety that promotes resident learning about error prevention and about improved communication during handovers.

Economic Implications

Further adjustments to duty hour limits or the content of resident work will have economic implications for institutions with resident training programs just as the 2003 rules did (Arora, 2007; Daschbach, 2008; Dyne, 2008; Hara, 2008; Liekweg, 2008; Noah, 2008; Opas, 2008). Addition of other personnel to substitute for resident work time was a frequently used but costly strategy employed by academic medical centers in response to the ACGME mandate (ACGME, 2007a). On some services, especially surgical services, resident duty hours were reduced by 20-25 percent as they adapted from workweeks of 100 hours to 80 hours (ACS, 2008).

More modifications to the 2003 duty hour limits raised questions of where the additional workforce would come from to substitute for resident-delivered care, and who would bear these costs. Without additional outside funds, many institutions will have a difficult time adapting and continuing to provide care to the same number of patients (AAMC, 2008c; Opas, 2008). The economic benefit that society has derived from the long duty hours of residents working at relatively low wages is substantial, and speakers felt that additional funding would be necessary to implement further changes to duty hours or workload (CIR/SEIU, 2007; Daschbach, 2008; Liekweg, 2008; Noah, 2008; Opas, 2008).

Chapter 9 concludes the report with recommendations for funding and evaluation. The chapter includes a summary of an economic analysis commissioned by the committee to estimate the order of magnitude of costs for substituting current resident duty hours with those of other personnel or additional residents according to various scenarios for changes in the duty hour and workload requirements of residents.

Finally, the committee is aware of the possibility that even well-considered recommendations might have unintended consequences, some of which will be discovered only after they are implemented. It urges an evaluation strategy incorporating data gathering and analysis from initiation of the recommended changes with the aim of detecting and minimizing unintended consequences on patient safety or resident education.

Limits on Scope of the Study

Some important issues necessarily fall outside the purview of this study because of time and other resource constraints. Thus, the committee's report does not attempt to reorganize the whole healthcare workforce or reform the healthcare system relative to the issue of patient and resident safety but examines the system as it works now. Specifically, while Medicare funding for GME flows to hospitals rather than other training sites or medical schools and reimbursement rates are highly variable from institution to institution despite a shared teaching mission, it was beyond the scope of this committee to analyze the effects and offer explicit alternative funding strategies. Similarly, the committee does not make a judgment on the concentration of training programs in certain geographic areas or address the development of integrated delivery systems or other models of care. Furthermore, the committee's focus is on residents, as charged, not on students in medical school or physicians in practice. Fellows (doctors who have completed their specialty residency and are continuing training in a subspecialty) are included only to the extent that data concerning them are embedded within the studies of residents. Nonetheless, much of the discussion and recommendations may be applicable to fellows as well as residents. The study also does not address the fatigue problems of physicians who continue to work long hours well after their training or of nurses or other healthcare providers who work long hours, because they are not covered by the ACGME duty hour limits. The statement of task for the project did not require detailed cost estimates of every recommendation. Establishing conditions that promote patient safety involves many interacting variables. Duty hours are but one element. The committee necessarily has narrowed its focus with respect to patient safety given the limits of its charge and restricted its exploration to related contextual issues such as supervision, workload, handovers, and teamwork.

REFERENCES

AAMC (Association of American Medical Colleges). 2004. *The handbook of academic medicine: How medical schools and teaching hospitals work*, edited by W. T. Mallon et al. Washington, DC: Association of American Medical Colleges.

———. 2007a. *AAMC survey of housestaff stipends, benefits and funding: Autumn 2007 report*. Washington, DC: Association of American Medical Colleges.

———. 2007b. *Medical school graduation and attrition rates*. Washington, DC: Association of American Medical Colleges.

———. 2008a. *AAMC data book: Medical school and teaching hospitals by the numbers*, edited by K. Brandenburg, S. Gaillard, W. Geraci, P. Vassev, and J. Youngclaus. Washington, DC: Association of American Medical Colleges.

———. 2008b. *Help wanted: More U.S. doctors projections indicate America will face shortage of M.D.s by 2020*. http://www.aamc.org/workforce/helpwanted.pdf (accessed June 4, 2008).

———. 2008c. Resident duty schedules and resident education. Presentation by Debra Weinstein to the Committee on Optimizing Graduate Medical Trainee (Resident) Hours and Work Schedules to Improve Patient Safety, May 8, 2008, Washington, DC.

———. 2008d. *U.S. medical school enrollment projected to rise 21 percent by 2012: Both new and existing schools will fuel growth.* http://www.aamc.org/newsroom/pressrel/2008/080501.htm (accessed May 2, 2008).

———. 2008e. *U.S. medical school seniors enjoy most successful "match day" in 30 years.* http://www.aamc.org/newsroom/pressrel/2008/080320.htm (accessed March 21, 2008).

ABMS (American Board of Medical Specialties). 2008. Presentation by Kevin Weiss to the Committee on Optimizing Graduate Medical Trainee (Resident) Hours and Work Schedules to Improve Patient Safety, May 8, 2008, Washington, DC.

ABNS (American Board of Neurological Surgery). 2008. Presentation by H. Hunt Batjer to the Committee on Optimizing Graduate Medical Trainee (Resident) Hours and Work Schedules to Improve Patient Safety, March 4, 2008, Irvine, CA.

ACGME (Accreditation Council for Graduate Medical Education). 2007a. The ACGME's approach to limiting resident duty hours. Presentation by Paul Friedmann to the Committee on Optimizing Graduate Medical Trainee (Resident) Hours and Work Schedules to Improve Patient Safety, December 3, 2007, Washington, DC.

———. 2007b. *Common program requirements.* http://acgme.org/acWebsite/dutyHours/dh_dutyhoursCommonPR07012007.pdf (accessed March 23, 2008).

———. 2008. The ACGME duty hour standards: One element for promoting good learning, safe patient care and resident well-being: Data on compliance and effect. Presentation by Ingrid Philibert and Thomas Nasca to the Committee on Optimizing Graduate Medical Trainee (Resident) Hours and Work Schedules to Improve Patient Safety, March 4, 2008, Irvine, CA.

ACS (American College of Surgeons). 2008. Presentation by Gerald Healy to the Committee on Optimizing Graduate Medical Trainee (Resident) Hours and Work Schedules to Improve Patient Safety, March 4, 2008, Irvine, CA.

AMA (American Medical Association). 2008. *State medical licensure requirements and statistics, 2008.* Chicago, IL: American Medical Association.

AMA Resident/Fellow Section. 2007. Duty hours: The resident perspective. Presentation by Sunny Ramchandani to the Committee on Optimizing Graduate Medical Trainee (Resident) Hours and Work Schedules to Improve Patient Safety, December 3, 2007, Washington, DC.

American Medical Student Association. 2007. Presentation by Michael Ehlert to the Committee on Optimizing Graduate Medical Trainee (Resident) Hours and Work Schedules to Improve Patient Safety, December 3, 2007, Washington, DC.

American Orthopaedic Association. 2008. Executive summary of the American Orthopaedic Association (AOA) resident work hours survey. Rosemont, IL: American Orthopaedic Association.

ANSR (Americans for Nursing Shortage Relief). 2008. *Testimony regarding FY 2009 appropriations for Title VIII—nursing workforce development programs.* https://www.ncsbn.org/FY09_Senate_Testimony.pdf (accessed August 26, 2008).

Arora, S. 2007. Duty hours: Past, present and future: The University of Chicago internal medicine experience. Presentation to the Committee on Optimizing Graduate Medical Trainee (Resident) Hours and Work Schedules to Improve Patient Safety, December 3, 2007, Washington, DC.

Baldwin, D. C., Jr., S. R. Daugherty, R. Tsai, and M. J. Scotti, Jr. 2003. A national survey of residents' self-reported work hours: Thinking beyond specialty. *Academic Medicine* 78(11):1154-1163.

Bellini, L. 2008. Beyond duty hour reform: Redefining the learning environment. Presentation to the Committee on Optimizing Graduate Medical Trainee (Resident) Hours and Work Schedules to Improve Patient Safety, May 8, 2008, Washington, DC.

Brotherton, S. E., and S. I. Etzel. 2007. Graduate medical education, 2006-2007. *JAMA* 298(9):1081-1096.

CIR/SEIU (Committee of Interns and Residents-Service Employees International Union). 2007. Healthcare. Presentation by L. Toni Lewis to the Committee on Optimizing Graduate Medical Trainee (Resident) Hours and Work Schedules to Improve Patient Safety December 3, 2007, Washington, DC.

Clancy, C. 2007. Remarks by Carolyn M. Clancy, M.D., director, Agency for Healthcare Research and Quality (AHRQ). Presentation to the Committee on Optimizing Graduate Medical Trainee (Resident) Hours and Work Schedules to Improve Patient Safety, December 3, 2007, Washington, DC.

Cohn, F. 2008. Ethical decision making: A framework for considering resident work hours. Presentation to the Committee on Optimizing Graduate Medical Trainee (Resident) Hours and Work Schedules to Improve Patient Safety, May 8, 2008, Irvine, CA.

Consumers Union of the United States. 2008. Presentation by Elizabeth Imholz to Committee on Optimizing Graduate Medical Trainee (Resident) Hours and Work Schedules to Improve Patient Safety, March 4, 2008, Irvine, CA.

Council of Teaching Hospitals and Health Systems. 2008. *AAMC teaching hospitals and health systems*. Washington, DC: Association of American Medical Colleges.

Czeisler, C. A. 2007. Presentation to the Committee on Optimizing Graduate Medical Trainee (Resident) Hours and Work Schedules to Improve Patient Safety, December 3, 2007, Washington, DC.

Daschbach, C. C. 2008. IOM panel: Resident duty hours cost impact estimates 2003-2008. Presentation to the Committee on Optimizing Graduate Medical Trainee (Resident) Hours and Work Schedules to Improve Patient Safety, March 4, 2008, Irvine, CA.

de Virgilio, C. 2008. The 80-hour resident workweek: A perspective from a university-affiliated Los Angeles County surgical training program. Presentation to the Committee on Optimizing Graduate Medical Trainee (Resident) Hours and Work Schedules to Improve Patient Safety, May 8, 2008, Irvine, CA.

Dingell, J. D., J. Barton, B. Stupak, and E. Whitfield. 2007. Letter to William Munier, acting director, Agency for Healthcare Research and Quality. Washington, DC: U.S. House of Representatives Committee on Energy and Commerce.

Dyne, P. L. 2008. Emergency medicine: Shift workers and safety net. Presentation to the Committee on Optimizing Graduate Medical Trainee (Resident) Hours and Work Schedules to Improve Patient Safety, March 4, 2008, Irvine, CA.

Fletcher, K. E., F. C. Wiest, L. Halasyamani, J. Lin, V. Nelson, S. R. Kaufman, S. Saint, and M. Schapira. 2007. How do hospitalized patients feel about resident work hours, fatigue, and discontinuity of care? *Journal of General Internal Medicine* 23(5):623-628.

FSMB (Federation of State Medical Boards). 2007. *State-specific requirements for initial medical licensure*. http://fsmb.org/usmle_eliinitial.html (accessed March 17, 2008).

Hara, J. 2008. Presentation to the Committee on Optimizing Graduate Medical Trainee (Resident) Hours and Work Schedules to Improve Patient Safety, March 4, 2008, Irvine, CA.

HHS (U.S. Department of Health and Human Services). 2006. *Physician supply and demand: Projections to 2020*. http://bhpr.hrsa.gov/healthworkforce/reports/physiciansupplydemand/physiciansupplyprojections.htm (accessed March 23, 2008).

IPRO. 2007. Working hours and conditions post-graduate trainees annual compliance assessment: Contract year 6 10/1/06-9/30/07.

Landrigan, C. P. 2007. Effects of traditional 24-hour work shifts on physician and patient safety. Presentation to the Committee on Optimizing Graduate Medical Trainee (Resident) Hours and Work Schedules to Improve Patient Safety, December 3, 2007, Washington, DC.

Landrigan, C. P., L. K. Barger, B. E. Cade, N. T. Ayas, and C. A. Czeisler. 2006. Interns' compliance with Accreditation Council for Graduate Medical Education work-hour limits. *JAMA* 296(9):1063-1070.

Liekweg, R. 2008. Resident work hours: A look at the University of California experience. Presentation to the Committee on Optimizing Graduate Medical Trainee (Resident) Hours and Work Schedules to Improve Patient Safety, March 4, 2008, Irvine, CA.

Ludmerer, K. M. 1999. *Time to heal: American medical education from the turn of the century to the era of managed care.* New York: Oxford University Press.

National Residency Match Program. 2008. *Advance data tables, 2008 main residency match.* http://www.nrmp.org/data/advancedatatables2008.pdf (accessed March 26, 2008).

National Sleep Foundation. 2002. *2002 "Sleep in America" poll.* http://www.kintera.org/atf/cf/%7BF6BF2668-A1B4-4FE8-8D1A-A5D39340D9CB%7D/2002SleepInAmericaPoll.pdf (accessed October 16, 2008).

Noah, M. 2008. Resident duty hour restrictions: Cost impact. Presentation to the Committee on Optimizing Graduate Medical Trainee (Resident) Hours and Work Schedules to Improve Patient Safety, March 4, 2008, Irvine, CA.

Opas, L. 2008. IOM workshop: Cost impact of duty hour change. Presentation to the Committee on Optimizing Graduate Medical Trainee (Resident) Hours and Work Schedules to Improve Patient Safety, March 4, 2008, Irvine, CA.

Public Citizen. 2007. Presentation by Peter Lurie to the Committee on Optimizing Graduate Medical Trainee (Resident) Hours and Work Schedules to Improve Patient Safety, December 3, 2008, Washington, DC.

Resident Panel. 2008. Presentation to some members of the Committee on Optimizing Graduate Medical Trainee (Resident) Hours and Work Schedules to Improve Patient Safety, March 4, 2008, Irvine, CA.

Rosekind, M. R. 2008. Managing alertness and performance in 24/7 settings: Lessons learned. Presentation to the Committee on Optimizing Graduate Medical Trainee (Resident) Hours and Work Schedules to Improve Patient Safety, March 4, 2008, Irvine, CA.

Shannon, S. C. 2007. *Graduate medical education.* http://www.aacom.org/advocacy/policy%20Statement/06-28-07%20-%20AACOM%20Policy%20on%20Graduate%20Medical%20Education.pdf (accessed March 21, 2008).

U.S. Department of Labor, Bureau of Labor Statistics. 2007. *Occupational outlook handbook, 2008-09 edition, physicians and surgeons, on the Internet.* http://www.bls.gov/oco/ocos074.htm (accessed March 17, 2008).

Vidyarthi, A. 2007. Findings and strategies from research literature. Presentation to the Committee on Optimizing Graduate Medical Trainee (Resident) Hours and Work Schedules to Improve Patient Safety, December 3, 2007, Washington, DC.

Volpp, K. G. 2007. Impact of work hour regulation on mortality in teaching hospitals: Some findings. Presentation to the Committee on Optimizing Graduate Medical Trainee (Resident) Hours and Work Schedules to Improve Patient Safety, December 3, 2007, Washington, DC.

Weiss, G. G. 2005. Exclusive survey: Productivity takes a dip. *Medical Economics* 82(22):86-87, 89, 91-93.

———. 2006. Exclusive survey—Productivity: Work hours up, patient visits down. *Medical Economics* 83(21):57-58, 60, 62-63.

2

Current Duty Hours and Monitoring Adherence

In 2003, the Accreditation Council for Graduate Medical Education (ACGME) established a maximum but not required 80-hour workweek for residents, averaged over 4 weeks. The best available national data show that first-year residents across various specialties reported working 66.6 hours a week on average during 2003-2004. Hours of work tend to be higher for residents in their first year of training, during rotations with overnight call responsibilities, and for certain specialty programs (e.g., general surgery). Lack of adherence to the ACGME limits remains an issue in some programs, particularly with respect to the limitations on the number of consecutive hours a resident can work and requirements for adequate time off for recovery sleep and personal activities. As a result, residents remain susceptible to acute and chronic sleep deprivation, despite the intent of the 2003 duty hour limits to prevent fatigue. The committee found the need to enhance monitoring of and adherence to duty hour rules. These changes should include (1) unannounced audits of duty hours by ACGME and strengthened whistle-blower protections at the local and national levels for better detection and resolution of violations; (2) oversight by the Centers for Medicare and Medicaid Services (CMS) and the Joint Commission of duty hours in relation to patient safety and quality improvement; and (3) evaluation of the hours worked and the frequency of violations, including documentation of their causes (such as patient needs or unnecessary workload) by specialty.

Residency is a unique career stage, a time to focus exclusively on training and professional development. Residents experience a sense of growing every day as they become more competent while performing important, meaningful work. The design of this training ideally maximizes rich educational experiences and has traditionally meant working long hours. In settings where camaraderie with faculty and fellow residents is strong, trainees experience a healthy learning environment and exponential growth as physicians, not just the fatigue associated with working long hours. However, concerns have been raised about the quality of care delivered by fatigued residents and the humane treatment of residents themselves. These concerns have led to duty hour reforms.

The committee's task has been to consider whether the current Accreditation Council for Graduate Medical Education (ACGME) duty hour limits are optimal for resident safety, patient safety, and education. Before determining this, it was important to understand how the current limits came to be, whether implementation of any of the limits has been problematic, how the limits have been monitored, and whether data on adherence to these limits are reliable. Additionally, the committee reviewed duty hour limits in safety-sensitive transportation industries, including the processes for establishing regulations in these industries. Based on this appraisal, the committee makes recommendations in this chapter for future duty hour monitoring practices at the local and national levels.

SETTING DUTY HOUR LIMITS

Resident work schedules first received widespread public attention in 1984 after the death of Libby Zion, an 18-year-old woman treated in the emergency department of a New York hospital for fever and earache. Her family charged that she died due to the poor care by overworked and undersupervised medical residents (no attending physician saw her, although one was consulted by phone). Upon investigation of her death, a grand jury in 1986 exonerated the doctors involved in the case (Ludmerer, 1999) and, instead, faulted the broader system of graduate medical education (New York Supreme Court, 1986):

> the underlying causes of the medical deficient care and treatment in this case might be prevalent in other Level One hospitals . . . the most serious deficiencies can be traced to the practice of permitting inexperienced physicians to staff emergency rooms and allowing interns and junior residents to practice medicine without supervision. . . . Moreover, those patients who are admitted into these hospitals for treatment are often cared for by interns and residents who are not required to have contemporaneous, in person consultations with senior physicians before they initiate a course of treatment. As a consequence, the most seriously ill patients may be cared for by the most inexperienced physicians.

The grand jury concluded that the long duty hours of residents at that time were counterproductive to patient care and to resident learning, and it recommended limitations on resident duty hours and enhanced supervision (Bell, 2003). During further court proceedings in 1995 related to Libby Zion's death, questions continued to be raised about whether duty hours and supervision were the only contributing factors (Andrews, 1995; Douglas, 1995). Regardless of the cause, concerns about the circumstances surrounding her death led to changes in resident scheduling in New York State and ultimately throughout the country. Table 2-1 outlines changes in duty hour limits over the past 25 years.

The New York State Commissioner of Health appointed the New York State Ad Hoc Advisory Committee on Emergency Services, commonly re-

TABLE 2-1 Evolution of Duty Hour Limits

1981	ACGME begins to require "time for rest"
1984	Patient (Libby Zion) dies in a New York teaching hospital, her father claiming she died due to poor care delivered by overworked, fatigued, and inadequately supervised residents
1986	Responding to Zion's death, a grand jury recommends patient care improvements including limits on resident duty hours in New York State
1987	The Bell Commission recommends specific limits on resident duty hours and increased supervision of their work
1989	New York State sets duty hour requirements[a]
1989-1990	ACGME sets an 80-hour limit in several specialties (internal medicine, dermatology, ophthalmology, and preventive medicine) and limits in-house call to every third night with 1 day off in 7 in all specialties, on average
1998	Surprise inspections in 12 New York hospitals find extensive violations of New York duty hour limits
2001	Public Citizen, Committee of Interns and Residents of the SEIU, and American Medical Student Association petition OSHA; Representative Conyers introduces the Patient and Physician Safety and Protection Act of 2001 (not enacted)
2001-2002	ACGME Work Group on Resident Duty Hours and the Learning Environment develops common duty hour limits for all specialties
2003	ACGME requires current common duty hour limits; Representative Conyers and Senator Corzine introduce the Patient and Physician Safety and Protection Act of 2003 (not enacted)
2005	Representative Conyers and Senator Corzine introduce the Patient and Physician Safety and Protection Act of 2005 (not enacted)

NOTE: OSHA = Occupational Safety and Health Administration; SEIU = Service Employees International Union.
[a]New York State Laws and Regulations: Title 10 NYCRR, § 405.4 (1998).
SOURCES: ACGME, 2007b; GovTrack.us, 2005a,b; IPRO, 2007a; Steinbrook, 2002.

ferred to as the Bell Commission, to investigate the conditions of residency. The Bell Commission, named after its chair Bertrand Bell, recommended changes in graduate medical education, including limiting duty hours and improving supervision of residents. In 1989, as a result of the recommendations, New York State implemented rules limiting total resident duty hours per week and the length of extended duty periods, as well as providing for time off from work to address resident fatigue (Table 2-2).[1] Table 2-2 compares New York State rules with subsequent proposals (Public Citizen petition and legislative proposals) and the 2003 ACGME limits established for all residencies nationwide.

Attempts to limit resident duty hours through regulatory or legislative bodies separate from the medical establishment have repeatedly been stopped both in the U.S. Congress and in state legislatures other than New York and Puerto Rico (IPRO, 2007a). The ACGME acknowledges that its "initiative to institute common minimum standards for duty hours unfolded against a political backdrop in which groups pursued federal intervention to regulate resident hours" (ACGME, 2003, p. 1; Steinbrook, 2002). Previously in 2001, Public Citizen petitioned the U.S. Occupational Safety and Health Administration (OSHA) to establish national duty hour limits for residents by arguing that long duty hours are physically and mentally harmful to medical residents (Table 2-2) (Public Citizen, 2001). OSHA turned down the group's petition in 2002 saying, "Because the issues involved with medical resident hours go well beyond job safety and affect hospital patient safety, because other knowledgeable groups are taking action to work on this problem . . . , the Agency has decided to deny your petition" (Department of Labor, 2002).

Several bills to legislate duty hour limits have come before the U.S. Congress since 2001—all have been referred to committee, and none has ever been called up for a vote. The most recent bills, introduced by Representative Conyers and Senator Corzine, are the Patient and Physician Safety and Protection Act of 2005 (H.R. 1228 and S. 1297, respectively). This legislation would have amended title XVIII (Medicare) of the Social Security Act and authorized the U.S. Secretary of Health and Human Services to establish regulations on resident duty hours, supervision, and whistle-blower protections (Table 2-2). The bills would have authorized funding for training facilities to help meet regulations and required fines for nonadherence (GovTrack.us, 2005a,b).

[1]New York State Laws and Regulations: Title 10 NYCRR, § 405.4 (1998).

ACGME 2003 Duty Hour Rules

Beginning July 1, 2003, the ACGME required that duty hours for residents "must" meet the following provisions:

- An 80-hour workweek averaged over a period of 4 weeks, including all in-house calls;
- 1 day in 7 off without any educational or clinical duties or call, averaged over 4 weeks;
- In-house overnight call frequency of no more than every third night, averaged over 4 weeks;
- A maximum onsite duty period of 24 hours with up to 6 additional hours available for didactic education as well as transfer of patients (residents may not take any new patient after 24 hours on duty); and
- Although at-home or pager calls do not count toward the every third night or 24 + 6 hour limit, they "must" not be utilized so frequently that the resident is unable to rest or to have a reasonable amount of personal time (ACGME, 2007b).

Additionally, residents "should" have 10 hours off between shifts for adequate rest (ACGME, 2003). The ACGME used "must" to designate mandatory requirements (ACGME, 2008a). These duty hour limits apply to trainees in their fellowship years as well.

The 80-hour workweek specified in the New York and ACGME rules was not empirically determined. The Bell Commission put forth the 80-hour week using the following heuristic: "There are 168 hours in a week. It is reasonable for residents to work a 10-hour day for 5 days a week. It is humane for people to work every fourth night. If you subtract the 50-hour week . . . from 168 hours, you end up with 118 hours. If you then divide 118 by 4 (every fourth night), it equals 30. If you then add 50 to 30, that equals an 80-hour week" (Bell, 2003, p. 40). Similarly, Dr. Paul Friedmann, co-chair of the later ACGME's duty hours working group said that 80 hours is "a number with some general acceptance, without much scientific underpinning" (Steinbrook, 2002, p. 1298). The ACGME adopted the 80-hour limit to help protect against the sleep loss associated with working long hours (ACGME, 2003). ACGME indicated that the +6 hours was added to the extended duty period to prevent residents from driving home at their circadian nadir and to provide time for learning activities and handover of patient information (ACGME, 2003).

TABLE 2-2 Comparison of Duty Hour Provisions

Duty Hour Provisions	New York Code 405 Rules (1989, updated in 1998)[a]
Maximum hours per week	80 hours, averaged over 4 weeks
Maximum shift length	24 hours + 3 hours for transitional activities
Maximum in-hospital on-call frequency	Every third night, with averaging
Minimum time off between scheduled shifts	8 hours
Mandatory time off duty	24 hours off per week, no averaging
Emergency room limits	12-hour limits in hospitals with more than 15,000 unscheduled visits
Whistle-blower protections	
Enforcement	Civil penalties issued by the state; originally $2,000 per violation; in 2000 raised to $6,000 per item, plan of correction within 30 days; $25,000 penalty for noncompliance with correction plan; additional $50,000 penalty for subsequent noncompliance with correction plan[e]
Funding (for additional staff to make up for loss of resident duty time)	Initially, yes; also funding for compliance monitoring

SOURCES: [a]New York State Laws and Regulations: Title 10 NYCRR, § 405.4 (1998).
[b]Public Citizen, 2001.
[c]ACGME, 2003.
[d]GovTrack.us, 2005a,b.
[e]New York State Department of Health, 2002.

Public Citizen Petition to OSHA (2001)[b]	ACGME Accreditation Standards (2003)[c]	H.R. 1228 and S. 1297 (2005)[d]
80 hours, no averaging	80 hours, averaged over 4 weeks; 88 hours for select programs for a sound educational rationale	80 hours, no averaging
24 hours	24 hours + 6 hours for transitional activities	24 hours + 3 hours for transitional activities (exception for patient emergencies)
Every third night, no averaging	Every third night, with averaging	Every third night, no averaging
10 hours	10 hours	10 hours
24 hours off per week, no averaging	24 hours off per week, averaged over 4 weeks	24 hours off per week, one full weekend off per month; no averaging
	12-hour shift limit, at least an equivalent period of time off between shifts; 60-hour workweek with additional 12 hours for education	12 hours
Yes	Some confidentiality protection in complaint procedure	Yes
Civil penalties sufficiently large to deter violations; unannounced inspections	Potential loss of accreditation; plan of correction	Civil penalties, not to exceed $100,000 per training program in a hospital, with corrective action plans to the Secretary of Health and Human Services; public disclosure on a hospital and residency training program-specific basis
No	No	Yes, to cover hospital incremental costs to comply with regulations

Definition of Terms

ACGME's definition of graduate medical trainee duty hours includes all time spent in "clinical and academic activities related to the program, that is: patient care (both inpatient and outpatient), administrative duties relative to patient care, provision for transfer of patient care, time spent in-house during call activities, and scheduled activities, such as conferences. Duty hours do not include reading and preparation time spent away from the duty site" (ACGME, 2008a). The term *extended duty period* (also known as "long call") is used in this report to refer to the 30-hour (24 + 6) maximum continuous duty period allowed under the 2003 limits (ACGME, 2008a). The term *shift* is applied to any other scheduled period of work, whether during the day, evening, or night. Residents may or may not have any time to sleep during extended duty periods, depending on how busy their service is and the presence or absence of mechanisms for distributing responsibilities to other residents in a "night float" system or to a hospitalist service who will admit new patients or respond to the needs of patients already in the hospital. On a day or night float schedule, residents are not assigned to a single service but float across services or teams to help with admissions and follow-up (PAIRO, 2008). Cross-coverage means being available to care for patients admitted by other residents when the resident who has had primary care responsibility for these patients is not at the hospital or is otherwise unavailable. Other definitions of terms are available in a glossary in Appendix D.

Suggested Refinements to Duty Hours

The committee heard testimony from organizations representing the graduate medical education community, which favored continuation of the current rules over any further reductions in duty hours (AAMC, 2008; ABNS, 2008; ACGME, 2007b, 2008c; ACS, 2008; AMA Resident/Fellow Section, 2007). Other speakers, however, pointed out that resident training in different countries used much shorter workweeks and suggested that the committee consider shortening the total number of hours allowed (Landrigan, 2007; Public Citizen, 2007). Speakers from the surgical community described implementation problems encountered in the United Kingdom as it has reduced hours substantially. Their perception was that fewer hours did not ensure sufficient operative experience to attain the level of competence required for independent practice (ABMS, 2008; ACS, 2008; Royal College of Surgeons of England, 2008). Appendix C contains an expanded discussion of the lessons from the international experience.

The main objection to the 2003 rules raised in testimony concerned the extended duty period of 24 + 6 hours. Public Citizen, the Committee of Interns and Residents (CIR) of Service Employees International Union (SEIU) Healthcare, and Drs. Landrigan and Czeisler advocated that residents work no more than 16 hours straight, including time for transfer of patient care and resident education (American Medical Student Association, 2007; CIR/SEIU, 2007; Czeisler, 2007; Landrigan, 2007). The CIR/SEIU Healthcare also recommended that averaging of duty hours no longer be allowed in order to prevent large variations in the number of duty hours from week to week (CIR/SEIU, 2007).

ADAPTING TO 2003 DUTY HOURS

Have duty hours changed in response to 2003 duty hour limits? The best available evidence shows that mean hours have been reduced over time (Baldwin et al., 2003; Landrigan et al., 2006). Despite progress in reducing overall hours, residents and their residency programs do not always adhere to every aspect of the 2003 ACGME limits (ACGME, 2004; IPRO, 2007b; Landrigan et al., 2006). Certain elements of the rules are more problematic than others, and certain specialties have more problems with adherence than others.

Change in Mean Duty Hours

Mean hours are useful information but they alone do not capture the variation within and across institutions or specialties. Mean duty hours for first-year residents (interns) appear to have declined nationwide in response to duty hour reforms from approximately 83 hours per week in 1998-1999 to 66.6 hours after the 2003 limits. Table 2-3 shows the results from a national survey of a randomly drawn sample of residents (n = 3,493) in the 1998-1999 academic year, with interns from two-thirds of specialties working more than 80 hours (83.3 hours) on average (Baldwin et al., 2003). Second-year residents averaged 76.2 hours per week, with one-third of specialties (all but one a surgical specialty) over 80 hours (Baldwin et al., 2003). Landrigan and colleagues (2006) looked at the duty hours of a national self-selected sample of first-year residents (n = 1,278) from at least 13 different specialties (e.g., internal medicine, pediatrics, psychiatry, general surgery, obstetrics-gynecology [OB/GYN]) using a monthly web-based survey to track duty hours. They found a decline from a mean of 70.7 hours before duty hour reform (2002-2003) to 66.6 hours in the first year of implementation (2003-2004) (Landrigan et al., 2006). The difference in pre-duty hour levels between Baldwin's 83 and Landrigan's 70.7 hours may be due to many programs starting to transition to expected limits even before the ACGME limits became official, although it may also reflect

TABLE 2-3 Average Reported Weekly Work Hours and Percentage of PGY-1 and PGY-2 Residents Working Over Proposed 80-Hour Limit by Specialty, 1998-1999 National Survey

	PGY-1		PGY-2	
Specialty	Mean (SD)	Percent of Residents Working Over 80-Hour Limit	Mean (SD)	Percent of Residents Working Over 80-Hour Limit
Anesthesiology	78.1 (18.1)	44.1	77.7 (16.2)	42.3
Dermatology			59.9 (16.7)	6.9
Emergency medicine	80.1 (17.5)	41.2	71.0 (15.2)	14.3
Family practice	78.1 (16.1)	39.1	67.6 (17.1)	17.9
Internal medicine	83.7 (15.5)	52.4	77.1 (17.3)	35.6
Internal medicine/pediatrics	81.6 (13.9)	46.7	77.5 (16.5)	37.3
Neurological surgery			110.6 (14.4)	100.0
Neurology			82.4 (17.5)	41.0
Obstetrics/gynecology	90.5 (13.7)	69.5	90.8 (17.1)	71.1
Ophthalmology			72.4 (18.9)	23.1
Orthopedic surgery	94.5 (19.1)	75.9	93.8 (16.3)	70.7
Otolaryngology			88.6 (16.5)	57.7
Pathology	60.8 (16.2)	9.4	56.7 (11.2)	5.1
Pediatrics	81.3 (14.9)	43.6	78.1 (14.9)	36.2
Physical medicine/rehabilitation			64.2 (18.6)	27.3
Preventive medicine			58.7 (18.9)	0.0
Psychiatry	69.7 (16.5)	20.4	59.2 (14.9)	7.1
Radiation oncology			67.4 (10.8)	9.1
Radiology			66.5 (14.7)	20.4
Surgery (general)	102.0 (16.1)	89.0	105.7 (13.6)	93.3
Transitional year	80.1 (17.2)	38.2		
Urology			98.5 (19.3)	66.7
Overall	83.0 (17.7)	49.7	76.2 (19.9)	35.1

NOTE: PGY-1 = postgraduate year 1; PGY-2 = postgraduate year 2.
Reprinted with permission by Academic Medicine. Baldwin, D. C., Jr., S. R. Daugherty, R. Tsai, and M. J. Scotti, Jr. 2003. A national survey of residents' self-reported work hours: Thinking beyond specialty. *Academic Medicine* 78(11):1154-1163.

methodological differences in sampling and recall period for hours worked. No other national study is available since the 2003 reforms on mean duty hours across such a spectrum of specialties; studies tend to be specialty or institution specific.

Landrigan et al. (2006) also noted a significant decline in the mean length of extended duty periods from 32.1 to 29.9 hours for interns in multiple specialties. In a subsequent analysis of the change of pediatric residents' extended duty hours in three institutions, Landrigan et al. (2008)

found a significant decline in the mean from 29.3 ± 3.2 hours before the 2003 rules were implemented to 28.5 ± 2.4 hours afterward.

Degree of Compliance with Current Limits

The responsibility for ensuring that residents and institutions adhere to the 2003 duty hour standards falls to institutions themselves and the ACGME as part of its announced accreditation visits. The ACGME accreditation review occurs once every 1-5 years, or once every 3.7 years on average (ACGME, 2008b). After 10 years of experience with the 80-hour workweek, New York State mandated yearly, unannounced audits of its training institutions because surprise inspections in 1998 found widespread violations: 94 percent of residents in New York City and 37 percent of those throughout the rest of the state worked more than 85 hours per week, and 77 percent of surgical residents in New York City and 60 percent in the rest of the state worked more than 95 hours per week. Further 38 percent of all residents and 67 percent of surgical residents worked more than 24 consecutive hours. Emergency room residents were in compliance with their 12-consecutive-hour limits (DeBuono and Osten, 1998; Kennedy, 1998). To achieve annual review, the New York Department of Health has contracted with IPRO since 2001 to focus solely on monitoring duty hour compliance (IPRO, 2008). IPRO, an independent, not-for-profit healthcare and quality improvement organization, is the New York Medicare Quality Improvement Organization and its Medicaid Utilization Review/Quality Assurance Agent. All 124 New York teaching hospitals have monitoring by both IPRO and ACGME.

Substantial Compliance Versus Counting Each Violation

ACGME and IPRO both use what they term a "substantial compliance model" for monitoring duty hours. Using this model, a program will not receive a citation for single or isolated violations of duty hour rules. The violations need to be more systemic. ACGME examines programs more closely if 15 percent or more residents report violations on three or more standards through ACGME's yearly resident survey. IPRO also uses a threshold of 15 percent for resident nonadherence to a single duty hour rule before giving a citation. IPRO does not trigger a violation at precisely the hour limit. Instead 15 percent of residents would have to be over an 85-hour week average, for example, to trigger a violation or 15 percent would have to be over 28 hours on an extended duty period (New York limit is 24 + 3 hours).[2] Various accreditation bodies (e.g., The Hague Accreditation

[2]Personal communication, V. Wilbur, IPRO, June 9, 2008.

and Approval Standards) use such substantial compliance models (Council on Accreditation, 2008).

Both the ACGME and IPRO monitor residents and fellows in all years of graduate medical training and annually report duty hour violations based on substantial compliance. Annually, ACGME surveys residents across the country by asking about their compliance with duty hours within the past week. When on site, ACGME looks back at a longer period, and IPRO reviews the previous 3 months of records. Their reported rates are not directly comparable because ACGME reports violations by specialty program and by residents while IPRO reports by facility. In academic year 2006-2007, 8.8 percent of the residency *programs* reviewed by ACGME received one or more citations for a violation of "any rule" related to duty hour limits (Table 2-4A) (ACGME, 2007a). In its 2006-2007 contract year, IPRO found that 16 percent of 124 New York teaching *facilities* had violated "any rule" (IPRO, 2007b). Since initiating duty hour monitoring, 46 percent of sponsoring *facilities* have received a duty hour citation from ACGME for one or more of their *programs* (ACGME, 2008c).

Most research studies on duty hours include counts of every violation reported by residents and do not use a substantial compliance threshold. Thus, these studies can be expected to report higher levels of duty hour violations than the two monitoring organizations. One study found that for the first year of duty hour rule implementation (academic year [AY] 2003-2004), 83.6 percent of interns, 85.4 percent of residency programs, and 90.8 percent of teaching facilities had a violation of "any rule" during at least 1 month of the year (Landrigan et al., 2006). While it is clear from this work that duty hour violations are common, these especially high rates of nonadherence might be accounted for in several ways: respondents were first-year residents (interns typically work more hours than residents in other years of training), the data are from the first year that duty hour rules were implemented, the data were collected monthly covering 11 months rather than a more limited period examined during an accreditation or audit visit, and no threshold is applied.

Violations of Specific Duty Hour Rules

Certain duty hour rules have been more difficult to adhere to than others, chiefly those limiting duty periods to 30 hours and requiring 1 day off in 7. Tables 2-4A and 2-4B present data on violation rates for breaking "any rule" as well as each of the components. These data are most useful in determining which duty hour limits are most difficult to adhere to across multiple monitoring systems rather than determining which monitoring system discovers the most violations. The basis for reporting differs by source, with violation rates reported by "facility" (IPRO, 2007b; Landrigan et al.,

TABLE 2-4A Comparison of Reported Duty Hour Violation Rates by Facility and Program

Rule and Source of Data	Pre-2003[a] (%)	2003-2004 (%)	2004-2005 (%)	2005-2006 (%)	2006-2007 (%)
Rates Reported for Facilities (% of facilities)					
Any rule violation					
IPRO	46, 42	21.0	13.0	17.0	16.0
Landrigan et al. (2006)		90.8			
80-hour rule violation					
IPRO	28, 10	0.0	2.0	0.0	0.0
Landrigan et al. (2006)		81.8			
Extended continuous hour shift					
IPRO	45, 32	15.0	5.0	18.0	12.0
Landrigan et al. (2006)		79.8			
Average 1 day off in 7					
IPRO	14, 5	3.0	4.0	8.0	10.0
Landrigan et al. (2006)		63.6			
Separation between shifts					
IPRO	18, 13	4.0	2.0	1.0	1.0
Rates Reported for Programs (% of programs)					
Any rule violation					
ACGME[b]		5.0	7.3	7.9	8.8
Landrigan et al. (2006)		85.4			
80-hour rule violation					
ACGME (15% of residents)				2.0	
ACGME (any resident)				9.8	
Landrigan et al. (2006)		69.7			
Extended continuous hour shift					
ACGME (15% of residents)				9.5	
ACGME (any resident)				20.9	
Landrigan et al. (2006)		70.2			
Average 1 day off in 7					
ACGME (15% of residents)				33.4	
ACGME (any resident)				45.9	
Landrigan et al. (2006)		50.9			
Separation between shifts					
ACGME (15% of residents)				12.0	
ACGME (any resident)				27.5	
Average call no more than every third night					
ACGME (15% of residents)				10.5	
ACGME (any resident)				25.0	

[a]IPRO percentages in this column represent values for the first 2 years of its contract with New York State.
[b]Percentage of ACGME site-reviewed programs; other ACGME data come from resident surveys.
SOURCES: ACGME, 2008f; IPRO, 2007b; Landrigan et al., 2006.

TABLE 2-4B Comparison of Reported Duty Hour Violation Rates by Residents

Rule and Source of Data	Pre-2003 (%)	2003-2004 (%)	2004-2005 (%)	2005-2006 (%)
Rates Reported for Residents (% of Residents)				
Violation of any rule by residents				
ACGME				
Landrigan et al. (2006)		83.6		
Violation of the 80-hour rule				
Baldwin et al. (2003), PGY-1s	49.7			
Baldwin et al. (2003), PGY-2s	35.1			
ACGME				2.2
Landrigan et al. (2006)		43.0		
Extended continuous hour shift				
ACGME				7.6
Landrigan et al. (2006)		67.4		
Average 1 day off in 7				
ACGME				15.3
Landrigan et al. (2006)		43.7		
Separation between shifts				
ACGME				5.2
Average call no more than every third night				
ACGME				5.7

SOURCES: ACGME, 2008f; Landrigan et al., 2006.

2006), "program" (ACGME, 2004, 2006, 2007a; Landrigan et al., 2006), or "resident" (ACGME, 2004, 2006, 2007a; Baldwin et al., 2003; Landrigan et al., 2006). In addition, ACGME released data for AY 2005-2006 on what the "program" violation rate would be if it counted every resident-reported violation or applied a 15 percent threshold (Table 2-4A). Where data are available, Table 2-4A includes compliance rates for each component by "facility" or "program" and Table 2-4B by "residents."

The 80-hour rule is more often adhered to than other limits, based on ACGME and IPRO reporting (Tables 2-4A and 2-4B). IPRO reports that it seldom finds excessive violations of the 80-hour rule any more in New York, although some flexibility is factored into its monitoring as noted earlier.[3] Similarly, ACGME data show few program or resident violations of the 80-hour limit when the substantial compliance threshold of 15 percent is applied. Even when every resident is counted, the 9.8 percent program

[3]Personal communication, V. Wilbur, IPRO, January 30, 2008.

violation rate is lower than the violation rates for other limits (Tables 2-4A and 2-4B). Only 2.2 percent of residents report violating the 80-hour rule in the past month on the ACGME resident survey, less than reported violations for other limits (Table 2-4B). Landrigan et al. (2006) reported that 43 percent of first-year residents violated the 80-hour limit in the first year of implementation, and other institution-specific and specialty-specific reports, that also do not use a substantial compliance threshold, show variable rates of violation for the 80-hour week including some that were quite high (e.g., 16-94 percent of residents) (Carpenter et al., 2006; Jagsi et al., 2008; Lin et al., 2006; Reiter and Wong, 2005). Two studies that examined trends over time found decreasing levels of violations of the 80-hour week (Jagsi et al., 2008; Landrigan et al., 2006).

The elements of the 2003 duty hour limits that provide opportunities for recovery from fatigue (days off per week, separation between shifts, limiting frequency of call) and limit consecutive hours on duty have had higher violation rates than the 80-hour limit by all measures (Tables 2-4A and 2-4B). In 2005-2006 according to ACGME data, 15.3 percent of residents went without 4 days off in a month and 45.9 percent of programs had at least one resident without the required days off (Tables 2-4A and 2-4B). Even in closely monitored New York, 8-10 percent of facilities failed to always deliver the days off (Table 2-4A).

Over time, the extended duty period has had the highest violation rate of any limit in New York facilities. ACGME found fewer, but still frequent, violations of the long duty period than of providing mandatory days off (Tables 2-4A and 2-4B). Other institution-specific or specialty-specific studies show greater adherence problems relative to other limits than either the ACGME or the IPRO data indicate for the 30-hour extended duty period limit. For example, at one major training center in 2005, 85 percent of medical and general surgery residents reported violations of the 24 + 6 hour limit compared with 65 percent in violation of the 80-hour limit and 28 percent in violation of the 1 day off in 7 rule (Carpenter et al., 2006). When asked if they "always" comply with a rule, 50 percent of otolaryngology surgery residents surveyed across the country reported violating the 30-hour limit, 39.5 percent reported violating the 80-hour workweek (averaged over 4 weeks), 30 percent reported missing their 1 day off in 7, and 66 percent reported not having the proper separation between shifts (Reiter and Wong, 2005). Rates of violation of 30-hour extended duty periods were more likely on inpatient wards, intensive care rotations, and surgical rotations (Cull et al., 2006; IPRO, 2007b; Landrigan et al., 2006).

Jagsi et al. (2008) reported improved compliance with the 30-hour limit after the 2003 limits in 76 programs at two institutions. Their analysis compared programs that made substantial reductions in their total workweek hours (reduced-hours programs) to those that did not (other programs).

Nonadherence with the extended duty period of 30 hours was reduced from 40.8 percent to 11.4 percent in the reduced-hours programs and from 12.6 percent to 5.0 percent for other programs. As noted earlier, Landrigan and colleagues have found the mean length of the extended duty period to decrease since the 2003 rules (Landrigan et al., 2006, 2008).

Provisions for time off between shifts and call frequency have similar violation rates according to ACGME data. IPRO reports fewer problems maintaining the proper separation between shifts, perhaps because New York requires that time off "must be" 8 hours long, whereas ACGME rules recommend that time off "should be" 10 hours. This is illustrated by a report from one surgical program in New York reporting 98 percent adherence to IPRO regulations but only 88 percent adherence to ACGME limits (Goldstein et al., 2005). The use of "should" with respect to this provision of ACGME rules while its other duty hour rules use "must" has caused confusion in the extent to which it must be followed. ACGME is conducting pilot tests on whether to change this rest requirement to "must be 8 hours" (ACGME, 2008f).

Compliance by Year of Training and Specialty

Year of training and type of specialty both influence duty hours worked and the potential for violation of duty hour limits. These observations raise questions about whether the same limits should apply across all specialties or years of training.

Compliance Across Specialties

Before implementation of the 2003 duty hour limits, there was great variability in total hours worked by different specialties, but even when mean duty hours were less than 80, on average, for a given specialty, a large percentage of its residents would still have been in violation of that limit at some point in the year (with the exception of those in pathology, dermatology, psychiatry, and preventive medicine) (Table 2-3) (Baldwin et al., 2003). In 2002, program directors anticipated greater relative difficulty for surgical programs in adhering to the duty hour changes, and since 2003, surgery programs have had to reduce the duty hours of residents by 20 percent or more (e.g., neurosurgery second-year residents averaged 110.6 hours per week before 2003) to meet the 80-hour limit (Baldwin et al., 2003; Brotherton et al., 2002; Lieberman et al., 2005). Only half (49 percent) of surgical residents compared with three-fourths of medical residents (73 percent) were expected to be in compliance with the proposed 2003 limits (Lieberman et al., 2005).

Studies on adherence since initiation of the 2003 rules vary whether

TABLE 2-5 Duty Hour Violations in New York State by Specialty (2006-2007)

	Violation of Any Duty Hour Limit (%)	Violation of Extended Duty Period of 24 + 3 Hours (%)	Violation of 1 Day Off in 7 (%)
Statewide	16	12	10
Specialty			
Anesthesiology	0	1	2
Emergency department	0	0	1
Family practice	5	4	4
Internal medicine	7	37	33
OB/GYN	0	5	9
Pediatrics	7	10	10
Surgery	13	44	42

SOURCE: IPRO, 2007b, Tables 36 and 40.

they identify worse duty hour violations among surgery programs than other specialties. IPRO data show that surgery, pediatrics, internal medicine, and family practice programs in FY 2006-2007 still had violations of some aspect of duty hour rules even after years of intensive monitoring, with surgery having almost twice the rate of nonadherence to "any rule" as the others (Table 2-5) (IPRO, 2007b). Landrigan and colleagues (2006) reported that programs with interns in internal medicine, pediatric programs, emergency medicine, and even psychiatry were equally likely to violate some duty hour rule as surgical programs for at least 1 month in the year.

In New York, IPRO has found that internal medicine and surgery programs are almost equally noncompliant with the 24 + 3 hour extended duty period and the day off per week, at rates three or more times greater than the other specialties (Table 2-5). ACGME data over time show a high portion of surgical programs cited for extended duty period violations, but frequent citations are also found in family medicine, internal medicine, pediatrics, and transitional year programs (ACGME, 2004, 2006, 2007a).

ACGME grants exemptions to the 80-hour workweek rule that allow up to a maximum workweek of 88 hours, after determining a program has a "sound educational rationale." The number of programs receiving exemptions decreased from 68 in 2004-2005 to 40 in 2007-2008: 34 of the 40 in 2007-2008 are in neurological surgery, 6 in thoracic surgery, and 1 in general surgery (ACGME, 2007a). Thus, approximately 40 percent of neurosurgery programs have an exemption from the 80-hour-a-week limit. Neurological surgery programs still have programs cited, although ACGME citation reports do not distinguish between programs that have the 88-hour limit and those that do not (ACGME, 2007a).

Compliance by Year of Training

First-year residents typically work longer hours than residents in other years, according to a 2007 ACGME resident survey and other studies (ACGME, 2008c; Baldwin et al., 2003; Carpenter et al., 2006). Interns have the most to learn and take longer to accomplish tasks while they are learning not only new medical information but how to work efficiently in the training environment. Learning how to manage one's time is an important part of the first-year experience. Among interns, 9 percent are not "always or usually" in compliance with the 80-hour workweek, compared to 4.6 percent of residents in later years, and 11 percent of interns are not "always or usually" in compliance with the 24 + 6 shift length compared to 6 percent of residents in later years (ACGME, 2008f).

Underreporting of Violations by Residents

Residency programs monitor resident duty hours in a variety of ways, including by self-report on time sheets, telephone or computer log-in/log-outs in the hospital, badge readers at entries and exits of hospitals or parking garages, and personal monitoring by program directors to ensure that residents are not still on the floor outside of duty hour limits (Asad et al., 2006; Chao and Wallack, 2004; Goldstein et al., 2005; Landrigan et al., 2008). The degree of resident participation determines the quality of information from any system, and the ease of use and the degree of monitoring and enforcement activities all influence resident participation (Chang et al., 2006; Chao and Wallack, 2004).

Testimony before the committee and other reports revealed that some residents have underreported the extent of their duty hours (American Medical Student Association, 2007; Arora et al., 2006; CIR/SEIU, 2007). Residents give multiple reasons for failing to report duty hours accurately. One rationale given for residents being disinclined to call ACGME's attention to duty hour violations is because this could lead to probationary accreditation for their program or loss of its accreditation altogether (AMA Resident/Fellow Section, 2007; CIR/SEIU, 2007). Residents perceive that graduation from a program that is on probation or without accreditation may hurt their chances when they seek employment. An American Medical Association (AMA) survey of residents in 2005 revealed that 7 in 10 residents know how to report excessive duty hours but half would be uncomfortable actually reporting them (AMA Resident/Fellow Section, 2005). A subsequent AMA survey (2006) found that some residents experience intimidation from attending physicians, senior residents, and fellows (AMA, 2006). These AMA survey data are not based on a representative response from residents and fellows. Such intimidation may be subtle—or at times not so subtle—cultural expectations that long hours are necessary and should be borne without complaint (Arora et al., 2008; CIR/SEIU, 2007).

Residents give other reasons for underreporting hours (Lamberg, 2002), such as wanting to experience more cases, not wanting to call attention to the fact that they do not work as fast as others, or knowing that everyone is overburdened by workload demands. Others attribute underreporting to the desire to be responsive to patient care needs and not wanting to be considered unprofessional if not following through in the care of a sick or unstable patient (Associated Press, 2003) or missing other patient-related activities such as meetings with families (Fletcher et al., 2008). A survey of internal medicine, pediatric, and general surgical residents at one major teaching center found that 85 percent exceeded duty hour limits at least once in the previous 3 months and 48 percent admitted underreporting their hours (Carpenter et al., 2006). Eighty percent of the residents noted that their concern for patient care was the greatest motivation in working the extra hours. Similar sentiments of not wanting to leave their patients' care to someone else were echoed in statements of residents interviewed in closed session by some members of the committee (Resident Panel, 2008). Carpenter et al. (2006) expressed concern for the ethical dilemma facing residents: professionalism and care for their patients drive residents to exceed duty hour limits, but they then must act unprofessionally by falsely reporting their time to avoid negative consequences for their program.

DUTY HOURS IN SELECTED INDUSTRIES

This section presents a brief overview of the hours of service and the nature of rule making in certain safety-sensitive transportation industries, comparing the length of duty hours for residents and other workers. There is considerable variation among transport modes in federal work and rest requirements. A more extensive review of this topic can be found in an article by Rogers in the Institute of Medicine (IOM, 2003) report *Keeping Patients Safe: Transforming the Work Environment of Nurses.*

Hours of Service per Week and per Shift

The 80-hour workweek limit for residents, although long, is not substantially different from those of some transportation industries that also have worker and public safety concerns. In examining the history of setting weekly limits on work time, one finds that the older the industry (e.g., trains and maritime vs. commercial trucking and aviation), the longer are its allowable hours in a workweek (Table 2-6). There is no limit on the total workweek for railroad conductors, and shipboard personnel on tank-

TABLE 2-6 Federally Mandated Work Hour Limitations for U.S. Transportation Modes and ACGME Duty Hours

Industry	Weekly Limitations (h)	Limits of Single Shift Duration (h)	Minimum Rest Between Shifts (h)	Minimum Rest Period Given Weekly (h)	Regulatory Agency	Enforcement
Airline pilots[a,b]	30 h of flying time in any 7 consecutive days[a]; 32 h of flying time in any 7 consecutive days[b]	8 h of flight time per 24 h	11 h of continuous rest in the 24 h prior to 9+ h of scheduled flight time[c]	None	Federal Aviation Administration (FAA)	FAA
Shipboard personnel on tankers[d]	84 h per week	15 h per every 24 h and 36 h per 72 h	None	None		
Railroad conductors[e,f]	None	12 h	10 consecutive hours after a 12 h shift *and* 8 consecutive hours during the 24 h prior to any shift	None	Federal Railroad Administration (FRA)	Currently hours are recorded by hand; 4 major railroads have upgraded to electronic recordkeeping
Long-haul truck drivers[g]	60 or 70 h driving time per 7- or 8-day shift	14 h on duty with a maximum of 11 h spent driving	10 consecutive hours, drivers with sleeper berth must spend minimum of 8 consecutive hours in berth and 2 h in berth or off-duty in any combination	34 h continuous rest period prior to any 7- or 8-day working period	Federal Motor Carrier Safety Administration (FMCSA)	Drivers are required to record a log of hours for each 24 h period, including a record of the prior 7 days. Record can be electronic or handwritten, depends on motor carrier[h,i]

Medical residents[j,k]	80 h per week averaged over 4 weeks	24 h + 6 h transition time	10 consecutive hours (recommended but not required)	One continuous 24 h rest period per week	ACGME	ACGME

[a]Federal Aviation Administration. 2008. *Title 14: Aeronautics and Space, Part 121,* Subpart Q, § 121.471 (b):3 and § 121.471 (a):3. This regulation applies to domestic flights that occur within the contiguous 48 states of the United States or the District of Columbia, or entirely within any State, territory, or possession of the United States.

[b]Federal Aviation Administration. 2008. *Title 14: Aeronautics and Space, Part 121,* Subpart R, § 121.481 (d) and § 121.483 (a). This regulation applies to flights that occur between Alaska, Hawaii, or any territory or possession of the United States and any point outside these locations; or between any point within the 48 contiguous States of the United States or the District of Columbia and any point outside these locations; between any two points outside the United States.

[c]If a pilot receives less than 9 h of rest during a 24 h period, he or she must be compensated with at least 10 continuous hours of rest that begins no later than 24 h after the commencement of the reduced rest period.

[d]United States Coast Guard Marine Safety Manual, Chapter 24, Section C.

[e]Federal Railroad Administration. 2008. Title 49: Transportation, Subtitle V—Rail programs, Part A, Chapter 211—hours of service, § 21103.

[f]Federal Railroad Administration, 2008.

[g]Federal Motor Carrier Safety Administration. 2005. Federal motor carrier safety regulations 49 CFR, Part 395: Hours-of-service change.

[h]Federal Motor Carrier Safety Administration. 2008. Part 395: Hours of service of drivers, § 395.8.

[i]Federal Motor Carrier Safety Administration. 2008. Part 395: Hours of service of drivers, § 395.15.

[j]ACGME, 2007a.

[k]Not a federal mandate; guidelines are set by ACGME.

ers have an 84-hour week. Only airline pilots have a significantly shorter workweek, but the limit is based on only one type of work (e.g., pilots' work on non-flying activities is not counted against the 30-hour weekly limit). Statutes and regulations often establish different duty hour limits for different types of work (e.g., truckers can work on non-driving activities for 3 hours beyond their daytime driving time limits). Paperwork is frequently done in all transportation modes outside duty hour limits.

The allowable length of a single shift is considerably shorter in these other industries than the 24 + 6 hour extended duty period for residents (Table 2-6). Some medical and surgical specialties view the extended duty period as necessary to obtain unique patient care learning experiences. To preserve this aspect of residency training, while acknowledging that residents have the same physiological needs for rest as other human beings, the committee examined ways to prevent and mitigate acute sleep deprivation when residents have extended duty periods (see Chapter 7).

Need for Modernization Based on Sleep Science

Although federally mandated hours of service still rely on a model that assumes the length of work time is the factor most relevant to fatigue, this is only one component of the relationship of fatigue to risk. Other factors can include the time of day work occurs in relation to one's circadian rhythm, the volume and intensity of work, and the amount of sleep obtained (Dinges, 1995; Drake et al., 2004; Folkard et al., 2005; Rosa, 2001; Van Dongen and Dinges, 2005). These factors are detailed in Chapter 7.

Over the past century, federally mandated hours of service (HOS) for aviation, trucking, railroad, and marine workers have not kept pace with the extensive science on the biological causes, consequences, and prevention of fatigue, prompting the National Transportation Safety Board to urge repeatedly that the relevant regulatory agencies set working hour limits based on fatigue research, circadian rhythms, and sleep and rest requirements (NTSB, 2007). Despite this, federal HOS in these industries have remained largely unchanged for decades and are seen as either "antiquated" (e.g., railroad; Boardman, 2007) or inadequate as prescriptive rules because they do not permit the operational flexibility increasingly required in modern systems (e.g., commercial aviation; Gilligan, 2007).

Once promulgated as either statutes or regulations, hours of service in transportation modes have proven remarkably difficult to revise to incorporate new scientific evidence on the biological causes of fatigue, due to lack of political consensus and legal challenges to even the smallest changes

(Boardman, 2007).[4] As a result, federally mandated hours of service are often seen as a barrier to a modern evidenced-based approach to preventing fatigue in many industries. Moreover, it is increasingly recognized that even meeting the nominal requirements of current HOS rules is not sufficient to effectively manage fatigue. Additional efforts involving organizational commitment and allocation of resources for establishing and sustaining fatigue management are necessary (McCallum et al., 2003). Thus, the committee concludes that one goal for its recommendations is to combine scientifically based duty hour limits with adequate adherence.

MONITORING DUTY HOURS

The level of adherence to resident duty hour limits has raised questions about the current approach to monitoring duty hours and whether the culture of expectation, if not overt intimidation, results in pressure on residents to work more than their assigned hours (AMA, 2006; Arora et al., 2008; CIR/SEIU, 2007). All hospitals, including teaching hospitals, are under pressure to increase revenue and manage their costs (Weissman et al., 2007). In a teaching environment, residents are relatively low-cost personnel available to handle increasing admissions (AAMC, 2007; Kozak et al., 2006). Achieving the correct balance between providing service and meeting educational goals has been a long-standing issue since the first report on graduate medical education was issued in the 1940s (Ludmerer, 1999).

ACGME Review of Adherence to Duty Hour Limits

The ACGME metric is that a program must have at least substantial compliance with accreditation standards for institutions and programs, including duty hour limits. A program once cited for deficient educational practices or duty hour violations can remain accredited while the problems are remediated if the ACGME judges that these do not immediately jeopardize the overall performance of the program. In AY 2006-2007, ACGME issued 8,804 citations for the 2,589 programs under accreditation review; most citations (54 percent) were for educational deficiencies and only 2.9 percent were for duty hour compliance issues (ACGME, 2007a). Serious educational deficiencies might include the following: (1) less than 50 percent of an internal medicine residency program's graduates pass the American Board of Internal Medicine exam on the first try, (2) the pediatric inpatient population lacks sufficient diversity and complexity for adequate training,

[4] *Public Citizen, et al., v. Federal Motor Carrier Safety Administration*, 374 F.3d 1209, 362 (U.S. App. D.C. 384).

or (3) residents perform an insufficient or excessive number of orthopedic procedures (ACGME, 2008b,e).

The sanctions that ACGME currently applies involve program probation or withdrawal of accreditation. The adverse action rate for ACGME is about 8 percent (this includes probationary accreditation, withdrawal of accreditation, withholding of requests for new program accreditation, and forced reductions in resident complement); most actions result in programs being placed on probation.[5] During AY 2007-2008, 10 programs had their accreditation withdrawn or were on track for expedited withdrawal; 3 of these had duty hour violations.[6]

Training facilities must maintain systems and documentation (e.g., rotation schedules, call rosters, sign-in/sign-out systems) to assure ACGME that staff hours are under the required limits. ACGME looks at this documentation during scheduled onsite visits and analyzes responses to its annual resident survey to see what residents report about compliance with duty hours, including whether there is any undue pressure to work more than the required hours. Typically, the annual survey involves half of the residents in the country. ACGME interviews another 12,000 or so residents during site visits (ACGME, 2008c). Complaints about training program quality including violations of duty hours can be made to ACGME, but those with a complaint are directed before filing a formal complaint to discuss the issues with the local program director or the supervising institutional official responsible for all graduate medical education (ACGME, 2004, 2007a). If the issue detailed in the complaint is egregious enough, there will be an immediate site visit (ACGME, 2007c).

Institutions pay ACGME $2,750-$3,500 per residency program for the accreditation review (including assessment of duty hours). If an academic medical institution has 20 programs, it could cost $54,000 to $70,000 (ACGME, 2008d).

New York State Monitoring

Unlike ACGME's scheduled accreditation visits, IPRO reviews are unannounced. The review team examines 3 months of schedules and further validates that residents are staying within the limits by collecting about 9 days of detailed data through interviews, direct observation and review of chart notes, operating room logs, clinic records, and test orders. This intensive audit is designed to ensure that residents are not coming in before their scheduled hours to prepare patient data for rounds or procedures, or staying after their hours. For example, a facility schedule may say that a

[5]Personal communication, Ingrid Philibert, ACGME, July 22, 2008.

[6]Personal communication, Ingrid Philibert, ACGME, August 2, 2008.

resident starts at 7 a.m. but he or she actually comes in at 5 a.m. to prepare for rounds, or the schedule says the resident is off on Saturday but other documents show the resident was in the operating room that day. An IPRO team will be onsite for 1 to 2 weeks. The IPRO overall sample size of a facility's resident trainees is 50-60 percent, with a nearly 100 percent sample in areas that have proven to have more violations over years of review—surgery and intensive care units.

After the audit, IPRO presents facilities with a detailed exit report listing any deficiencies to make sure violations are clearly documented by time and dates. Facilities must submit a plan for correction within 30 days of being notified of deficiencies. IPRO tries to work with facilities to resolve issues before the state gets to the stage of fining the hospital. New York State imposes fines for residency programs that persist in noncompliance ($6,000 for a first offense escalating to $50,000 for a third offense). When a deficiency persists, facilities are not able to receive any certificate-of-need approvals from the state (e.g., to expand capacity).[7]

In New York, the state, not the institution, pays for the reviews of duty hour compliance. For the IPRO contract year 2008-2009, the cost for duty hour review alone is $2.9 million annually, averaging about $24,000 for each of the 124 hospitals under review.[8] New York State has approximately 15 percent of the graduate medical trainees in the country (IPRO, 2007b). For 2006-2007, 16 percent of the facilities in New York State had violations of some duty hour rule (IPRO, 2007b).

Future Approach to Monitoring Hours

The committee concludes that violations of duty hours are frequent and underreported and that more intensified monitoring is necessary immediately to ensure adherence. Achieving adherence to existing duty hour rules is an established and essential first step to which stakeholders have already agreed. The next step would be adherence to the committee's recommended duty hour parameters. The committee considered carefully whether ACGME should remain the body that sets and monitors duty hour limits, or whether an alternative organization and approach are warranted. The goals of the committee are to have an effective monitoring process under a substantial compliance model and documentation of when and why violations occur in order to guide institutions in reconfiguring their scheduling and workload and provide a better understanding of the circumstances when exceptions to duty hour limits might be permitted.

The main monitoring alternatives include: (1) the status quo, con-

[7]Personal communication, V. Wilbur, IPRO, January 30, 2008.
[8]Personal communication, V. Wilbur, IPRO, January 30, 2008.

tinuing with the ACGME and its current approach, (2) a new agency to conduct the monitoring, or (3) the ACGME with changes. The committee concluded that neither of the first two options was acceptable. The committee preferred the third alternative, strengthening the ACGME procedures and providing additional oversight and evaluation by other organizations to ensure that duty hours are considered in the context of quality improvement and patient safety. Below the committee discusses the advantages and disadvantages of each option.

Alternative 1: The Status Quo

The ACGME's approach to monitoring duty hours through site visits and national surveys of residents is described earlier in this chapter. Several advantages have been cited supporting the continuation of ACGME's current monitoring procedures. ACGME would be able to respond more easily and quickly than a new organization to implement the committee's recommendations since it already has a monitoring process in place, trained field staff, and relationships with all the training institutions and programs. It would not require new legislation. Also, ACGME could adapt readily if future scientific evidence prompts fine tuning or adjustments to the duty hours that the committee now recommends without the time needed to enact legislation and develop governmental regulations.

ACGME's current monitoring of duty hours is a relatively inexpensive add-on because it is embedded in its overall accreditation and survey processes, onsite monitoring happens infrequently for individual residency programs, and it uses volunteers to a large extent, a financial advantage. Additionally, both the ACGME and the AAMC advocate a continued role for ACGME in monitoring as well as establishing duty hours, and training institutions seem more comfortable with the status quo, also (AAMC, 2008; ACGME, 2008c).

Since ACGME reviews the quality of the residency programs seeking their accreditation, it can readily integrate duty hour compliance data with assessments of educational programs. For example, when ACGME was reviewing a surgical program that wanted to expand the size of its training program, it discovered duty hour violations and would not allow expansion because of a pattern of such violations (Kowalczyk, 2008). Additionally, ACGME presented to the committee national data on the positive correlation between substantial violations of duty hours and other undesirable educational program characteristics (problems with faculty, teachers, resident intimidation, excessive service obligations) as reported by residents ($p = .0001$). Residents in these 115 outlier programs (3.8 percent) report that they are less likely to participate in scholarly activities (38 percent vs. 58 percent) and more likely to be required to provide support services (45 percent vs. 22 percent). Resident survey data from previous years show that 91 percent of programs with educational program citations are also in the most noncompliant quartile for duty hours (ACGME, 2008c).

ACGME now uses a substantial compliance threshold for assessing adherence to duty hour limits, so that no program is punished for isolated, individual incidents when there is not a pattern of abuse. This indicates an appropriate effort to focus monitoring attention on the more serious cases and an attempt to avoid unfairly punishing programs for isolated events.

There are disadvantages associated with continued ACGME monitoring. Its monitoring processes have not been effective in bringing adherence to the duty hour limits that have been in place since 2003. Suboptimal adherence to current limits means their expected positive effect on resident fatigue and patient safety may be less than anticipated to date, making it difficult to assess the national impact of the 2003 duty hour rule changes on patient outcomes, as discussed in Chapter 6.

ACGME assurances that reasonable duty hour limits have been set and are being followed and that ACGME can detect the full extent of violations have been met with skepticism. Violation data from multiple other sources find higher levels of violations although there are methodological differences in reporting and timeframe that may account for some of the differences. Some groups suggest the membership of ACGME[9] has been slow to accept duty hour limits and that some members may have a conflict of interest in enforcing limits on resident duty hours because of the costs of replacing resident labor with other personnel (CIR/SEIU, 2007; Sleep-deprived doctors, 2002).

The long time between accreditation visits (e.g., 3.35 years even for programs with citations) (ACGME, 2008c) leaves opportunity for duty hour violations to escalate between visits. The current average time between visits can be longer than a complete residency period for some trainees. When the monitoring visit does occur, it is announced by ACGME and expected by the training institution, which also diminishes its value in uncovering problems.

The current ACGME procedures for residents to report violations of the duty hour limits and undue pressure to work beyond the limits are a deterrent to whistle-blowing since the residents are expected to report through their residency program director first, before taking a complaint to the ACGME. Residents are concerned that the process will identify them to their program directors or senior colleagues, whose recommenda-

[9]American Board of Medical Specialties, American Hospital Association, American Medical Association, Association of American Medical Colleges, Council of Medical Specialty Societies.

tions can determine a resident's opportunities for fellowship training or employment.

Current ACGME monitoring data and studies that simply report violation rates are insufficient for policy purposes. They do not provide a complete picture of actual hours worked and when exceptions to the rules might be necessary for educational or patient safety reasons. Aside from limited national data on mean hours worked in a week and consecutively by interns in 2003-2004, the committee does not know for certain how many hours over or under the time limits residents are working nationally and by specialty, or the frequency of different reasons that might push residents over the time limits (e.g., unstable patient, lack of system supports, workload). Such data from a nationally representative sample of institutions and specialty programs would have helped the committee determine whether there are other reasonable adjustments to duty hours that would help achieve the training goals of each specialty and provide safe working conditions for residents and patients, beyond the committee's recommendations in Chapter 7.

The status quo is unacceptable to the committee, although ACGME is an attractive option because it links the education and duty hour policy development and monitoring. Many of the shortcomings of the ACGME monitoring process could be corrected with some additional effort and resources. Also, neutral organizations could provide oversight of the ACGME process to provide assurance to the public, patients, and residents.

Alternative 2: A New Organization

An organization unrelated to ACGME, such as a government or an independent agency, could take over the responsibility for duty hours monitoring. This might take the form of an existing organization new to resident hours monitoring, a newly formed and purpose built organization, or one with experience monitoring duty hours at the state level but not nationally. One existing organization that fits the latter category by already performing duty hour monitoring is IPRO in New York State. Its monitoring processes are discussed earlier in this chapter. IPRO has the expertise to perform duty hour monitoring as demonstrated over their multiple years of experience.

The advantage of having a different organization take on the functions of duty hour monitoring currently performed by ACGME would be that it could have a clean slate without the perceived biases of ACGME. There would be disadvantages to using a new organization or even having IPRO take on the function nationwide. These include a major expense to create any new agency from scratch, and whether a new agency or an expanded IPRO, major delays because of the need to get legislation or some other acceptable authority to delegate the responsibility to conduct such monitoring and determine a way to provide the necessary funds. There would be major disruptions to all parties while new working relationships are established. The new agency would need to establish credibility with the public, Congress, physicians, and residents. Another drawback of public agencies would be the difficulty of adapting quickly to changing circumstances, as evidenced by historical impediments to updating existing legislation and regulations on hours of service requirements in the transportation arena to use newer scientific evidence (see discussion earlier in this chapter) (Boardman, 2007; Gilligan, 2007; NTSB, 2007).

The committee decided that the expense and delays involved with creating a new organization were unacceptable. Among existing organizations, OSHA was an obvious option because it has responsibility for enforcing work hours in other industries. However, when Public Citizen requested OSHA to set duty hour limits for residents in 2001, it declined in favor of ACGME because the issues involved patient as well as worker safety and because others were taking action on duty hours (Department of Labor, 2002).

The committee also considered whether IPRO or other QIOs (Quality Improvement Organizations) could fulfill the monitoring role in a manner that would be acceptable to all the involved parties. The advantage of IPRO fulfilling this function nationwide would be its expertise in duty hour monitoring and providing education to facilities on how to better achieve compliance. The main disadvantage would be that duty hour monitoring would be separate from review of the educational program. Additionally, the IPRO approach is quite expensive per institution ($24,000 annually covering all programs in an institution) within New York State; more extensive travel requirements likely would mean even higher costs even if they could develop sufficient staff capacity to perform the function. Currently, IPRO's duty hour monitoring function falls under state authority and other than in New York and Puerto Rico there has been limited interest in passing duty hour limits at the state level. This implies it might not be a priority for funding in all states. In addition, New York State training facilities still have significant levels of violations despite years of intensive monitoring by IPRO, as noted earlier in this chapter. Possibly all QIOs or a selected few could have duty hours monitoring added to their portfolio of duties in the contracts they negotiate with CMS, but they would need time to develop staff expertise and procedures since such monitoring is not consistent with their other responsibilities and they would also need to establish working relationships with all the residency programs.

The committee did not find any of these options preferable to the third alternative, a change in ACGME practices along with oversight and evaluation.

Alternative 3: ACGME with Changes

The committee recommendation is to retain ACGME's current role in establishing duty hour limits and monitoring, in part because of the importance of maintaining the link between the residency programs' quality of education and duty hour compliance. It is essential to design educational programs in concert with duty hour schedules. The committee urges the ACGME to foster not only changes in scheduling and staffing patterns in response to duty hours but innovations in education and ways to measure competency while ensuring patient safety. Chapter 4 discusses educational considerations associated with duty hours in more depth. The changes to the ACGME monitoring process that are built into Alternative 3 are designed to correct some of the shortcomings identified in the discussion above.

The committee recommends:

- *Strengthening the ACGME monitoring process.* Increasing the frequency of duty hour audits from the more than 3 year interval for programs with citations and making unannounced visits would allow observation of operations under normal circumstances without advance preparation by the facility for a visit. This expansion of monitoring would require ACGME to raise additional funds to cover the review costs whether using its own staff, voluntary reviewers, or contractors for these additional reviews of duty hour adherence. The costs might be borne by the institutions to be audited through additional visit fees and by organizational members of ACGME. Being mindful of the potential costs, the committee does not expect that unannounced visits would be needed yearly to every institution as New York State requires. Currently, ACGME makes separate visits to institutions for each program review; a team on site for one specialty program's accreditation review could build in an unannounced look at some other program's scheduling practices and compliance in order to minimize transportation costs incurred for more frequent reviews. Given that institutions often have 30 or more residency programs, there would be numerous opportunities for unannounced visits to one or more programs.
- *Creating robust whistle-blower protections and alternative violations reporting procedures.* Encouraging residents (1) to complain about duty hour violations directly to the ACGME without first requiring them to go through their program director and the designated institutional officer for graduate medical education or (2) to complain to their local institutional compliance office could help address residents' concerns that by reporting onerous working

conditions they place their own career in jeopardy. Institutional compliance offices have a broader role than just graduate medical education and exist to ensure compliance with laws, regulations, and policies that govern medical facility operations (e.g., worker safety, Health Insurance Portability and Accountability Act, research subjects' protection, billing practices). They investigate complaints in a confidential manner and develop plans of corrective action. The institution's compliance office might be able to provide a more immediate response to a duty hours problem than a national organization even when direct reporting is allowed.

- *Gathering useful data to drive policy and evaluate progress.* Duty hour compliance audits by ACGME could be made more useful than they are now. They provide opportunities not only to ensure adherence to rules but also to gather data on how long residents are really working (by specialty and rotation), why they violate limits, and when they violate limits (e.g., night shifts but not day shifts). Such data may illuminate when exceptions might be permissible and how to target fatigue mitigation strategies and staffing. Some program directors and residents have complained that the limits are inflexible and sometimes interfere with professional obligations and important educational opportunities (Fletcher et al., 2008; Lin et al., 2006). Better collection and analysis of monitoring data could document such problems.

These changes recommended by the committee should greatly improve adherence to the duty hour limits. The committee, however, concluded that oversight of the ACGME process was also needed.

Future Approach to Providing Oversight of ACGME Monitoring

The committee considered different ways to provide assurance to the public and Congress that ACGME's discovery of the extent of duty hour violations is accurate, that residency programs move more quickly towards full adherence, and that residents can safely report violations when necessary. Oversight is recommended to provide such assurance. This function could be assigned to a new organization, a government agency or to an existing organization. The advantages and disadvantages of these options are discussed below.

Alternative 1: A New Organization for Oversight

Creating a new organization to provide oversight would have the advantage of independence from all the existing stakeholders, if structured appropriately, and a clean slate with no perceptions of bias. The main disadvantages to doing this are similar to those mentioned for a new organization to conduct monitoring: delays in getting authority and funds to create such an organization and the need to establish public credibility. It would also need to design and establish a mechanism and procedures for providing oversight. The committee concluded that the delays and expense involved with creating a new oversight organization were not justified.

Alternative 2: A Government Agency for Oversight

Some of the advantages of using a public federal agency are that it is less likely to be co-opted by the profession than are private bodies, it can be tough and authoritative, and may already enjoy a measure of public trust. The most obvious government agencies that might conduct oversight of duty hour monitoring are the Centers for Medicare and Medicaid Services (CMS) or the Agency for Healthcare Research and Quality (AHRQ). The specific pros and cons of using CMS and AHRQ are discussed below.

The role envisioned for CMS would be to help ensure the accuracy and reliability of ACGME procedures, data, and reports by supporting periodic evaluations of duty hours that would look not only at compliance but also examine the reasons behind violations and to suggest when exceptions to rules might be necessary to promote patient safety and under what circumstances (e.g., direct supervision) a resident might be able to stay beyond his or her hour limit to participate in an unusual learning opportunity. This overview of the exceptions process as well as duty hour adherence is important in light of the committee's recommendations on adjustments to duty hours and provisions for exceptions in Chapter 7.

CMS has an Office of Clinical Standards and Quality (OCSQ) that serves as a focal point for all quality and safety issues and it has direct access to funds from the Medicare Trust Fund that support contracts for research and evaluation related to quality and safety. A very small percentage of those funds could support periodic contracted evaluations of duty hours and their monitoring and their relationship to quality of care, patient safety, resident safety, and educational outcomes. CMS could either contract for studies of duty hour compliance and manage the contracts directly or it could support research managed by another federal agency, such as AHRQ. Alternatively, OCSQ also has ongoing contracts with private quality improvement organizations in each state, such as IPRO in New York, and could support one or more of them to conduct an evaluation of the outcomes of ACGME monitoring on adherence to rules.

There are several reasons why CMS would be the most appropriate agency to take on some of the necessary evaluative responsibilities, but also reasons why that might not be desirable. On the positive side, CMS has had an intimate relationship with teaching hospitals since 1965 concerning graduate medical education (GME) funding and resident education. It expends more than $8 billion in funds annually related to GME and associated patient care. CMS is the main federal agency responsible for assuring healthcare quality, paying for the care of millions of patients in teaching hospitals, as well as auditing the facilities. In addition, CMS has had ongoing relations with the VA and DOD health systems over the years concerning Medicare-eligible veterans and retirees, which might facilitate cooperative oversight of resident hours in those systems. CMS has the resources to conduct the evaluative studies envisioned for their oversight role.

Having CMS involved in the oversight of duty hours monitoring provides additional possibilities for increasing adherence to the rules. If the changes in ACGME monitoring practices and whistle-blower protections do not prove sufficient to have institutions comply, financial levers should be considered in addition to the threat of ACGME accreditation withdrawal or placing a residency program on probationary status. For example, in New York State, fines for duty hour violations are levied on institutions. The committee suggests that ACGME and CMS explore this and other options related to Medicare's program rules for institutions receiving direct or indirect GME funds. CMS would want to have confidence in the monitoring process before leveling such fines and having conducted an evaluation of the process would be critical. Additionally, ACGME through its Committee on Innovations may discover that certain carrot-and-stick approaches will foster adherence and these should also be considered (Volpp and Landrigan, 2008).

On the negative side, CMS is a large bureaucracy that has not done such oversight of the GME program in the past and that function may not be a top priority for funding and attention in the organization. Since it is a federal agency, its policies and staff could potentially change significantly from one administration to another. Some people might object to giving a government agency oversight over a private organization's monitoring of duty hours as has been evidenced in opposition to previous attempts to regulate duty hours in HHS through legislation.

While AHRQ might also be an appropriate option for the evaluative studies, it would likely have more difficulty obtaining needed funds than would CMS, and it does not have the leverage over training institutions that CMS has. The committee expects that AHRQ would play a significant role in implementing the recommendation concerning future research and evaluation, discussed in Chapter 9, and AHRQ would benefit from having a neutral role when working with other research-oriented parties planning a research agenda rather than direct oversight responsibilities. Additionally, the newly initiated AHRQ program on Patient Safety Organization reporting could yield complementary information on whether residents and/or fatigue contribute to reported events if resident status and fatigue are included in those reporting requirements (AHRQ, 2008).

The committee recommends that CMS should provide evaluative oversight of ACGME's monitoring of duty hours and the possible effects of violations on quality of care and patient safety. That oversight function would be enhanced by complementary oversight by an existing private organization as well.

Alternative 3: An Existing Private Organization for Oversight

An existing private organization conducting related functions could have certain advantages over both a new organization and a public one. It could move quickly and readily update its procedures, have stature and recognition among the profession and the public, and a focus on quality and safety. The Joint Commission, which currently accredits hospitals, could play a complementary role to CMS's oversight of the duty hours monitoring.

The oversight role for the Joint Commission would differ from that of CMS and should fit consistently with its own accreditation process, which focuses on patient safety and quality during periodic, unannounced visits to institutions by a team of surveyors. Testimony by the Joint Commission on its approach to monitoring quality of care and safety indicated that rather than monitoring whether resident duty hours meet ACGME limits within an institution, the Joint Commission's approach could be to determine whether residents or other staff were involved in patient safety events examined through patient-centered tracer cases and whether fatigue was a contributing cause (Joint Commission, 2008). For the Joint Commission to take on this systematic oversight function, it would likely need to adjust its policies and procedures to include a stronger focus and guidance on fatigue, safety, and work hours, although it already has raised the issue through its publications. Since their tracer case process as well as preliminary data analysis related to each hospital's accreditation visit include a wide variety of data and record checks, the marginal increase in work and costs based on cases with fatigued residents to assure proper monitoring policies and procedures would likely not be great. The surveyors would not have to check adherence documentation for all the residency programs in an institution, just those related to programs in a tracer case.

There are advantages to including the Joint Commission in the oversight process. The Joint Commission currently accredits 97.5 percent of major teaching hospitals and 93.6 percent of minor teaching hospitals (Joint Commission, 2008). An oversight role would place adherence to duty hours and prevention of fatigue within institution-specific quality and safety efforts and highlight the role of residents and their importance to patient safety. Joint Commission accreditation affects the entire hospital not just the educational programs and, through its recommendations for systems improvements, receives the attention of institutional administrators. The accreditation process is used to identify areas for correction and improvement in a hospital (i.e., Requirements for Improvement that facilities must address specified by surveyors based on findings of deficiencies), and if a problem concerning adherence to duty hours is uncovered, it should be treated as other similar violations by the Joint Commission. The loss of accreditation is rare and occurs because of large, serious, and persistent problems.

Disadvantages associated with the inclusion of the Joint Commission in the oversight process include the need for the Joint Commission to expand its survey process to include some specific attention to resident fatigue and to adjust its working relationship with other organizations to recognize the duty hour limits set by ACGME. The Joint Commission's priorities are on patient safety and quality issues. Their complaint process, which receives approximately 12,000 complaints per year, gets only 5-8 related to resident work hours and their voluntary sentinel events reporting system rarely finds resident fatigue mentioned in the root cause analyses, indicating that currently the resident fatigue issue does not demand much attention (Joint Commission, 2008). It could happen that the Joint Commission's use of its tracer case method will reveal very few patient events related to resident fatigue and duty hours, but that would not necessarily negate the value of its oversight role.

The committee concludes that the advantages of a strengthened ACGME monitoring process along with external oversight by *both* CMS and the Joint Commission would help assure the public that programs would be more likely to adhere to the rules, problems with duty hours compliance would be uncovered and dealt with properly, and there would be more rapid implementation of the committee's recommended adjustments to duty hours. CMS, the Joint Commission, and ACGME should discuss how their functions could complement each other and what information can be shared. The recommended oversight functions discussed for CMS and the Joint Commission are designed to be practical, derive from existing functions, and not be overly burdensome. Also, residents are more likely to report problems when they arise if probation of their educational program was not the only lever. The stature of existing relationships of both CMS and Joint Commission with teaching institutions would significantly add weight at the institutional level to ACGME processes.

Recommendation 2-1: ACGME and residency programs should ensure adherence to the current limits now, and to any new limits when implemented, by strengthening their current monitoring practices. To provide additional support, the Centers for Medicare and Medicaid Services and the Joint Commission should take an active oversight role:

- **ACGME should maintain responsibility for duty hour monitoring and should enhance its procedures by including unannounced visits for monitoring duty hours and regular collection of sufficient data to understand when and why limits are violated.**
- **Sponsoring institutions should provide for confidential, protected reporting of duty hour violations by residents through their compliance office or by an entity above the program level that does not have direct responsibility over the residency programs.**
- **ACGME should strengthen its complaint procedures to provide more confidentiality and protection to persons reporting violations of duty hours, as well as other violations of residency rules.**
- **The Centers for Medicare and Medicaid Services should assess the reliability of ACGME procedures and data and should sponsor periodic independent reviews of ACGME's duty hour monitoring to determine the characteristics of and reasons for violations.**
- **The Joint Commission should seek to ensure that duty hour monitoring is linked to broader activities to improve patient safety in hospitals, including the use of ACGME's adherence data as part of the Joint Commission's hospital surveys and accreditation actions.**

Service demands on residents and educational expectations can create pressures for longer hours of service than are necessary for achieving educational competence alone. In instituting the 2003 duty hour reforms, the ACGME indicated that training programs needed to "decouple notions of professionalism from the number of hours worked" (AAMC, 2003). Going forward, professionalism should not just mean staying long hours. Educational leaders, hospital administrators, and residents themselves should recognize that ensuring adequate sleep for residents is part of responsible behavior to promote safe conditions for both residents and patients. This chapter has focused on the need for increased monitoring of resident duty hours and increasing transparency of why rules are violated. The committee's ultimate intent is not to establish a burdensome and costly monitoring process that must be continued forever, but to ensure that there is a change in practice and that we learn from its implementation.

REFERENCES

AAMC (Association of American Medical Colleges). 2003. *Adapting to the new duty hour requirements: GME directors meet to compare notes.* http://www.aamc.org/newsroom/reporter/nov03/dutyhours.htm (accessed September 9, 2008).

———. 2007. AAMC Survey of house staff stipends, benefits and funding: Autumn 2007 report. Washington, DC: Association of American Medical Colleges.

———. 2008. Resident duty schedules and resident education. Presentation by Debra Weinstein to the Committee on Optimizing Graduate Medical Trainee (Resident) Hours and Work Schedules to Improve Patient Safety, May 8, 2008, Washington, DC.

ABMS (American Board of Medical Specialties). 2008. Presentation by Kevin Weiss to the Committee on Optimizing Graduate Medical Trainee (Resident) Hours and Work Schedules to Improve Patient Safety, May 8, 2008, Washington, DC.

ABNS (American Board of Neurological Surgeons). 2008. Presentation by H. Hunt Batjer to the Committee on Optimizing Graduate Medical Trainee (Resident) Hours and Work Schedules to Improve Patient Safety, March 4, 2008, Irvine, CA.

ACGME (Accreditation Council for Graduate Medical Education). 2003. *The ACGME's approach to limit resident duty hours: The common standards and activities to promote adherence.* http://acgme.org/acWebsite/GME_infor/history/GME.pdf (accessed December 27, 2008).

———. 2004. *The ACGME's approach to limit resident duty hours 12 months after implementation: A summary of achievements.* http://www.acgme.org/acWebsite/dutyHours/dh_dutyhoursummary2003-04.pdf (accessed December 27, 2007).

———. 2006. *The ACGME's approach to limit resident duty hours 2005-06: A summary of achievements for the third year under the common requirements.* http://www.acgme.org/acWebsite/dutyHours/dh_achieveSum05-06.pdf (accessed December 27, 2007).

———. 2007a. *The ACGME's approach to limit resident duty hours 2006-07: A summary of achievements for the fourth year under the common requirements.* http://www.acgme.org/acWebsite/dutyhours/dh_achieve0607.pdf (accessed December 27, 2008).

———. 2007b. The ACGME's approach to limiting resident duty hours. Presentation by Paul Friedmann to the Committee on Optimizing Graduate Medical Trainee (Resident) Hours and Work Schedules to Improve Patient Safety, December 3, 2007, Washington, DC.

———. 2007c. *Procedures for addressing complaints against residency programs and sponsoring institutions.* http://www.acgme.org/acWebsite/resIfo/ri_complaint.asp (accessed March 29, 2008).

———. 2008a. *Accreditation Council for Graduate Medical Education: Glossary of terms.* http://www.acgme.org/acWebsite/about/ab_ACGMEglossary.pdf (accessed July 17, 2008).

———. 2008b. *The accreditation process and the role of the ACGME site visitor.* http://www.acgme.org/acWebsite/fieldStaff.fs_siteRole.asp (accessed March 26, 2008).

———. 2008c. The ACGME duty hour standards: One element for promoting good learning, safe patient care and resident well-being: Data on compliance and effect. Presentation by Ingrid Philibert and Thomas Nasca to the Committee on Optimizing Graduate Medical Trainee (Resident) Hours and Work Schedules to Improve Patient Safety, March 4, 2008, Irvine, CA.

———. 2008d. *Fees for evaluation and accreditation.* http://www.acgme.org/acWebsite/GME_info/gme_feesAccred.asp (accessed March 30, 2008).

———. 2008e. Frequent citations of orthopaedic surgery residencies. *Orthopaedic RRC Newsletter* (Winter 2008):1-2.

———. 2008f (unpublished). The ACGME limits on resident duty hours: Standards, promoting compliance and effect. May 2008.

ACS (American College of Surgeons). 2008. Presentation by Gerald Healy to the Committee on Optimizing Graduate Medical Trainee (Resident) Hours and Work Schedules to Improve Patient Safety, March 4, 2008, Irvine, CA.

AHRQ (Agency for Healthcare Research and Quality). 2008. *The patient safety and quality improvement act of 2005.* http://www.ahrq.gov/qual/psoact.htm (accessed November 13, 2008).

AMA (American Medical Association). 2005. Medical students and residents work-hour survey. Chicago, IL: American Medical Association, Division of Market Research and Analysis.

———. 2006. *Issues of importance to resident physicians and fellows highlights.* Chicago, IL: American Medical Association, Division of Market Research and Analysis.

———. 2007. Duty hours: The resident perspective. Presentation by Sunny Ramchandani to the Committee on Optimizing Graduate Medical Trainee (Resident) Hours and Work Schedules to Improve Patient Safety, December 3, 2007, Washington, DC.

American Medical Student Association. 2007. Presentation by Michael Ehlert to the Committee on Optimizing Graduate Medical Trainee (Resident) Hours and Work Schedules to Improve Patient Safety, December 3, 2007, Washington, DC.

———. 2008. *Principles regarding resident and student work hours.* http://www.amsa.org/about/ppp/rwh.cfm (accessed February 10, 2008).

Andrews, A. 1995. I was juror no. 6, the lone dissenter in the Libby Zion case. *New York Times*, February 21.

Arora, V., C. Dunphy, V. Y. Chang, F. Ahmad, H. J. Humphrey, and D. Meltzer. 2006. The effects of on-duty napping on intern sleep time and fatigue. *Annals of Internal Medicine* 144(11):792-798.

Arora, V., D. B. Wayne, R. A. Anderson, A. Didwania, and H. J. Humphrey. 2008. Participation in and perceptions of unprofessional behaviors among incoming internal medicine interns. *JAMA* 300(10):1132-1134.

Asad, M., A. K. Shah, and F. Ehrlich. 2006. Electronic submission of resident work hours: A convenient way of exchange of information. *American Surgeon* 72(8):735-738.

Associated Press. 2003. *Hospitals face limit on resident hours.* http://www.paworld.net/pasmeetneed.htm (accessed August 27, 2008).

Baldwin, D. C., Jr., S. R. Daugherty, R. Tsai, and M. J. Scotti, Jr. 2003. A national survey of residents' self-reported work hours: Thinking beyond specialty. *Academic Medicine* 78(11):1154-1163.

Bell, B. M. 2003. Reconsideration of the New York State laws rationalizing the supervision and the working conditions of residents. *Einstein Journal of Biological Medicine* 20(1):36-40.

Boardman, J. H. 2007. *Written statement to the Subcommittee on Surface Transportation and Merchant Marine Infrastructure, Safety, and Security, Committee on Commerce, Science, and Transportation, U.S. Senate.* Administrator, Federal Railroad Administration, U.S. Department of Transportation.

Brotherton, S. E., F. A. Simon, and S. I. Etzel. 2002. U.S. graduate medical education, 2001-2002: Changing dynamics. *JAMA* 288(9):1073-1078.

Carpenter, R. O., J. Spooner, P. G. Arbogast, J. L. Tarpley, M. R. Griffin, and K. D. Lomis. 2006. Work hours restrictions as an ethical dilemma for residents: A descriptive survey of violation types and frequency. *Current Surgery* 63(6):448-455.

Chang, L. W., A. R. Vidyarthi, and R. J. Kohlwes. 2006. Baseline duty hours recorded with time-cards: A pre-regulation study of internal medicine residents. *Medical Education* 40(7):662-666.

Chao, L., and M. K. Wallack. 2004. Changes in resident training affect what you can expect from your next partner. *Bulletin of the American College of Surgeons* 89(9):12-15.

CIR/SEIU (Committee of Interns and Residents-Service Employees International Union Healthcare). 2007. Presentation by L. Toni Lewis to the Committee on Optimizing Graduate Medical Trainee (Resident) Hours and Work Schedules to Improve Patient Safety, December 3, 2007, Washington, DC.

Council on Accreditation. 2008. *Understanding the Hague accreditation and approval standards and the substantial compliance system.* http://coanet.org/front3/page.cfm?sect=54&cont=4251 (accessed March 24, 2008).

Cull, W. L., H. J. Mulvey, E. A. Jewett, E. L. Zalneraitis, C. E. Allen, and R. J. Pan. 2006. Pediatric residency duty hours before and after limitations. *Pediatrics* 118(6): e1805-e1811.

Czeisler, C. A. 2007. Presentation to the Committee on Optimizing Graduate Medical Trainee (Resident) Hours and Work Schedules to Improve Patient Safety, December 3, 2007, Washington, DC.

DeBuono, B. A., and W. M. Osten. 1998. The medical resident workload: The case of New York State. *JAMA* 280(21):1882-1883.

Department of Labor-Occupational Safety and Health. 2002. Letter from John L. Henshaw, Assistant Secretary for Occupational Safety and Health to Sidney M. Wolfe, Director, Health Research Group, Public Citizen. http://www.citizen.org/documents/ACFC2B.pdf.

Dinges, D. F. 1995. Sleep deprivation, impairment of resident performance, and medical error. *Journal of Bone and Joint Surgery* 87a(11):2581-2582.

Douglas, R. G. 1995. Zion case verdict vindicates training system. *New York Times*, February 14.

Drake, C. L., T. Roehrs, G. Richardson, J. K. Walsh, and T. Roth. 2004. Shift work sleep disorder: Prevalence and consequences beyond that of symptomatic day workers. *Sleep* 27(8):1453-1462.

Federal Railroad Administration. 2008. *National Rail Safety Action Plan final report 2005-2008.* Cambridge, MA: U.S. Department of Transportation.

Fletcher, K. E., V. Parekh, L. Halasyamani, S. R. Kaufman, M. Schapira, K. Ertl, and S. Saint. 2008. Work hour rules and contributors to patient care mistakes: A focus group study with internal medicine residents. *Journal of Hospital Medicine* 3(3):228-237.

Folkard, S., D. A. Lombardi, and P. T. Tucker. 2005. Shiftwork: Safety, sleepiness and sleep. *Industrial Health* 43(1):201-223.

Gilligan, P. 2007. *Statement of Peggy Gilligan, Deputy Associate Administrator for Aviation Safety. Testimony before the Congressional Committee on Transportation and Infrastructure, Subcommittee on Aviation on the Most Wanted List of the National Transportation Safety Board.* Washington, DC: Federal Aviation Administration, U.S. Department of Transportation.

Goldstein, M. J., B. Samstein, A. Ude, W. D. Widmann, and M. A. Hardy. 2005. Work Hours Assessment and Monitoring Initiative (WHAMI) under resident direction: A strategy for working within limitations. *Current Surgery* 62(1):132-137.

GovTrack.us. 2005a. *H.R. 1228—109th Congress: Patient and Physician Safety and Protection Act of 2005.* http://www.govtrack.us/congress/bill.xpd?bill=h109-1228&tab=summary (accessed March 29, 2008).

———. 2005b. *S. 1297—109th Congress: Patient and Physician Safety and Protection Act of 2005.* http://www.govtrack.us/congress/bill.xpd?bill=h109-1297&tab=summary (accessed March 29, 2008).

IOM (Institute of Medicine). 2003. *Keeping patients safe: Transforming the work environment of nurses.* Appendix C. Washington, DC: The National Academies Press.

IPRO. 2007a. Graduate medical education—Past, present and future. *Resident Times* 1(1):2.

———. 2007b. *Working hours and conditions post-graduate trainees annual compliance assessment: Contract year 6 10/1/06-9/30/07.* Albany, NY: IPRO.

———. 2008. *IPRO milestones—20 years of quality improvement.* http://www.ipro.org/index/milestones (accessed January 29, 2008).

Jagsi, R., D. F. Weinstein, J. Shapiro, B. T. Kitch, D. Dorer, and J. S. Weissman. 2008. The Accreditation Council for Graduate Medical Education's limits on residents' work hours and patient safety: A study of resident experiences and perceptions before and after hours reductions. *Archives of Internal Medicine* 168(5):493-500.

Joint Commission. 2008. Presentation by Paul Schyve to the Committee on Optimizing Graduate Medical Trainee (Resident) Hours and Work Schedules to Improve Patient Safety, May 8, 2008, Washington, DC.

Kennedy, R. 1998. Residents' work hours termed excessive in hospital study. *New York Times*, May 1, p. 1.

Kowalczyk, L. 2008. Beth Israel cited for residents' long hours. *Boston Globe*, September 7.

Kozak, L. J., C. J. DeFrances, and M. J. Hall. 2006. *National Hospital Discharge Survey: 2004 annual summary with detailed diagnosis and procedure data.* Hyattsville, MD: National Center for Health Statistics.

Lamberg, L. 2002. Long hours, little sleep: Bad medicine for physicians-in-training? *JAMA* 287(3):303-306.

Landrigan, C. P. 2007. Effects of traditional 24-hour work shifts on physician and patient safety. Presentation to the Committee on Optimizing Graduate Medical Trainee (Resident) Hours and Work Schedules to Improve Patient Safety, December 3, 2007, Washington, DC.

Landrigan, C. P., L. K. Barger, B. E. Cade, N. T. Ayas, and C. A. Czeisler. 2006. Interns' compliance with Accreditation Council for Graduate Medical Education work-hour limits. *JAMA* 296(9):1063-1070.

Landrigan, C. P., A. M. Fahrenkopf, D. Lewin, P. J. Sharek, L. K. Barger, M. Eisner, S. Edwards, V. W. Chiang, B. L. Wiedermann, and T. C. Sectish. 2008. Effects of the Accreditation Council for Graduate Medical Education duty hour limits on sleep, work hours, and safety. *Pediatrics* 122:250-258.

Lieberman, J. D., J. A. Olenwine, W. Finley, and G. G. Nicholas. 2005. Residency reform: Anticipated effects of ACGME guidelines on general surgery and internal medicine residency programs. *Current Surgery* 62(2):231-236.

Lin, G. A., D. C. Beck, and J. M. Garbutt. 2006. Residents' perceptions of the effects of work hour limitations at a large teaching hospital. *Academic Medicine* 81(1):63-67.

Ludmerer, K. M. 1999. *Time to heal: American medical education from the turn of the century to the era of managed care.* New York: Oxford University Press.

McCallum, M., T. Sanquist, M. Mitler, and G. Krueger. 2003. *Commercial transportation operator fatigue management reference.* Washington, DC: U.S. Department of Transportation Human Factors Coordinating Committee.

New York State Department of Health. 2002. *State health department cites 54 teaching hospitals for resident working hours violations.* http://www.health.state.ny.us/press/releases/2002/resident_working_hours.htm (accessed May 26, 2008).

New York Supreme Court. 1986. Report of the fourth grand jury for the April/May term of 1986 concerning the care and treatment of a patient and the supervision of interns and junior residents at a hospital in New York County.

NTSB (National Transportation Safety Board). 2007. *NSTB most wanted list—Transportation safety improvements: 2008*. Washington, DC: National Transportation Safety Board.

PAIRO (Professional Association of Interns and Residents of Ontario). 2008. *Principles of cross-coverage*. http://www.pairo.org/Content/Default.aspx?pg=1177 (accessed July 21, 2008).

Public Citizen. 2001. Petition to the Occupational Safety and Health Administration requesting that limits be placed on hours worked by medical residents. Washington, DC: Public Citizen.

———. 2007. Presentation by Peter Lurie to the Committee on Optimizing Graduate Medical Trainee (Resident) Hours and Work Schedules to Improve Patient Safety, December 3, 2008, Washington, DC.

Reiter, E. R., and D. R. Wong. 2005. Impact of duty hour limits on resident training in otolaryngology. *Laryngoscope* 115(5):773-779.

Resident Panel. 2008. Presentation to some members of the Committee on Optimizing Graduate Medical Trainee (Resident) Hours and Work Schedules to Improve Patient Safety, March 4, 2008, Irvine, CA.

Rosa, R. R. 2001. Examining work schedules for fatigue: It's not just hours of work. In *Stress, workload, and fatigue*, edited by P. A. Hancock and P. A. Desmond. Mahwah, NJ: Lawrence Erlbaum Associates, Inc. Pp. 513-528.

Royal College of Surgeons of England. 2008. Lessons from surgery. Presentation by Bernard Ribeiro to Committee on Optimizing Graduate Medical Trainee (Resident) Hours and Work Schedules to Improve Patient Safety, March 4, 2008, Irvine, CA.

Sleep-deprived doctors. 2002. *New York Times*, June 14.

Steinbrook, R. 2002. The debate over residents' work hours. *New England Journal of Medicine* 347(16):1296-1302.

Van Dongen, H. P. A., and D. F. Dinges. 2005. Circadian rhythm in sleepiness, alertness and performance. In *Principles and practice of sleep medicine*, 4th ed., edited by M. H. Kryger, T. Roth, and W. C. Dement. Philadelphia, PA: Elsevier Saunders. Pp. 435-443.

Volpp, K. G., and C. P. Landrigan. 2008. Building physician work hour regulations from first principles and best evidence. *JAMA* 300(10):1197-1199.

Weissman, J. S., J. M. Rothschild, E. Bendavid, P. Sprivulis, E. F. Cook, R. S. Evans, Y. Kaganova, M. Bender, J. David-Kasdan, P. Haug, J. Lloyd, L. G. Selbovitz, H. J. Murff, and D. W. Bates. 2007. Hospital workload and adverse events. *Medical Care* 45(5):448-455.

3

Adapting the Resident Educational and Work Environment to Duty Hour Limits

Numerous factors in the learning and work environments contribute to the content of work and the caseload that residents can manage. Trends over time have shown that patients admitted to the hospital are less stable and have more complex diagnostic and treatment needs than in past decades, yet their hospital stays are shorter. These changes have compressed the time residents have available to complete work and to learn from individual patients. The intensity of resident work appears to have increased for some specialties and rotations since the 2003 duty hour limits when they are expected to admit and manage the same caseload in fewer hours on duty. The committee recommends the development of specialty-specific workload guidelines by Residency Review Committees and continued reduction of noneducational work to support both learning and patient safety goals. The committee also concludes, based on a review of adaptations since 2003, that there is not a single approach to scheduling duty hours that fits all training facilities or specialties.

The Accreditation Council for Graduate Medical Education (ACGME) announced new duty hour limits in February 2003, with a required start date of July 1, 2003 (ACGME, 2003). Many programs anticipated the changes and had started to adapt a year or two prior to ACGME's announcement. Sponsoring institutions and their program directors responded by redesigning schedules, strengthening duty hour monitoring practices, assigning some tasks usually performed by residents to other health professionals and support staff, trying new educational approaches, and altering the work environment. No national funding allocation was dedicated to these program adaptations, and teaching institutions report that the changes were costly.

First, this chapter looks at how residents fit within a complex and changing work and learning environment, with particular attention to the influence of the content of residents' work and workload on their ability to meet duty hour limits. The committee makes recommendations with respect to the content of resident work and caseload. Next, the chapter provides examples of how programs responded to the challenges of the 2003 duty hour limits, the variety of scheduling practices adopted, and the committee's comments on scheduling preferences. Finally, the chapter examines how duty hour changes have required hiring substitutes for lost resident time, resulting in additional costs.

RESIDENT EDUCATIONAL AND WORK SYSTEMS

Resident education takes place on a daily basis through the delivery of direct care to patients, supplemented by lectures, conferences, and daily review of their patients with attending physicians. Most of the education occurs through the many conversations about their patients that residents have with attending physicians, consultants, and fellow residents. Residency can be thought of as on-the-job training since very little is detached from direct patient care. Yet educationally valuable work has not always been given priority over the service needs of institutions (Cohen, 1999; Ludmerer, 1999). The 2003 reduction in duty hours reemphasized the need to find the right balance between education and service because compressing unaltered workload into fewer hours can put pressure on residents to violate duty hour limits or rush through their work, perhaps leading to patient harm (e.g., forgetting to order a test, which delays the diagnosis and care a patient receives, or forgetting to convey critical information during handovers).

Residency programs and their sponsoring institutions needed to take many workplace factors into account when they redesigned resident work schedules in response to the 2003 limits, and these will remain considerations as additional duty hour adjustments are implemented. Ideally, the redesign took into account the ultimate outcomes of patient safety, resident safety, and educational attainment not just compliance with duty hours. A useful framework when redesigning healthcare operations in the context of patient safety is an adaptation by Vincent and colleagues of Reason's taxonomy of factors that contribute to accidents and adverse events in clinical environments. These include patient characteristics, task factors (including the content of work and workload), team factors, work and learning environment, and organizational and management factors (Reason, 1990; Vincent et al., 1998). Change in one area is not without repercussions in others, because change in the "work situation can alter substantially the individual's level of performance or decrease the probability that the performance will be maintained at a satisfactory level" (Chiles, 1982). These other factors affect whether residents can comply with duty hour limits, maximize their learning, and care for patients under the safest conditions. Duty hours are not the only factor driving resident performance.

Patient Characteristics

The growing number of admissions to hospitals of complex patients, and the availability of ever-expanding advanced technologies for diagnosis and treatment have increased the intensity of the inpatient care experience in hospitals in general and thus for residents in training (Anderson and Horvath, 2004; Bodenheimer, 2005; Lawler et al., 2001; Vogeli et al., 2007). A declining overall number of acute care beds due to cost containment measures, the shifting of many formerly hospitalized patients to outpatient care (e.g., increased use of ambulatory surgery), and reduced length of stay have meant that the inpatient population that residents care for today is sicker and more equivalent to the patients in intensive care units (ICUs) 20 to 30 years ago (Carayon and Gurses, 2005; Ludmerer, 1999; Oransky, 2003). As the U.S. population ages, some experts have advocated for new bed capacity; if demand increases without growth in beds, or better management of existing beds, there will be pressure to turn over beds sooner increasing throughput (Bazzoli et al., 2003; The Chartis Group, 2007).

The average length of stay over the past 25 years has decreased dramatically: 7.3 days in 1980, 6.4 in 1990, 4.9 in 2000, and 4.8 in 2004 (Kozak et al., 2006). In fact, many hospital stays are shorter than 4.8 days. One university-teaching hospital reported that patients with 29 of their 88 most frequent diagnostic codes in 1986 were out of the hospital in less than 2 days, and if residents were to obtain the same breadth of experience as 1980, they would need more ambulatory care experiences (Rosevear and Gary, 1989).

Brief intense patient stays in the hospital today also mean that residents have less time to get to know their patients and observe the progression of a patient's illness or injury and recovery than they did 25 years ago. Duty hour limits implemented in 2003 may have further eroded the time for interacting with individual hospitalized patients. For example, Horwitz et al. (2006a) reported that the primary admitting resident team covers approximately 47 percent of an average inpatient's 4-day hospitalization on an internal medicine service compared with 70 percent reported prior to the 2003 duty hour limits (Petersen et al., 1998). Depending on how training programs schedule their residents under duty hour limits, a resident who admits a patient may or may not be available the next day to evaluate the patient's progress before discharge (Gilsdorf, 2008).

Teaching hospitals usually have a more complex inpatient case mix

TABLE 3-1 Case Mix Index by Teaching Status for FY 2007

Teaching Status	Number of Hospitals	% of Total	Case Mix Index (CMI) FY 2007 Mean	Median	Minimum
Major teaching	303	8.24	1.60	1.59	0.84
Other teaching	795	21.62	1.49	1.49	0.80
Non-teaching	2,579	70.14	1.28	1.24	0.41
All	3,677	100.00	1.36	1.31	0.41

NOTE: CMIs are transfer adjusted and based on Medicare Grouper Version 24. Major teaching is defined as having an intern and resident-to-bed ratio greater than or equal to 0.25.
SOURCE: Inpatient Prospective Payment System (IPPS) Final Rule FY 2007; data analyzed and provided by AAMC (September 30, 2008).

than other hospitals, and the case mix index (CMI) is often used as a proxy for the relative severity of illness. The CMI measures the amount of services provided to patients with different diagnoses. The higher the case mix average, the greater the severity of illness in that institution's patient population tends to be, and therefore, more resources are used, on average, to care for them (Andrews et al., 2007). The Centers for Medicare and Medicaid Services (CMS) takes this more complex caseload and the greater number of services that may be delivered in the course of teaching into account when determining payments for teaching hospitals through its indirect medical expenditure payment for graduate medical education. As illustrated in Table 3-1, the mean and the median CMIs for teaching hospitals are higher than for non-teaching hospitals, although there is considerable variation within each category, reflecting the diversity of specialized services (e.g., transplantation, burn units) offered to their patient populations (COTH, 2008). The mean and median CMIs have not changed much from fiscal year 2000 to the present. Comparisons of the CMI over a longer period of time to assess changes in the severity of patients and the services provided are of questionable validity because there have been changes in the classification of certain illnesses within the relative diagnosis-related group weights established by CMS, on which the CMI is based.[1]

Since patients differ in terms of severity of illness and length of stay from specialty to specialty (e.g., obstetrics vs. other types of surgery) and even among rotations within specialties (e.g., ICU rotation vs. ambulatory care), patient factors must be considered when determining what type of resident work schedule will best provide continuity of patient care and

[1]Personal communication, Erika Steinmetz and Karen Fisher, Association of American Medical Colleges, April 18, 2008.

high-quality learning experiences. Patient severity is a key factor in determining the number of cases that a resident might manage within his or her duty hours.

Task Factors

Noneducational Activities

With reduced duty hours, it is critical to assess not only the number of hours that residents spend in the hospital but also the educational value of that time. A review of the literature on how residents spend their time, covering studies from the time of the Bell Commission to 2003, found that residents spent up to 36 percent of their time learning while delivering patient care services, an additional 15 percent was spent in formalized teaching activities (e.g., conferences, grand rounds), but up to 35 percent of the day was spent in non- or marginally educational patient-related activities (Boex and Leahy, 2003). Although more limited duty hours and ACGME guidance have encouraged the transfer of some tasks with marginal educational value (e.g., transport, phlebotomy) to others, residents typically still spend a substantial amount of time searching for test results and supplies, completing paperwork, obtaining and transporting specimens for laboratory tests, moving patients, making appointments, and completing paperwork for patient discharges (Gabow et al., 2006). The content of residents' work and the amount of time residents spend on different tasks have received little analysis since the 2003 change in duty hour regulations, but a few limited studies indicate that a considerable amount of noneducational work remains: from 8 to 24 percent for residents in one surgical program, with the highest values for PGY-1s, -2s, and -5s (Brasel et al., 2004), and 10 to 30 percent for another institution's residents across multiple specialties and training years (Dola et al., 2006). In a national survey of internal medicine programs, only 9 percent reported that ancillary services were more available now to help with these tasks than prior to duty hour reduction (Horwitz et al., 2006b). Addressing this issue now is a way to add to the number of resident hours available for direct patient care, enhancing both their ability to meet patient care needs and their learning.

Currently, ACGME requires that sponsoring institutions "must provide services and develop health care delivery systems to minimize residents' work that is extraneous to their GME [graduate medical education] programs' educational goals and objectives." These services and systems must include patient support services: Peripheral intravenous access placement, phlebotomy, and laboratory and transporter services must be provided in a manner appropriate to and consistent with educational objectives and quality patient care (ACGME, 2007b). As noted, such practices are not always followed. The committee concludes that ACGME should expand the protections for residents by monitoring and assessing these practices as well as broadening the current definition of support services to include administrative and secretarial support in order to reduce resident time unnecessarily spent on those tasks (e.g., making appointments, tracking down paperwork). Later, in this chapter's discussion of support services adaptations, the experiences of several programs in transferring these tasks to others are illustrated.

Reducing the amount of time residents spend on these marginally educational activities is not meant to undermine multidisciplinary team-based approaches to medical care or to establish silos of work effort ("that's not my job") and cause delays in care delivery. There may be times when a resident might be able to do such tasks in a manner that is more timely, accurate, and complete, than others can, thereby accelerating care delivery to the patient or better coordinating care by assisting in the navigation of hospital systems.

Resident Caseload

The reduction of duty hours in 2003 was not typically accompanied by a reduction in the caseload that residents manage. Workload has been implicated as a factor in resident error, delays in patient care, and possible effects on patient outcomes (Jagsi et al., 2008; Ong et al., 2007; Vidyarthi et al., 2007). Working beyond shift length because of workload contributes to violations in duty hour limits and is observed in the practice of residents as noted in Chapter 2 and of nurses as well (Rogers et al., 2004; Scott et al., 2006; Tucker and Spear, 2006). Reports on nurses find that heavy workload (e.g., nurse-patient ratios), time pressures due to work system factors (e.g., patient severity, having to perform nonnursing tasks; spending time tracking down patients' charts), and reduced supervision can contribute to poorer patient care (e.g., delays in care, complications), increased mortality, and a climate for error (Aiken et al., 2001; Carayon and Gurses, 2008; Lang et al., 2004; Tarnow-Mordi et al., 2000; Tibby et al., 2004). These are of concern in the resident work and learning environment as well. Specialty-specific and rotation-specific workload guidelines should take into account the number and severity of patients as well as the number of procedures required to determine the intensity of the experience and its effect on promoting safe conditions for residents and patients. The contribution of residents' workload to error and patient safety has not received the same investigative or public attention as their duty hours (Parshuram et al., 2004). Common sense indicates that an excessive workload might result in cutting corners that could affect patient safety (e.g., forgetting to transmit vital information during a handover or to order a needed diagnostic test leading to delays in care). Additionally, Chapter 4 examines the impact of excessive workload on learning.

Although the daily patient census for residents may have remained the same or even decreased over the past 20 to 30 years, the number of admissions and discharges has increased due to shorter lengths of stay. For example, one institution reported that although its average length of stay decreased by 13 percent and daily census decreased by 5 percent, the daily number of admissions and discharges for residents increased by 15 percent (Dellit et al., 2001). From a financial perspective, this is just what hospitals want—greater productivity, higher throughput, and faster turnover of beds, all of which maximize hospital revenues to address their costs (The Chartis Group, 2007; Gregory et al., 2003; Larson, 2003).

After the 2003 reduction in duty hours, it appears that there has not been a significant reduction in the number of patients a resident admits, manages, or cross-covers based on reports across a variety of residency specialties. Maintenance of the same caseload may lead to increased work compression or intensity during work hours (Bellini, 2008; Dawson and Zee, 2005; Horwitz et al., 2006b; Jagsi et al., 2008). A national survey of ACGME-accredited programs in internal medicine found that only 28 percent reduced the average daily census for interns in response to duty hour limits (Horwitz et al., 2006b). Numerous reports from general and other surgical programs report that they also have maintained pre-2003 surgical volume despite the reduction in duty hours (e.g., Baskies et al., 2008; Bland et al., 2005; Ferguson et al., 2005; Shin et al., 2008).

Admissions and discharges are among the most time-consuming and complex tasks that residents must complete (Dellit et al., 2001), and for many specialties these activities are limiting factors in the caseload that can be managed thoroughly within allotted duty hours. For example, Ong et al. (2007) found that increased resident workload for an internal medicine service on admission days (i.e., each additional team admission) was associated with increases in average length of stay, total costs, and risk of mortality, with the risk even higher when more than nine patients were admitted to a team on their admitting day. The authors suggested that the increased workload may have led to residents' making an "inaccurate initial clinical assessment or pushing workup activity onto subsequent days, leading to longer lengths of stay," thereby increasing the costs per patient and potentially having a detrimental impact on patient mortality. This study examined the care experiences of more than 5,000 patients over 3 years, but the authors recommend additional trials to increase the statistical power to detect changes in mortality (Ong et al., 2007). Teams were able to make some short-term adjustments to respond to increasing workload; the authors suggest that these short-term adjustments might mean that residents skip offered didactics or stay overtime to catch up on work in order to reduce the overall patient census. However, it appears that fatigue may accumulate as the team once again becomes less efficient if the census remains high over the month-long rotation.

Ong et al. (2007) also make a business case for reducing workload per team and using the savings to support additional physician-level staff or midlevel providers. Earlier studies have also found effects for the number and timing of admissions on length of stay and total charges (Griffith et al., 1997; Hillson et al., 1992). An additional approach to workload management is having teams admit a few patients each day rather than in boluses of a large number of cases every third to fourth night (Volpp and Landrigan, 2008).

Maintenance of the same caseload can affect the time available for conference attendance, educational activities other than direct patient care, adherence to duty hour limits, and on-call sleep (Arora et al., 2008a; Horwitz et al., 2006b). These effects may not be static over the training year and may differ according to various measures of caseload (e.g., new admissions vs. overall census). For example, Arora et al. (2008a) found that interns early in the training year (July-October) had 10.5 minutes less sleep for each additional on-call admission, and this declined to 1.9 minutes less sleep per admission later in the year (March-June) on extended duty periods (30-hour shifts). Thus, workload measures should recognize the growth in competence of residents over time. The study also showed that each additional patient added extra time to shift duration (e.g., 13.2 and 15.5 minutes per patient, respectively); approximately 30 percent of extended duty periods on this internal medicine service were found to be noncompliant (i.e., more than 30.5 hours in length). Reduction in workload can assist in greater adherence to duty hours.

It appears that efforts to maintain caseload have not been supported by sufficient reductions in noneducational tasks that consume large amounts of resident time. This likely leads to a smaller proportion of available time for educational activity as throughput increases and sacrifices depth of learning for greater exposure to learning episodes that are more brief and may be less rich depending on the specialty and resident rotation. A well-designed caseload of the right variety and number of patients can enhance learning, while too much work can overwhelm cognitive processing and lessen learning (Chewning and Harrell, 1990; Choo, 1995; Wiener et al., 1984). A reduction in duty hours suggests that the number of patients a resident can care for at one time, especially in the first year of residency, and the amount of noneducational work need to be reduced so that resident time and workload are maximally attuned to the higher-yield learning events of a patient stay. The amount of time spent in daily care of patients varies from specialty to specialty, necessitating specialty-specific workload guidance; for example, the most time-demanding portion of the work day for surgical residents may be the time spent in preparation and performance of procedures rather than admissions and discharges.

Team Factors

One of the key elements of the residency experience is working as part of teams, both resident teams and a larger interprofessional team (e.g., nurses, physicians, pharmacists). Each resident team is made up of several levels of residents and medical students, with those in each advancing year of training having increasing levels of responsibility. Even when working in teams, the ultimate responsibility for patient care resides with each individual patient's attending physician. Reducing resident duty hours has meant changing team dynamics and potentially affecting teaching, learning, and performance. A few studies indicate that some work has shifted within existing team structures, particularly from interns to more senior residents and from residents to faculty; the volume of work and/or its intensity, as noted above, appears to have remained the same or even increased for some training years but not others (Coverdill et al., 2006a,b; Hutter et al., 2006; Parekh et al., 2005). Reorganization of team structures has been necessary to enhance patient continuity (Mathis et al., 2006) and provide coverage of services around the clock.

The effects of reduced hours on resident team dynamics vary according to specialty and the size of programs. Neurosurgery programs that often only have one resident per training year, very long operations, and patients who need close observation after surgery have had difficulty meeting the 80-hour limit (ACGME, 2007a; Cohen-Gadol et al., 2005). Having to cover duty hours with just a few residents per year makes it difficult to sustain traditional hierarchical relationships and progressively increasing training and experience from intern to second year to third year and up until the attending; these programs may have to match individual residents with attending physicians. This diminishes the traditional involvement of senior residents in teaching junior residents (Cohen-Gadol et al., 2005). Training programs, regardless of specialty, that have just a few residents will have a harder time adapting to reduced duty hours than those with more residents; these programs will need to find alternatives to resident coverage and redesign their approaches to care, or they might be unable to maintain accreditation.

Under duty hour restrictions, an excessive workload (i.e., numbers of patients, complexity of caseload, amount of noneducational work) for the given time is one of the obstacles that residents, their mentors, and other professionals must overcome in providing quality care to patients. There is extensive research in other fields that indicates the detrimental effects on individual performance of excessive workload (Gonzalez, 2005; Hancock et al., 1995; Rahman and Haque, 1992), but teams can help buffer these detrimental effects and even increase productivity by distributing work and workload among team members (Jung et al., 2002). Individuals in teams that work as a unit have a shared idea of how to accomplish a task and therefore provide assistance to one another (e.g., by providing backup or monitoring the situation for work to be done or to prevent errors). Teamwork has been shown to improve performance even under conditions of sleep deprivation (Baranski et al., 2007; Vander Wood et al., 2007).

In some programs, staff members have been added to the care team to help complete the work formerly done by residents; these include hospitalists, physician assistants, and nurse practitioners. This is discussed in more detail later in the chapter. Sometimes these additional staff members are well integrated into the resident team with good communication to provide continuity of care; for example, they might all have rounds together. However, others have more of a stopgap function to fill uncovered hours (e.g., moonlighting physicians) (Horwitz et al., 2006a).[2]

Work and Learning Environment

The philosophy of the sponsoring organization and the residency program director determines whether the balance of resident work is tilted toward service or education. The size and scope of residency programs vary greatly from site to site. Sponsoring institutions may have a handful of specialty residency programs, while others might have more than 100 different programs; medical school sponsors tend to have the most programs, an average of 35.5. There are numerous types of sponsoring organization (e.g., for-profit and nonprofit groups including government, church, or private ownership) with the majority being nonprofit (ACGME, 2007d).

Some organizations, regardless of philosophy, have limited resources and thus may have trouble providing supplementary services or hiring replacements for residents even if they would like to do so. The committee is cognizant of this and in Chapter 9 recommends additional funding to implement changes in workload and hours, with special consideration for safety net teaching hospitals so that they can maintain robust training programs while providing desirable community service.

Some educators and residents have expressed concern that educational opportunities are diminished for today's residents. Studies report decreased attendance at formal didactics, less availability for ambulatory care clinics, less opportunity for residents to discuss their cases thoroughly with attendings, and fewer other educational opportunities since 2003 (Arora et al., 2008b; Parekh et al., 2005; Reed et al., 2007). On the other hand, orga-

[2]Personal communication, D. Meltzer, University of Chicago, August 12, 2008.

nizations report on how they redesigned their programs to preserve these elements and how they maintained or even improved educational outcomes (Basu et al., 2004; de Virgilio et al., 2006; Horwitz et al., 2007). More information from reports on education after resident duty hours reform is contained in Chapter 4.

Many of the work processes in the system as a whole are inefficient, affect residents' performance, and inhibit their ability to complete their work in a timely fashion. Gabow and others have found that the workflow of residents is fragmented by frequent interruptions and changes in focus that interfere with task completion and cognitive processing, and that often (e.g., 25-26 percent) these interruptions are rated as being for unimportant reasons (Blum and Lieu, 1992; Gabow et al., 2006). Specifically, Gabow and her colleagues (2006) found that residents performed 5.0 to 11.3 different activities per hour of non-sleeping time. Residents "experienced frequent interruptions and changes in focus"; interruptions can lead to errors, and sleepy residents will have more trouble recovering from interruptions to focus on their tasks (Gabow et al., 2006). Research in other environments finds that as interruptions increase, the frequency of error also increases (Hirst and Kalmar, 1987; Speier et al., 1997). Such interruptions have been implicated as contributing to pilot error (Dismukes et al., 1998) and to medication-dispensing errors by nurses and pharmacists (Flynn et al., 1994; Gladstone, 1995; Peterson et al., 1999).

Human factors and systems engineering approaches help programs analyze their current work practices to determine the amount of time residents spend on key activities and how they interact with others in the work environment (Barach and Johnson, 2006). These could reveal ways to reorganize work processes and resident work time to increase efficiency and decrease interruptions (Chung and Ahmed, 2007; Gabow et al., 2006). For example, one surgical program's self-study revealed a need to reorganize morning and evening rounds to make them more efficient. These activities designed to improve both time and team management also resulted in other improvements (e.g., greater punctuality at conferences, clinics, and operations). Resident satisfaction improved as well because they did not perceive that time was wasted when they were on duty (Chung and Ahmed, 2007). Other programs have drawn up schedules and shift changes to match patient admission flow, thus reducing the amount of time residents spend waiting for patients to arrive (Levin et al., 2007; Ogden et al., 2006). Rethinking and reengineering how residents spend their time might help reduce the hours needed to complete the desired tasks; with increased efficiency, they could spend more time at the bedside caring for patients and in other learning activities, and when on night call they would have additional time for sleep (Lamberg, 2004; Morton et al., 2004; Viney, 2008).

Organizational and Management Factors

Although patient acuity is a factor in the number of patients that a resident can handle, organizational and management factors in the microsystem and macrosystem surrounding a resident can decrease or increase this number (Carayon and Gurses, 2008; Gurses and Carayon, 2007). Resident education is beginning to incorporate a greater understanding of the effect of system issues, not just science-based medical care, on patient outcomes through its focus on core competencies. Quality improvement is being viewed as an essential element of professional development (ACGME, 2007c; Batalden and Davidoff, 2007a,b).

The term *microsystem* has been applied to "a small, organized patient care unit with a specific clinical purpose, set of patients, technologies and practitioners who work directly with these patients" (e.g., neonatal intensive care, surgical care team, outpatient clinic) (Mohr et al., 2004). Resident teams operate within an interdisciplinary microsystem on the front lines of the overall complex macrosystem of a hospital. Residents do not operate in isolation without affecting the work of others and vice versa. Effective microsystems have been characterized as having extensive cooperation and teamwork with better communication and interdependence (e.g., use of multidisciplinary rounds, better use of information technology). Analyzing and mapping the processes within microsystems and the overall hospital are viewed as a way to reveal disorganization and inefficiencies that can compromise patient safety and contribute to wasting resources (IOM, 2001; Mohr et al., 2004). Chapter 8 examines ways to improve communication in handovers and teamwork with the aim of improving the conditions for safety through error prevention, detection, and feedback.

Redesign of workflow, paging practices, and having residents treating inpatients that are in close geographic proximity are a few suggestions for facilitating higher caseloads and improving care:

- Having inpatients grouped in a geographically cohesive area reduces the amount of time spent by residents traveling around the hospital (Bellini, 2008).
- Reliable, user-friendly computer order entry and electronic health records can reduce the time spent hunting for records, tracking down lab results, and deciphering illegible handwriting (saving 1-2 hours a day according to one report) and can reduce harmful drug interactions through e-prescribing (Armitage and Rathod, 2003; Henry Ford Health System, 2008).
- Health information systems can allow improved scheduling, faster utilization of laboratory and radiology tests (e.g., results obtained

more quickly, less duplication of testing), and access to electronic imaging in real time (Henry Ford Health System, 2008; Hillestad, 2008). These can enhance communication during team transitions and oversight by supervisors of resident activity.

- Services to handle transport, phlebotomy, intravenous (IV) services, and clerical support for appointments and discharges will free up resident time for educationally valuable patient care.
- Space and time set aside for residents to take naps help prevent or mitigate fatigue, reducing the propensity for error after many hours on duty (Arora et al., 2006; Flynn et al., 1999; Weinger et al., 2004).

Performance obstacles that exist for residents (e.g., wasting time trying to locate supplies or track down charts, ineffective communications) are likely contributing to inefficiencies and errors of other staff as well (Gurses and Carayon, 2007). System redesign efforts made for other purposes can also help residents access desired educational opportunities during reduced duty hours. For example, MetroHealth in Cleveland and Denver Health have made strides in improving operating room efficiency in teaching settings by reducing the time spent on nonoperative tasks and reducing nonclinical delays and interruptions in shorter-duration surgeries (e.g., less than 2 hours) (Gabow, 2008; Harders et al., 2006). Denver Health increased its operating room efficiency from 70 percent to almost 85 percent. Such efficiencies can allow residents to have the desired case experiences with less time wasted so that they can attend to their remaining duties. Organizational factors can also have an effect on patient outcomes (Volpp, 2008).

Understanding and resolving performance obstacles requires an investment in time and resources, but removing barriers and streamlining practices can yield increased revenues and improve quality, while residents learn state-of-the-science evidence-based medicine and quality improvement strategies. Institutions should, to the extent possible, redesign their systems; for example, if patients can be admitted or discharged earlier in the day and other efficiencies are in place, there will be less pressure for residents to work long into the evenings and nights.

REDESIGNING RESIDENT WORK AND WORKLOAD

The 2003 reduction of duty hours did not always translate into a reduced caseload for residents because of both perceived educational needs and institutional economic pressures for patient care and "throughput" (e.g., patients transferred and discharged per day), adding to work intensity due to compression of the same caseload into fewer hours and/or violation of duty hour limits. On the other hand, reports indicate that some resident work is inefficient or has little or no educational value, which suggests that time was and is available during duty hours that could be used to greater advantage from a learning perspective. The committee maintains that ACGME's Residency Review Committees (RRCs), sponsoring institutions, and residency programs need to study and rectify the issue of resident workload so that residents are able to comply with desired duty hours. It is possible that a reduction in noneducational work (e.g., the 8 to 30 percent observed in recent studies; Brasel et al., 2004; Dola et al., 2006) and overall caseload will still accommodate the number of patient educational experiences necessary to achieve competence without extending training time although, for some specialties, additional time may be necessary. It is unlikely that reducing noneducational tasks alone will resolve issues related to having a sufficient resident workforce to provide 24-hour resident coverage under further reduced duty hours, but removing these tasks could be a partial solution and could be achieved more quickly than the time required to generate additional medical school graduates (Jeon and Hurley, 2007). The committee also believes that the often high workload of residents and the compression of work into fewer hours are unrecognized contributors to risks for patient safety and resident well-being; for example, less information gets transferred during handovers when residents are rushed for time, and a workload that is overly heavy for the time allotted adds to stress. Chapter 4 includes information from studies examining the impact of excess workload on educational attainment and the importance of having adequate time for thorough evaluation of patients and reflection.

There is a dearth of information on the resident caseload that provides an optimal learning experience, promotes patient safety or enhances resident well-being. However, as clearly illustrated in Chapter 2, residents continue to violate duty hours and patient needs appear paramount in their decisions to do so, indicating that the current caseload combined with other required tasks exceeds the time available. Chapter 5 includes discussion of the effect of workload on residents. The committee could not obtain uniform information across resident specialties on how many patients residents are admitting or how many patients they are following or cross-covering under current schedules. Such information is necessary to establish caseload guidelines by specialty and rotation. There is no transparency in the number of patients or their complexity that would allow judgment of the appropriate balance between service and education. Additionally, the committee tried to obtain data from specialty certifying boards to determine the impact of the 2003 duty hour rules on educational outcomes, but data are just beginning to be analyzed systematically to determine whether the duty hours have had a significant impact on competency and preparation for board certification (ABMS, 2008).

The committee believes that workload limits should reflect the practices of each specialty, and this is its reason for recommending the ACGME RRCs's role in determining the limits. Here is an opportunity for and a challenge to the RRCs to determine the number and characteristics of patients that are optimal for both resident learning and patient safety on a day-to-day basis and for developing competence over the long term. ACGME already has a process in place as part of its accreditation evaluation to collect information on institutions and residency programs through PIFs (program information forms). This vehicle could be adapted to collect information to make a determination of maximum caseload for residents by specialty (e.g., patient census, number of admissions, number of cases and time as a surgical assistant during the day, cross-coverage). Depending on specialty, some data are already available through this process (e.g., number of residents, average annual caseload of operations, surgical volume, procedures as primary or assistant, admissions, average daily census, average number of patients per resident per shift). The committee believes that RRCs are in the best position to determine the appropriate guidance by training year for the specificity of data collection, maximum daily caseload and admissions, and overall census as well as the diversity of patient cases necessary.

The committee suggests that RRCs do this in consultation with certifying boards to ensure an adequate number and distribution of patient experiences for applicable board certification. RRCs may find specialty society databases (e.g., Society of Thoracic Surgeons national database on outcomes for cardiothoracic patients) (Society of Thoracic Surgeons, 2008) or research on physicians in practice that would inform workload guidelines. For example, a survey of the members of the National Association of Inpatient Physicians (hospitalists) showed that the average daily patient census for these *fully trained* physicians most often fell between 11 and 15 patients, with an average of 6 admissions per admission period, and varied depending on case complexity (Lurie and Wachter, 1999). The authors note that these represent averages, not necessarily optimal workloads, and that caseloads will have to vary according to the complexity of cases managed (Lurie and Wachter, 1999). Currently, only the internal medicine (IM) RRC has established a caseload cap, and there are efforts through the Association of Program Directors in Internal Medicine to reduce the existing cap further and to take case complexity and other factors into account (APDIM, 2008). The current IM cap varies by training year and whether the resident is providing a supervisory function; first-year residents on an inpatient rotation are not to have more than five new patients per admitting day or eight new patients in a 48-hour period, although they may accept other patient transfers. For ongoing care they are not to be responsible for more than 12 patients (ACGME, 2007d).

The committee believes there is a range in the number of patients that a resident may be able to manage depending on the supports available in the training environment, the complexity of the cases, the number of patients needing treatment, and the capability of the individual resident. When establishing specialty-specific recommendations on caseloads, mitigating factors could allow an incremental raising or lowering of the number, such as patient severity and availability of electronic medical records. The complexity of the resident caseload should be monitored after implementation of the RRC guidance and the caseload adjusted accordingly, especially if all straightforward admissions go to non-resident teams leaving only the most complex cases for residents.

The committee concludes that minimizing the time that residents spend in noneducational activities would help programs achieve compliance with duty hours by reducing unnecessary workload and would provide more time for sleep on long shifts and/or for enhancing educational content (Brasel et al., 2004; Schwartz et al., 1992). Knowledge of patient caseload metrics by specialty would have informed the committee's work, and the committee believes such data should be gathered by RRCs to establish guidelines. Residents are in training, and although having a sufficient volume and diversity of patient experiences is necessary to develop compentence, time for daily reflection strengthens learning whether considering various diagnoses, choosing the right intervention, or examining patient outcomes after surgery.

As a result of these findings and complementary findings in other chapters (4 and 5) on workload, the committee recommends the following:

Recommendation 3-1: To ensure that residency programs fulfill their core educational mission, ACGME should require that institutions sponsoring residency programs appropriately adjust resident workload by

- **Providing support services and redesigning healthcare delivery systems to minimize the current level of residents' work that is of limited or no educational value, is extraneous to their graduate medical education program's educational goals and objectives, and can be done well by others; and**
- **Providing residents with adequate time to conduct thorough evaluations of patients and for reflective learning based on their clinical experiences.**

ACGME should require each Residency Review Committee to define and then require appropriate limits on the caseload (e.g., patient census, number of admissions, number of surgical cases to assist per day, cross-coverage) that can be assigned to a resident at a given time, taking into consideration the severity and complexity of patient illness and the level of residents' competency.

CHANGES IN RESPONSE TO DUTY HOUR LIMITS

Programs use a mix of strategies to address the limits of the 2003 ACGME guidelines. Studies mainly document multiple layers of changes necessary to comply with the limits in an attempt to share best practices with others struggling with change. They also report that their first attempts did not always achieve their scheduling goal of being in compliance with the duty hour limits (e.g., Ogden et al., 2006; Yoon, 2007). Programs employed strategies such as making schedule changes, hiring individuals for certain routine support tasks, hiring midlevel providers and physicians to pick up clinical duties, adding residents or fellows to specific services, shifting work within the existing resident team and to faculty, and switching patients to non-teaching units. These reports give guidance on some options that residency programs will have as duty hours are adjusted further.

The frequency of using different strategies appears to vary both by specialty and by programs within specialties. Examples include the following:

- A survey of neurology programs found the following changes, in order of frequency: 75 percent reformed team structures; 42 percent increased staff, attending, or consultant responsibilities and/or coverage; 42 percent rescheduled educational activities; 25 percent added more residents; 25 percent eliminated some elective time and some previously required rotations; 17 percent added physician extenders (e.g., physician assistants, nurse practitioners); and 17 percent instituted night float (Watson, 2005).
- A neurosurgery program survey reported 68 percent of programs added ancillary healthcare professionals and that change did not limit residents' clinical exposure (Cohen-Gadol et al., 2005).
- Family medicine program directors reported that 50 percent of programs increased the patient care responsibilities of attendings, 60 percent eliminated post-call clinics, about 40 percent added night float, and 20 percent added more staff (Peterson et al., 2006).
- A survey of orthopedic surgery residents indicated that 82 percent of their programs used physician assistants, night float systems, and/or home-call to comply (Kusuma et al., 2007).
- A survey of internal medicine chief residents found that 34 percent increased float time (night or day) and decreased elective time (Horwitz et al., 2006b). Overall 76 percent of programs had night float, with a mean of 2.4 months of night float during residency. Non-university-based teaching facilities used night float more often (Wallach et al., 2006).

Thus, depending on numbers of residents, patient complexity, and service needs of individual institutions or programs, various approaches have been implemented to seek to achieve compliance with duty hours and maintain educational quality. Few studies report patient outcomes, and those single-site studies that do tend to be underpowered statistically to determine changes in mortality.

Changes to Scheduling

Hospitals operate on a 24-hour basis and need staffing around the clock 7 days a week. In most training institutions, residents are the first medical doctors to be called to admit and monitor patients. So when residency programs started to redraw their schedules, they sought to assess how and if they could continue to provide the same level of coverage. There are no national estimates of the frequency of specific scheduling strategies used, although some authors have commented that certain strategies (e.g., shift scheduling, night float) have become more common (Kaushal et al., 2004; Yoon, 2007). The purpose of reviewing these reports on scheduling is to examine the variations used in response to duty hour reform and to determine whether there is consensus on scheduling practices. One traditional approach that has continued to be used to schedule resident teams has been to combine daily "short call" (e.g., an 8- to 10-hour day shift) with "long call" (the extended duty period of 24 + 6 hours) every third to fourth day. On short call, residents begin working in the morning and usually admit patients until some designated time in the afternoon; they do not stay overnight (ACGME, 2008). Residents have time during the day to hand over their patients to another team or to a member of their own team that will be on overnight (Carey and Fishburne, 1989). The benefits of this combination of short call and extended duty are seen as having the resident present for daytime educational experiences of rounds and formal didactics, combined with the learning experience of following patients continuously from the time of admission overnight until they have stabilized. A drawback of this approach is the acute sleep deprivation that residents experience if they obtain little or no sleep during the extended duty period. In order to meet the 80-hour limit, programs have generally scheduled extended duty periods every fourth night on average, rather than every third as allowed under ACGME rules (Barden et al., 2002; Mendoza, 2003; Steinbrook, 2002).

Shift schedules (e.g., 8- to 16-hour day or night duty shifts) without using the extended 30-hour duty periods have been suggested as one way to reduce the acute sleep loss associated with extended duty periods, but there are other drawbacks. Regardless of whether using a shift schedule or a combination of short and long call, a period of overlap between schedules facili-

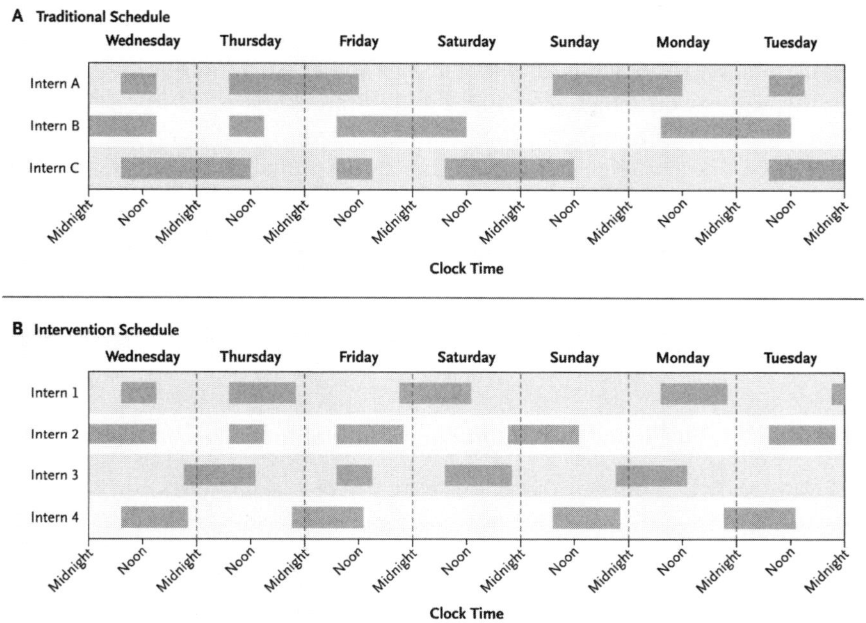

FIGURE 3-1 Representative work hours during a single week for the whole team of interns during the traditional schedule (Panel A) and the intervention schedule (Panel B).
SOURCE: Landrigan et al., 2004.

tates handovers. A drawback of multiple consecutive nights of shift work is that it has negative effects on residents' well-being, alertness, and ability to function (Cavallo et al., 2002). The number of transfers of patient care increases as the number of shifts increase, as Figure 3-1 clearly illustrates (Landrigan et al., 2004). An increased risk of error has been associated with poor communication, and increased numbers of transfers offer a greater opportunity for communication errors (Afessa et al., 2005; Landrigan et al., 2004; Petersen et al., 1994). However, studies have shown that good handover procedures can mitigate communication problems and concerns about continuity of information (Goldstein et al., 2004). Chapter 8 examines ways to improve handovers during these transitions in care.

Examples of Schedule Changes

Many schedule changes represent relatively no-cost strategies to meet duty hour rules. Schedule changes might include having fewer team members on call at a given time, scheduling an extra day off, matching call to patient flow, increasing cross-coverage responsibilities, or shortening call length (Cockerham et al., 2004; Dillingham et al., 2004; Yoon, 2007). Usually, reports on schedule changes make statements about the program's perceived ability or inability to maintain continuity of care, educational experience, and patient outcomes, rather than specific measures. Sometimes a program's first attempt did not achieve compliance with duty hours or created problems that were not anticipated. For example, one IM program used to send its interns home early in the evening to come under the duty hour limits, but then found that interns were not getting enough exposure to admitting their own patients and following those cases. As a result, the hospital reinstated extended duty periods for interns (Yoon, 2007).

Day-Night Shift Models

Although 12-hour shift schedules are the norm in emergency medicine programs, it is only recently that such day-night shifts have been more widely embraced by other specialties as a way to conform to the 80-hour workweek limit. Working a shift at night increases fatigue levels and decreases performance more quickly than working during the day, and because of our innate circadian propensity to fall asleep at night, the night shift worker finds it more difficult to make up a sleep deficit during the day (Akerstedt, 2003; Rosa, 2001). To provide access to care 24 hours a day, some staff will have to work at night. The effects of night shift work and approaches to minimizing sleep deficit through appropriate scheduling are discussed more fully in Chapter 7.

Several studies of limited duration or of small numbers of residents compared shift schedules to those incorporating extended duty periods and assessed compliance with duty hours, and improvements in patient and resident outcomes (Afessa et al., 2005; Goldstein et al., 2004; Landrigan et al., 2004). Each program had its own unique mix of hours and changes to staffing, but they all achieved fewer work hours per week under the 80-hour limit. Because of the increased transfers in care experienced, they scheduled an overlap in shifts to better conduct handovers, and the two programs based in an ICU setting each required the addition of more residents (Afessa et al., 2005; Goldstein et al., 2004; Landrigan et al., 2004). With fewer resident hours available, some programs have chosen to have fewer residents on at night and thereby increased cross-coverage responsibilities (e.g., Cockerham et al., 2004). Otherwise, additional staff would have to be added.

Night Float

Night float is a scheduling strategy that programs in different specialties have now embraced when they did not want to go to a straight shift schedule (ACGME, 2004; Bell, 2005; Calverley, 2003; Darosa et al., 2003; Lieu et al., 1992; Lin et al., 2006; Rosenfeld, 2003; Sanfey et al., 2003). Many internal medicine programs have used night float for years, while others had preferred to provide their own night coverage rather than sign out to another team. However, the limit on duty hours has induced its adoption (Horwitz et al., 2006a,b; Vaughn et al., 2008; Whang et al., 2003).

The perceived benefits of the use of night float are that it improves continuity of care at night for longer-stay patients, allows a more regular sleep cycle for residents rather than alternating days and nights, and gives them more autonomy to gain confidence in their skills. Its perceived drawbacks like any night shift are that it limits access to didactic teaching and rounds with attendings, decreases time with resident's family, eliminates availability for daytime continuity clinics, and decreases operative experience for surgeons (Bell, 2005; Calverley, 2003; Darosa et al., 2003; Lieu et al., 1992; Moore et al., 2000). Programs are working to increase the educational value of night float (Lefrak et al., 2005).

Cavallo disputes the idea that weeks of night float or any night shift configuration are a safer option for residents or for patients than an extended duty period, given the known consequences for workers of night shift work in disrupting the sleep-wake cycle: impaired alertness, more irritability, greater risk of depressive symptoms (Cavallo, 2004; Cavallo et al., 2002, 2003). Since it is necessary to have 24/7 coverage, Cavallo advocates integrating changes to the physical environment, educating residents on the body's adjustment to different work shifts, and encouraging the use of naps, including making sure that residents actually sleep and do not use the time to catch up on other work (Cavallo et al., 2002). Afessa et al. (2006) note that there should be limits to the consecutive nights that a resident works (e.g., four nights) to prevent fatigue. Such strategies are discussed further in Chapter 7.

Day Float

Day float is less controversial than night float. Residents enjoy the quality-of-life benefits of fewer hours during this rotation, but they do not necessarily have more time for conferences or teaching (Roey, 2006; Wong et al., 2004). Their duties might include independently admitting new patients and completing work left from those who stayed overnight (checking on consults, arranging needed studies, writing progress notes, completing the discharge process) (Roey, 2006).

Preferred Scheduling Practices

The committee finds that there is no comprehensive national information on how programs adjusted to the 2003 rules. The past 5 years have been a period of experimentation for programs. Models adopted in the initial year have been replaced with different models and they continue to be refined to improve educational value, the quality of patient care delivered, and service coverage. One of the consequences of the 2003 duty hour rules may have been an increase in shift work schedules, and the number of days and months of night work in a year, but this is not documented nationally and should be part of the ACGME analysis of duty hours and monitoring compliance called for in Chapter 2.

Based on the collective field experiences of programs adapting to the 2003 duty hour rules, the committee concludes that no single scheduling model appears to fit all training facilities or specialties, or even training programs within a particular program or specialty, and that some flexibility will have to be retained. There are advantages and disadvantages to each approach from the perspective of complying with duty hour limits, patient continuity, and potentially, patient safety. Studies tend to report on individual institution-specific adaptations, but there is little rigorous analysis of the effects of specific models across a wider spectrum of sites. The international picture is the same; a variety of call schedules are utilized as noted in Appendix C. In making recommendations for adjustments to current duty hour rules that govern resident scheduling (Chapter 7), the committee draws on the scientific evidence that fatigue is an unsafe condition that can occur relative to the timing and duration of work and sleep opportunities.

Transferring Resident Work to Other Personnel

Earlier in this chapter, work within the resident day that is not educationally valuable is discussed, as well as the need for someone else to substitute for them when residents are not available to provide patient care. Determining who should replace residents depends on the tasks that the substitutes will perform (i.e., task-tailored substitutes). Knickman estimated in 1988 that 20 percent of a medicine resident's time could be replaced only by another physician, 35 percent by midlevel practitioners, 3.4 percent by nurses, 1.2 percent by laboratory technicians, and 6 percent by unskilled workers such as messengers and transporters. The remainder of the time was personal (13.4 percent) or education time (20.8 percent) (Knickman et al., 1992). This mix may be different for other specialties and may have changed in the intervening 20 years. In the New York City system where residents made up 13 percent of the total in the United States at that time, Green and Johnson (1995) projected that a 25 percent decrease in residents would require 2,389 midlevel providers, 1,280 physician full-time equivalents (FTEs), 232 nurse FTEs, 82 laboratory technicians, and 410 unskilled workers (Green and Johnson, 1995). Besides making scheduling changes, in response to the 2003 duty hour limits, programs have added personnel from various categories. The committee commissioned a paper to develop an estimate of the cost of replacing resident hours under different future scenarios. A summary of this work is found in Chapter 9 (Nuckols and Escarce, 2008). Future changes in duty hours will likely require a mix of substitutes different from those used in the past. Since some of the more routine and less complex tasks and patient cases have already been absorbed by others, more physician-level substitutes may be necessary.

Support Services

ACGME has encouraged institutions to transfer tasks such as patient transportation and blood drawing away from residents, as mentioned earlier for mitigating workload. A case can be made that by having someone who has supplies readily available for routine tasks such as blood drawing, electrocardiograms, or IV access, and whose skills are maintained through regular use, there will be efficiencies and perhaps patient safety benefits (Herbertson et al., 2007; Rogers et al., 2005). More recently, there has been a push to relieve residents of the time spent arranging appointments, scheduling tests and procedures, tracking down test results, or completing the paperwork associated with discharging patients. A few very preliminary studies with limited resident pools aim to make the business case as well as a quality argument for transferring these tasks to others; similar approaches warrant further follow-up studies. Moriarty and colleagues (2008) found that their internal medicine interns were spending an average of 187 minutes a day on the phone performing tasks such as arranging for diagnostic tests, making discharge follow-up appointments, obtaining records, and so forth. Adding a medical team assistant reduced resident phone time to 41 minutes. Another program substituted a centralized inpatient appointment service to handle the post-discharge follow-up process to limit intern time on the phone; there were neither duty hour violations nor apparent differences in no-shows, cancellations, readmission rates, or emergency department visits (Bellini, 2008). Last, a health technician on a surgical service picked up approximately 20 nonclinical tasks of interns per day, resulting in interns working 2 to 4 hours less each day and increasing their time in the operating room by 6.5 hours per week (Podnos et al., 2003). To ensure that residents maximize their patient care learning opportunities within their duty hours, the transfer of support service functions will continue to be required.

Physician Extenders

Nurse practitioners (NPs) and physician assistants (PAs), like residents, work under the supervision of a physician. They replace residents in many places, particularly in surgical residencies (Buch et al., 2008; Kirton et al., 2007; Reines et al., 2006; Todd et al., 2004). These physician extenders, also known as "midlevel" providers, have been hired to relieve residents after overnight call so that they could adhere to duty hour limits, to reduce workload by taking on more routine patients with little educational value for residents, and to prevent excess resident work from shifting to faculty (Abrass et al., 2001; Lundberg et al., 2006; Schneider et al., 2007). Replacing residents with non-physician providers can be an expensive option, although not as expensive as replacing them with physicians (Pisetsky et al., 1998).

These practitioners frequently know more than interns about day-to-day operational aspects of patient care, especially early in the academic year, by virtue of their experience and familiarity with the routines of the unit in which they work (Karlowicz and McMurray, 2000; Kirton et al., 2007; Silver and McAtee, 1988). However, if they are new to a unit themselves, these physician extenders may also be learning, and it may take 6 months to 1 year to understand the practices and procedures and develop the necessary skills (Mathur et al., 2005). Staff turnover becomes a key issue.

Few studies give practical guidance on hiring physician extenders. A study by Rudy et al. (1998) in two academic medical centers illustrates some differences in how physician extenders are utilized, and these have implications for determining substitution ratios. NPs or PAs took on traditional tasks of medicine, not nursing tasks, but they cared for 4.9 patients on average compared with 8.7 patients per resident. There were other differences: residents' patients were older and sicker, and residents did more invasive procedures (Kirton et al., 2007; Rudy et al., 1998). Thus, more than one midlevel practitioner would be needed to do the work of one resident. Green and Johnson (1995) estimated that three midlevel practitioners would be required to replace one resident's work hours. A more recent study looking at substitution strategies compared coverage of a pediatric intensive care unit (PICU) by four residents (two during the day and two at night) to a matched PICU with two residents (both on during the day; night call coverage by resident on a subspecialty rotation) paired with PAs. The PAs worked three to four 12-hour shifts per week. To have enough PAs available to replace the two full-time residents, the hospital had to hire 5.5 PAs. Recruitment, training, and turnover of the PAs in a competitive market have been continuing issues for the 5 years the program has been in operation (Mathur et al., 2005). Substitution of other clinical personnel for residents has cost implications as outlined in Chapter 9.

Hiring Additional Physician-Level Staff

Not all resident tasks, especially those of more advanced residents, can be delegated to mid levels; at some point, attending-level physicians (e.g., fellows, faculty, other attendings) must be utilized. The hospitalist movement has grown at the same time that there were requirements for resident replacement (Meltzer et al., 2002; Wachter, 2006). This confluence has allowed some institutions to develop a non-teaching service to prevent the resident service from becoming overloaded (Bellini, 2008). Hiring these physicians is more costly as an initial investment than using residents, although these costs might be recouped since they may independently bill for services. Hiring moonlighting physicians is another stopgap measure to fill resident shoes,[3] but these do not provide as much patient continuity (Horwitz et al., 2006a).

To date, much of the physician-level burden has fallen on staff that is already present in the training facility, although the evidence is mixed on how much the workload of faculty has increased (Klingensmith et al., 2006; Ladd, 2006). When more work shifts to chief residents and to attendings, there is less time for these physicians to teach. The shift in workload to attending-level physician staff at teaching institutions has raised questions about the future attractiveness of becoming a faculty member in an academic medical center (Reed et al., 2007).

Having Additional Residents and Fellows

The number of residency positions (including fellowship positions) in the country that qualifies for Medicare GME funding has not increased since the Balanced Budget Act of 1997 froze the number of positions at 1996 levels. This action was taken because of projections at that time of a physician surplus as well as pressure to reduce federal spending on graduate medical education. Thus, adding more residents to fill in the gap left by reduced residency hours has not been a widely available option since 2003. Some programs added positions through other means such as private funding, closure of another residency program, or transfer of positions among programs (Ladd, 2006). Duty hour reduction has also meant changes in relationships among institutions; for example, one teaching facility had to withdraw all of its medicine residents from the Department of Veterans Affairs hospital after a 35-year relationship in order to have a sufficient number of residents at its main facility (Daschbach, 2008). Even if a training program wanted to add more residents to fill in hours left open due to duty hour constraints, it could not always do so because it had to have the

[3]Personal communication, D. Meltzer, University of Chicago, August 12, 2008.

educational capacity (sufficient cases and faculty) and resources other than Medicare to fund the position.[4]

COSTS OF ADAPTING TO THE 2003 DUTY HOUR LIMITS

The replacement of resident duties and coverage of hours of work by other personnel required as a result of the 2003 ACGME rules came with no dedicated funding from outside sources. Such help for transition funding had been included in House and Senate legislative proposals to regulate resident duty hours; these proposals have not been called up for a vote in either body (GovTrack.us, 2005a,b).

Individual Program Costs

A few programs have reported in the literature and in testimony to the committee that adjusting to duty hours has carried substantial annual recurring costs. These estimates run from $1 million for a single specialty program to $7 million for all residencies across several hospitals (Knapp, 2002; Liekweg, 2008; Noah, 2008; Opas, 2008; Oransky, 2003). These funds primarily went to pay for hiring physician extenders, moonlighting physicians, and hospitalists and to privately fund additional residency positions.

Teasing apart the costs associated with duty hour reduction is difficult. The addition of personnel is the most visible component. Other expenses might include costs of monitoring such as electronic duty hour verification, capital investment in rooms for napping, additional office space for the residents to work in, and hidden costs of additional faculty work (Daschbach, 2008; Opas, 2008).

Medicare is a principal source of payments for graduate medical education, about $8.5 billion in 2007. For graduate medical training facilities without a Medicare population, other sources must suffice. For example, the Los Angeles hospital system received little of its funding from a dedicated source for graduate medical training (e.g., 0.6 percent of its funds came from Medicare because it does not have a large Medicare population). The hospital system and its graduate medical education programs must compete with other county and state needs for appropriations; 70.8 percent of their support comes from state and other federal funds (e.g., Medicaid), 25 percent from county taxes, and 0.6 percent from the DSH (disproportionate share hospital) program (Opas, 2008).

There is a clear need to address the costs of supplementing reduced

[4]Personal communication, S. Hamlin, Cincinnati Children's Hospital Medical Center, February 20, 2008.

resident work hours by hiring other medical providers as well as other expenses. These costs will be a factor in the ability of some but not all residency programs to adapt to future duty hour adjustments. See further discussion of costs in Chapter 9.

REFERENCES

ABMS (American Board of Medical Specialties). 2008. Presentation by Kevin Weiss to the Committee on Optimizing Graduate Medical Trainee (Resident) Hours and Work Schedules to Improve Patient Safety, May 8, 2008, Washington, DC.

Abrass, C. K., R. Ballweg, M. Gilshannon, and J. B. Coombs. 2001. A process for reducing workload and enhancing residents' education at an academic medical center. *Academic Medicine* 76(8):798-805.

ACGME (Accreditation Council for Graduate Medical Education). 2003. *ACGME board of directors approved resident duty hour standards at February meeting*. http://www.acgme.org/acWebsite/newsReleases/newsRel_02_17_03.asp (accessed August 10, 2008).

———. 2004. *2003-2004 annual report*. http://www.acgme.org/acWebsite/annRep/an_2003-04AnnRep.pdf (accessed August 12, 2008).

———. 2007a. *The ACGME's approach to limit resident duty hours 2006-07: A summary of achievements for the fourth year under the common requirements*. http://www.acgme.org/acWebsite/dutyhours/dh_achieve0607.pdf (accessed December 27, 2008).

———. 2007b. *ACGME institutional requirements: Checklist, effective: July 1, 2007*. http://www.acgme.org/acWebsite/irc/IRCheckList-07.pdf (accessed February 28, 2008).

———. 2007c. *Common program requirements: General competencies*. Chicago, IL: ACGME.

———. 2007d. *Data resource book academic year 2005-06*. Chicago, IL: ACGME.

———. 2008. *Accreditation Council for Graduate Medical Education: Glossary of terms*. http://www.acgme.org/acWebsite/about/ab_ACGMEglossary.pdf (accessed July 17, 2008).

Afessa, B., C. C. Kennedy, K. W. Klarich, T. R. Aksamit, J. C. Kolars, and R. D. Hubmayr. 2005. Introduction of a 14-hour work shift model for housestaff in the medical ICU. *Chest* 128(6):3910-3915.

Afessa, B., J. C. Kolars, and R. D. Hubmayr. 2006. Editorial correspondence. *Chest* 130(2): 626.

Aiken, L. H., S. P. Clarke, D. M. Sloane, J. A. Sochalski, R. Busse, H. Clarke, P. Giovannetti, J. Hunt, A. M. Rafferty, and J. Shamian. 2001. Nurses' reports on hospital care in five countries. *Health Affairs* 20(3):43-53.

Akerstedt, T. 2003. Shift work and disturbed sleep/wakefulness. *Occupational Medicine* 53(2):89-94.

Anderson, G., and J. Horvath. 2004. The growing burden of chronic disease in America. *Public Health Reports* 119(3):263-270.

Andrews, R. M., D. E. Stull, I. Fraser, B. Friedman, and R. L. Houchens. 2007. *Serving the uninsured: Safety-net hospitals, 2003*. Rockville, MD: Agency for Healthcare Research and Quality.

APDIM (Association of Program Directors in Internal Medicine). 2008. *APDIM learning environment task force*. http://www.im.org/meetings/Past/2008/Documents/2008%20APDIM%20Spring%20Meeting/Plenary%20Session%20V,%20Bellini.pdf (accessed March 10, 2009).

Armitage, K. B., and R. Rathod. 2003. *Implementation of a resident created electronic medical record (EMR) to improve compliance with duty hours and patient care*. http://www.acgme.org/acWebsite/dutyHours/dh_EMRCleveland.pdf (accessed October 17, 2008).

Arora, V., C. Dunphy, V. Y. Chang, F. Ahmad, H. J. Humphrey, and D. Meltzer. 2006. The effects of on-duty napping on intern sleep time and fatigue. *Annals of Internal Medicine* 144(11):792-798.

Arora, V., E. Georgitis, J. Siddique, B. Vekhter, J. N. Woodruff, H. J. Humphrey, and D. O. Meltzer. 2008a. Association of workload of on-call medical interns with on-call sleep duration, shift duration, and participation in educational activities. *JAMA* 300(10):1146-1153.

Arora, V. M., J. K. Johnson, D. O. Meltzer, and H. J. Humphrey. 2008b. A theoretical framework and competency-based approach to improving handoffs. *Quality and Safety in Health Care* 17(1):11-14.

Barach, P., and J. K. Johnson. 2006. Understanding the complexity of redesigning care around the clinical microsystem. *Quality & Safety in Health Care* 15(Suppl 1):i10-i16.

Baranski, J. V., M. M. Thompson, F. M. J. Lichacz, C. McCann, V. Gil, L. Pasto, and R. A. Pigeau. 2007. Effects of sleep loss on team decision making: Motivational loss or motivational gain? *Human Factors* 49(4):646-660.

Barden, C. B., M. C. Specht, M. D. McCarter, J. M. Daly, and T. J. Fahey III. 2002. Effects of limited work hours on surgical training. *Journal of the American College of Surgeons* 195(4):531-538.

Baskies, M. A., D. E. Ruchelsman, C. M. Capeci, J. D. Zuckerman, and K. A. Egol. 2008. Operative experience in an orthopaedic surgery residency program: The effect of work-hour restrictions. *Journal of Bone & Joint Surgery—American Volume* 90(4):924-927.

Basu, C. B., L. M. Chen, L. H. Hollier, Jr., and S. M. Shenaq. 2004. The effect of the Accreditation Council for Graduate Medical Education duty hours policy on plastic surgery resident education and patient care: An outcomes study. *Plastic & Reconstructive Surgery* 114(7):1878-1886.

Batalden, P. B., and F. Davidoff. 2007a. What is "quality improvement" and how can it transform healthcare? *Quality & Safety in Health Care* 16(1):2-3.

Batalden, P., and F. Davidoff. 2007b. Teaching quality improvement: The devil is in the details. *JAMA* 298(9):1059-1061.

Bazzoli, G. J., L. R. Brewster, G. Liu, and S. Kuo. 2003. Does U.S. hospital capacity need to be expanded? *Health Affairs* 22(6):40-54.

Bell, R. H., Jr. 2005. *How to integrate team, night float, and apprenticeship models into a surgery residency program.* www.facs.org/education/rap/bell2.html (accessed February 16, 2008).

Bellini, L. 2008. Beyond duty hour reform: Redefining the learning environment. Presentation to the Committee on Optimizing Graduate Medical Trainee (Resident) Hours and Work Schedules to Improve Patient Safety, May 8, 2008, Washington, DC.

Bland, K. I., D. A. Stoll, J. D. Richardson, L. D. Britt, and Committee-Surgery Members of the Residency Review. 2005. Brief communication of the Residency Review Committee-Surgery (RRC-S) on residents' surgical volume in general surgery. *American Journal of Surgery* 190(3):345-350.

Blum, N. J., and T. A. Lieu. 1992. Interrupted care: The effects of paging on pediatric resident activities. *American Journal of Diseases of Children* 146(7):806-808.

Bodenheimer, T. 2005. High and rising health care costs. Part 2: Technologic innovation. *Annals of Internal Medicine* 142(11):932-937.

Boex, J. R., and P. J. Leahy. 2003. Understanding residents' work: Moving beyond counting hours to assessing educational value. *Academic Medicine* 78(9):939-944.

Brasel, K. J., A. L. Pierre, and J. A. Weigelt. 2004. Resident work hours: What they are really doing. *Archives of Surgery* 139(5):490-494.

Buch, K. E., M. Y. Genovese, J. L. Conigliaro, S. Q. Nguyen, J. C. Byrn, C. L. Novembre, and C. M. Divino. 2008. Non-physician practitioners' overall enhancement to a surgical resident's experience. *Journal of Surgical Education* 65(1):50-53.

Calverley, J. 2003. *The "Galveston Plan" for resident duty hours.* https://www.acgme.org/acWebsite/ci/resident_duty_hours_PDFs/NightFloat3.pdf (accessed September 5, 2008).

Carayon, P., and A. P. Gurses. 2005. A human factors engineering conceptual framework of nursing workload and patient safety in intensive care units. *Intensive & Critical Care Nursing* 21(5):284-301.

———. 2008. Nursing workload and patient safety—A human factors engineering perspective. In *Patient safety and quality: An evidence-based handbook for nurses.* Rockville, MD: Agency for Healthcare Research and Quality. Original edition, AHRQ Publication No. 08-0043.

Carey, J. C., and J. I. Fishburne. 1989. A method to limit working hours and reduce sleep deprivation in an obstetrics and gynecology residency program. *Obstetrics & Gynecology* 74(4):668-672.

Cavallo, A. 2004. *Night float can affect resident performance and patient safety.* http://www.cincinnatichildrens.org/health/subscribe/horizons/archives/2004/2004-1/nightfloat.htm (accessed August 13, 2008).

Cavallo, A., J. Jaskiewicz, and M. D. Ris. 2002. Impact of night-float rotation on sleep, mood, and alertness: The resident's perception. *Chronobiology International: The Journal of Biological & Medical Rhythm Research* 19(5):893.

Cavallo, A., M. D. Ris, and P. Succop. 2003. The night float paradigm to decrease sleep deprivation: Good solution or a new problem? *Ergonomics* 46(7):653-663.

The Chartis Group. 2007. *White paper: Patient throughput: A critical strategy for success.* New York: The Chartis Group.

Chewning, E. G., and A. M. Harrell. 1990. The effect of information load on decision makers' cue utilization levels and decision quality in a financial distress decision task. *Accounting, Organizations and Society* 15(6):527-542.

Chiles, W. D. 1982. Workload, task, and situational factors as modifiers of complex human performance. In *Human Performance and Productivity*, edited by E. A. Alluisi and E. A. Fleishman. Hillsdale, NJ: Lawrence Erlbaum Associates.

Choo, F. 1995. Auditors' judgment performance under stress: A test of the predicted relationship by three theoretical models. *Journal of Accounting, Auditing, and Finance* 10(3):611-641.

Chung, R. S., and N. Ahmed. 2007. How surgical residents spend their training time: The effect of a goal-oriented work style on efficiency and work satisfaction. *Archives of Surgery* 142(3):249-252; discussion 252.

Cockerham, W. T., J. B. Cofer, P. L. Lewis, C. M. Scroggins, and R. P. Burns. 2004. Resident work hours: Can we meet the ACGME requirements? *American Surgeon* 70(8):687-690.

Cohen, J. J. 1999. Honoring the "E" in GME. *Academic Medicine* 74(2):108-113.

Cohen-Gadol, A. A., D. G. Piepgras, S. Krishnamurthy, and R. D. Fessler. 2005. Resident duty hours reform: Results of a national survey of the program directors and residents in neurosurgery training programs. *Neurosurgery* 56(2):398-403.

COTH (Council of Teaching Hospitals and Health Systems). 2008. *AAMC Teaching Hospitals and Health Systems.* Washington, DC: Association of American Medical Colleges.

Coverdill, J. E., G. L. Adrales, W. Finlay, J. D. Mellinger, K. D. Anderson, B. W. Bonnell, J. B. Cofer, D. B. Dorner, C. Haisch, K. L. Harold, P. M. Termuhlen, and A. L. Webb. 2006a. How surgical faculty and residents assess the first year of the Accreditation Council for Graduate Medical Education duty-hour restrictions: Results of a multi-institutional study. *American Journal of Surgery* 191(1):11-16.

Coverdill, J. E., W. Finlay, G. L. Adrales, J. D. Mellinger, K. D. Anderson, B. W. Bonnell, J. B. Cofer, D. B. Dorner, C. Haisch, K. L. Harold, P. M. Termuhlen, and A. L. B. Webb. 2006b. Duty-hour restrictions and the work of surgical faculty: Results of a multi-institutional study. *Academic Medicine* 81(1):50-56.

Darosa, D. A., R. H. Bell, Jr., and G. L. Dunnington. 2003. Residency program models, implications, and evaluation: Results of a think tank consortium on resident work hours. *Surgery* 133(1):13-23.

Daschbach, C. C. 2008. IOM panel: Resident duty hours cost impact estimates 2003-2008. Presentation to the Committee on Optimizing Graduate Medical Trainee (Resident) Hours and Work Schedules to Improve Patient Safety, March 4, 2008, Irvine, CA.

Dawson, D., and P. Zee. 2005. Work hours and reducing fatigue-related risk: Good research vs good policy. *JAMA* 294(9):1104-1106.

de Virgilio, C., A. Yaghoubian, R. J. Lewis, B. E. Stabile, and B. A. Putnam. 2006. The 80-hour resident workweek does not adversely affect patient outcomes or resident education. *APDS* 63(6):435-439.

Dellit, T. H., B. Armas-Loughran, G. J. Bosl, K. A. Sepkowitz, H. Thaler, and J. Blaskovich. 2001. A method for assessing house staff workload as a function of length of stay. *JAMA* 286(9):1023-1024.

Dillingham, R., J. S. Pope, D. Benson, and G. Donowitz. 2004. *A dynamic strategy for reducing internal medicine work hours.* http://www.acgme.org/outcome/implement/rsvpTemplate.asp?rsvpID=24 (accessed September 5, 2008).

Dismukes, K., G. Young, and R. Sumwalt. 1998. Cockpit interruptions and distractions: Effective management requires a careful balancing act. *ASRS Directline* (10). http://asrs.arc.nasa.gov/publications/directline/dl10_distract.htm (accessed September 5, 2008).

Dola, C., L. Nelson, J. Lauterbach, S. Degefu, and G. Pridjian. 2006. Eighty hour work reform: Faculty and resident perceptions. *American Journal of Obstetrics & Gynecology* 195(5):1450-1456.

Ferguson, C. M., K. C. Kellogg, M. M. Hutter, and A. L. Warshaw. 2005. Effect of work-hour reforms on operative case volume of surgical residents. *Current Surgery* 62(5):535-538.

Flynn, E. A., K. N. Barker, J. T. Gibson, R. E. Pearson, B. A. Berger, and L. A. Smith. 1994. *Impact of interruptions and distractions on dispensing errors in an ambulatory care pharmacy*. Paper presented at ASHP Midyear Clinical Meeting, Miami, FL.

———. 1999. Impact of interruptions and distractions on dispensing errors in an ambulatory care pharmacy. *American Journal of Health-System Pharmacy* 56(13):1319-1325.

Gabow, P. A. 2008. *Transforming delivery systems for quality and efficiency*. AHRQ 2008 Conference: Promoting Quality . . . Partnering for Change, Rockville, MD.

Gabow, P. A., A. Karkhanis, A. Knight, P. Dixon, S. Eisert, and R. K. Albert. 2006. Observations of residents' work activities for 24 consecutive hours: Implications for workflow redesign. *Academic Medicine* 81(8):766-775.

Gilsdorf, J. R. 2008. The disappearing doctors. *Health Affairs* 27(3):850-854.

Gladstone, J. 1995. Drug administration errors: A study into the factors underlying the occurrence and reporting of drug errors in a district general hospital. *Journal of Advanced Nursing* 22(4):628-637.

Goldstein, M. J., E. Kim, W. D. Widmann, and M. A. Hardy. 2004. A 360 degrees evaluation of a night-float system for general surgery: A response to mandated work-hours reduction. *Current Surgery* 61(5):445-451.

Gonzalez, C. 2005. Task workload and cognitive abilities in dynamic decision making. *Human Factors* 47(1):92-101.

GovTrack.us. 2005a. *H.R. 1228—109th Congress: Patient and Physician Safety and Protection Act of 2005.* http://www.govtrack.us/congress/bill.xpd?bill=h109-1228&tab=summary (accessed March 29, 2008).

———. 2005b. *S. 1297—109th congress: Patient and Physician Safety and Protection Act of 2005*. http://www.govtrack.us/congress/bill.xpd?bill=h109-1297&tab=summary (accessed March 29, 2008).

Green, B. A., and T. Johnson. 1995. Replacing residents with midlevel practitioners: A New York City-area analysis. *Health Affairs* 14(2):192-198.

Gregory, D., W. Baigelman, and I. B. Wilson. 2003. Hospital economics of the hospitalist. *Health Services Research* 38(3):905-918.

Griffith III, C. H., E. C. Rich, S. D. Hillson, and J. F. Wilson. 1997. Internal medicine residency training and outcomes. *Journal of General Internal Medicine* 12(6):390-396.

Gurses, A. P., and P. Carayon. 2007. Performance obstacles of intensive care nurses. *Nursing Research* 56(3):185-194.

Hancock, P. A., G. Williams, C. M. Manning, and S. Miyake. 1995. Influence of task demand characteristics on workload and performance. *International Journal of Aviation Psychology* 5(1):63-86.

Harders, M., M. A. Malangoni, S. Weight, and T. Sidhu. 2006. Improving operating room efficiency through process redesign. *Surgery* 140(4):509-516.

Henry Ford Health System. 2008. *Redesigning clinical care*. http://www.henryford.com/body.cfm?id=49541 (accessed September 29, 2008).

Herbertson, R., A. Blundell, and C. Bowman. 2007. The role of clinical support workers in reducing junior doctors hours and improving quality of patient care. *Journal of Evaluation in Clinical Practice* 13(3):449-452.

Hillestad, R. 2008. Statement of Richard Hillestad, Principal Researcher, RAND Corporation. Testimony before the Senate Finance Committee. *The potential benefits and costs of increased adoption of health information technology*. Washington, DC: RAND Corporation.

Hillson, S. D., B. Dowd, E. C. Rich, and M. G. Luxenberg. 1992. Call nights and patient care: Effects on inpatients at one teaching hospital. *Journal of General Internal Medicine* 7(4):405-410.

Hirst, W., and D. Kalmar. 1987. Characterizing attentional resources. *Journal of Experimental Psychology: General* 116(1):68-81.

Horwitz, L. I., H. M. Krumholz, M. L. Green, and S. J. Huot. 2006a. Transfers of patient care between house staff on internal medicine wards: A national survey. *Archives of Internal Medicine* 166(11):1173-1177.

Horwitz, L. I., H. M. Krumholz, S. J. Huot, and M. L. Green. 2006b. Internal medicine residents' clinical and didactic experiences after work hour regulation: A survey of chief residents. *Journal of General Internal Medicine* 21(9):961-965.

Horwitz, L. I., M. Kosiborod, Z. Lin, and H. M. Krumholz. 2007. Changes in outcomes for internal medicine inpatients after work-hour regulations. *Annals of Internal Medicine* 147(2):97-103.

Hutter, M. M., K. C. Kellogg, C. M. Ferguson, W. M. Abbott, and A. L. Warshaw. 2006. The impact of the 80-hour resident workweek on surgical residents and attending surgeons. *Annals of Surgery* 243(6):864-871; discussion 871-875.

IOM (Institute of Medicine). 2001. *Exploring innovation and quality improvement in health care micro-systems: A cross-case analysis*. Washington, DC: National Academy Press.

Jagsi, R., D. F. Weinstein, J. Shapiro, B. T. Kitch, D. Dorer, and J. S. Weissman. 2008. The Accreditation Council for Graduate Medical Education's limits on residents' work hours and patient safety: A study of resident experiences and perceptions before and after hours reductions. *Archives of Internal Medicine* 168(5):493-500.

Jeon, S., and J. Hurley. 2007. The relationship between physician hours of work, service volume and service intensity. *Canadian Public Policy* 33(Suppl):17-30.

Jung, D. I., J. J. Sosik, and K. B. Baik. 2002. Investigating work group characteristics and performance over time: A replication and cross-cultural extension. *Group Dynamics* 6(2):153-171.

Karlowicz, M. G., and J. L. McMurray. 2000. Comparison of neonatal nurse practitioners' and pediatric residents' care of extremely low-birth-weight infants. *Archives of Pediatrics & Adolescent Medicine* 154(11):1123-1126.

Kaushal, H., J. Fischer, H. Khurana, and A. Reddy. 2004. Does night float impact the quality of the admit history and physical. *Journal of General Internal Medicine* 19(Suppl 1):145.

Kirton, O. C., M. A. Folcik, M. E. Ivy, R. Calabrese, E. Dobkin, J. Pepe, J. Mah, K. Keating, and M. Palter. 2007. Midlevel practitioner workforce analysis at a university-affiliated teaching hospital. *Archives of Surgery* 142(4):336-341.

Klingensmith, M. E., E. R. Winslow, B. H. Hamilton, and B. L. Hall. 2006. Impact of resident duty-hour reform on faculty clinical productivity. *Current Surgery* 63(1):74-79.

Knapp, R. 2002. Financing graduate medical education and limiting resident work hours: A political assessment. *American Journal of Surgery* 184(3):187-195.

Knickman, J. R., M. Lipkin, Jr., S. A. Finkler, W. G. Thompson, and J. Kiel. 1992. The potential for using non-physicians to compensate for the reduced availability of residents. *Academic Medicine* 67(7):429-438.

Kozak, L. J., C. J. DeFrances, and M. J. Hall. 2006. *National hospital discharge survey: 2004 annual summary with detailed diagnosis and procedure data.* Hyattsville, MD: National Center for Health Statistics.

Kusuma, S. K., S. Mehta, M. Sirkin, A. J. Yates, T. Miclau, K. J. Templeton, and G. E. Friedlaender. 2007. Measuring the attitudes and impact of the eighty-hour workweek rules on orthopaedic surgery residents. *Journal of Bone & Joint Surgery—American Volume* 89(3):679-685.

Ladd, A. P. 2006. Pediatric surgery fellowship compliance to the 80-hour work week. *Journal of Pediatric Surgery* 41(4):687-692; discussion 691-692.

Lamberg, L. 2004. Impact of long working hours explored. *JAMA* 292(1):25-26.

Landrigan, C. P., J. M. Rothschild, J. W. Cronin, R. Kaushal, E. Burdick, J. T. Katz, C. M. Lilly, P. H. Stone, S. W. Lockley, D. W. Bates, and C. A. Czeisler. 2004. Effect of reducing interns' work hours on serious medical errors in intensive care units. *New England Journal of Medicine* 351(18):1838-1848.

Lang, T. A., M. Hodge, V. Olson, P. S. Romano, and R. L. Kravitz. 2004. Nurse-patient ratios: A systematic review on the effects of nurse staffing on patient, nurse employee, and hospital outcomes. *Journal of Nursing Administration* 34(7-8):326-337.

Larson, E. B. 2003. Twenty-first century hospitals: Intensification increases. *Health Services Research* 38(3):919-922.

Lawler, L. P., J. Fromke, R. G. Jost, and R. G. Evens. 2001. Results of and comments on the 2000 survey of the American Association of Academic Chief Residents in Radiology. *Academic Radiology* 8(8):777-781.

Lefrak, S., S. Miller, B. Schirmer, and H. Sanfey. 2005. The night float system: Ensuring educational benefit. *American Journal of Surgery* 189(6):639-642.

Levin, S., D. Aronsky, R. Hemphill, J. Han, J. Slagle, and D. J. France. 2007. Shifting toward balance: Measuring the distribution of workload among emergency physician teams. *Annals of Emergency Medicine* 50(4):419-423.

Liekweg, R. 2008. Resident work hours: A look at the University of California experience. Presentation to the Committee on Optimizing Graduate Medical Trainee (Resident) Hours and Work Schedules to Improve Patient Safety, March 4, 2008, Irvine, CA.

Lieu, T. A., C. B. Forrest, N. J. Blum, D. Cornfeld, and R. A. Polin. 1992. Effects of a night-float system on resident activities and parent satisfaction. *American Journal of Diseases of Children* 146(3):307-310.

Lin, G. A., D. C. Beck, and J. M. Garbutt. 2006. Residents' perceptions of the effects of work hour limitations at a large teaching hospital. *Academic Medicine* 81(1):63-67.

Ludmerer, K. M. 1999. *Time to heal: American medical education from the turn of the century to the era of managed care*. New York: Oxford University Press.

Lundberg, S., S. Wali, P. Thomas, and D. Cope. 2006. Attaining resident duty hours compliance: The acute care nurse practitioners program at Olive View-UCLA medical center. *Academic Medicine* 81(12):1021-1025.

Lurie, J. D., and R. M. Wachter. 1999. Hospitalist staffing requirements. *Effective Clinical Practice* 2(3):126-130.

Mathis, B. R., T. Diers, R. Hornung, M. Ho, and G. W. Rouan. 2006. Implementing duty-hour restrictions without diminishing patient care or education: Can it be done? *Academic Medicine* 81(1):68-75.

Mathur, M., A. Rampersad, K. Howard, and G. M. Goldman. 2005. Physician assistants as physician extenders in the pediatric intensive care unit setting—A 5-year experience. *Pediatric Critical Care Medicine* 6(1):14-19.

Meltzer, D., W. G. Manning, J. Morrison, M. N. Shah, L. Jin, T. Guth, and W. Levinson. 2002. Effects of physician experience on costs and outcomes on an academic general medicine service: Results of a trial of hospitalists. *Annals of Internal Medicine* 137(11):866-874.

Mendoza, B. 2003. Beat the clock: Challenges to training the new surgeon. *Current Surgery* 60(3):319-320.

Mohr, J., P. Batalden, and P. Barach. 2004. Integrating patient safety into the clinical microsystem. *Quality & Safety in Health Care* 13(Suppl 2):ii34-ii38.

Moore, S. S., M. D. Nettleman, S. Beyer, K. Chalasani, R. J. Fairbanks, M. Goyal, and M. Carter. 2000. How residents spend their nights on call. *Academic Medicine* 75(10):1021-1024.

Moriarty, J. P., G. Jenq, and S. J. Hout. 2008. Use of a medical team assistant to restructure tasks performed by interns. Paper read at 2008 APDIM Spring Meeting, Building an Effective Educational Environment, New Orleans, LA.

Morton, J. M., C. C. Baker, T. M. Farrell, M. E. Yohe, R. J. Kimple, D. C. Herman, P. Udekwu, J. A. Galanko, K. E. Behrns, and A. A. Meyer. 2004. What do surgery residents do on their call nights? *American Journal of Surgery* 188(3):225-229.

Noah, M. 2008. Resident duty hour restrictions: Cost impact. Presentation to the Committee on Optimizing Graduate Medical Trainee (Resident) Hours and Work Schedules to Improve Patient Safety, March 4, 2008, Irvine, CA.

Nuckols, T. K., and J. J. Escarce. 2008 (unpublished). Cost analysis for the Institute of Medicine project, "Optimizing Graduate Medical Trainee (Resident) Hours and Work Schedules to Improve Patient Safety." RAND Corporation. Available at www.iom.edu/residenthours or www.iom.edu/hcs.

Ogden, P. E., S. Sibbitt, M. Howell, D. Rice, J. O'Brien, R. Aguirre, P. Naus, and D. DiPette. 2006. Complying with ACGME resident duty hours restrictions: Restructuring the 80-hour workweek to enhance education and patient safety at Texas A & M Scott & White Memorial Hospital. *Academic Medicine* 81(12):1026-1031.

Ong, M., A. Bostrom, A. Vidyarthi, C. McCulloch, and A. Auerbach. 2007. House staff team workload and organization effects on patient outcomes in an academic general internal medicine inpatient service. *Archives of Internal Medicine* 167(1):47-52.

Opas, L. 2008. IOM workshop: Cost impact of duty hour change. Presentation to the Committee on Optimizing Graduate Medical Trainee (Resident) Hours and Work Schedules to Improve Patient Safety, March 4, 2008, Irvine, CA.

Oransky, I. 2003. Limits imposed on residents' work hours. Reduced workload said to increase patient—and resident—safety. *Lancet* 362(9381):378-379.

Parekh, V., B. Sharpe, A. Vidyarthi, K. E. Fletcher, and P. Cornia. 2005. The impact of resident work hour limits on internal medicine residents' continuity clinic experience. *Journal of General Internal Medicine* 20:165.

Parshuram, C. S., S. Dhanani, J. A. Kirsh, and P. N. Cox. 2004. Fellowship training, workload, fatigue and physical stress: A prospective observational study. *CMAJ: Canadian Medical Association Journal* 170(6):965-970.

Petersen, L. A., T. A. Brennan, A. C. O'Neil, E. F. Cook, and T. H. Lee. 1994. Does housestaff discontinuity of care increase the risk for preventable adverse events? *Annals of Internal Medicine* 121(11):866.

Petersen, L. A., E. J. Orav, J. M. Teich, A. C. O'Neil, and T. A. Brennan. 1998. Using a computerized sign-out program to improve continuity of inpatient care and prevent adverse events. *Joint Commission Journal on Quality Improvement* 24(2):77-87.

Peterson, G. M., M. S. Wu, and J. K. Bergin. 1999. Pharmacist's attitudes towards dispensing errors: Their causes and prevention. *Journal of Clinical Pharmacy & Therapeutics* 24(1):57-71.

Peterson, L. E., H. Johnson, P. A. Pugno, A. Bazemore, and R. L. Phillips, Jr. 2006. Training on the clock: Family medicine residency directors' responses to resident duty hours reform. *Academic Medicine* 81(12):1032-1037.

Pisetsky, M. A., D. A. Lubarsky, B. P. Capehart, C. K. Lineberger, and J. G. Reves. 1998. Valuing the work performed by anesthesiology residents and the financial impact on teaching hospitals in the United States of a reduced anesthesia residency program size. *Anesthesia & Analgesia* 87(2):245-254.

Podnos, Y. D., R. A. Williams, J. C. Jimenez, E. A. Stemmer, I. L. Gordon, and S. E. Wilson. 2003. Reducing the noneducational and nonclinical workload of the surgical resident; defining the role of the health technician. *Current Surgery* 60(5):529-532.

Rahman, A., and S. Haque. 1992. Perceived workload and performance of shift workers. *Journal of Human Ergology* 21(1):99-103.

Reason, J. 1990. *Human error*. Cambridge, UK: Cambridge University Press.

Reed, D. A., R. B. Levine, R. G. Miller, B. H. Ashar, E. B. Bass, T. N. Rice, and J. Cofrancesco, Jr. 2007. Effect of residency duty-hour limits: Views of key clinical faculty. *Archives of Internal Medicine* 167(14):1487-1492.

Reines, H. D., L. Robinson, M. Duggan, B. M. O'Brien, and K. Aulenbach. 2006. Integrating midlevel practitioners into a teaching service. *American Journal of Surgery* 192(1):119-124.

Roey, S. 2006. Medical education and the ACGME duty hour requirements: Assessing the effect of a day float system on educational activities. *Teaching & Learning in Medicine* 18(1):28-34.

Rogers, A. E., W. T. Hwang, L. D. Scott, L. H. Aiken, and D. F. Dinges. 2004. The working hours of hospital staff nurses and patient safety. *Health Affairs* 23(4):202-212.

Rogers, F., S. Shackford, S. Daniel, B. Crookes, K. Sartorelli, W. Charash, P. Igneri, F. A. Luchette, and S. Brotman. 2005. Workload redistribution: A new approach to the 80-hour workweek. *Journal of Trauma-Injury, Infection and Critical Care* 58(5):911-916.

Rosa, R. R. 2001. Examining work schedules for fatigue: It's not just hours of work. In *Stress, workload, and fatigue*, edited by P. A. Hancock and P. A. Desmond. Mahwah, NJ: Lawrence Erlbaum Associates, Inc. Pp. 513-528.

Rosenfeld, J. C. 2003. *From >110 to <80: More work for the program director a better educational experience for the residents.* http://www.acgme.org/outcome/implement/rsvp Template.asp?rsvpID=17 (accessed August 12, 2008).

Rosevear, G. C., and N. E. Gary. 1989. Changes in admissions, lengths of stay, and discharge diagnoses at a major university-affiliated teaching hospital: Implications for medical education. *Academic Medicine* 64(5):253-258.

Rudy, E. B., L. J. Davidson, B. Daly, J. M. Clochesy, S. Sereika, M. Baldisseri, M. Hravnak, T. Ross, and C. Ryan. 1998. Care activities and outcomes of patients cared for by acute care nurse practitioners, physician assistants, and resident physicians: A comparison. *American Journal of Critical Care* 7(4):267-281.

Sanfey, H., S. Lefrak, S. Miller, and B. Schirmer. 2003. *Night float at the University of Virginia's surgery program: Ensuring educational benefit.* Department of Surgery, University of Virginia Health System, Charlottesville, VA. http://www.acgme.org/acWebsite/dutyHours/dh_UVAsurgeryprogram.pdf (accessed September 5, 2008).

Schneider, J. R., J. J. Coyle, E. R. Ryan, R. H. Bell, Jr., and D. A. DaRosa. 2007. Implementation and evaluation of a new surgical residency model. *Journal of the American College of Surgeons* 205(3):393-404.

Schwartz, R. J., T. J. Dubrow, R. F. Rosso, R. A. Williams, J. A. Butler, and S. E. Wilson. 1992. Guidelines for surgical residents' working hours. Intent vs reality. *Archives of Surgery* 127(7):778-782.

Scott, L. D., A. E. Rogers, W. T. Hwang, and Y. Zhang. 2006. Effects of critical care nurses' work hours on vigilance and patients' safety. *American Journal of Critical Care* 15(1):30-37.

Shin, S., R. Britt, and L. D. Britt. 2008. Effect of the 80-hour work week on resident case coverage: Corrected article. *Journal of the American College of Surgeons* 207(1):148-150.

Silver, H. K., and P. McAtee. 1988. Speaking out. Should nurses substitute for house staff? *American Journal of Nursing* 88(12):1671-1673.

Society of Thoracic Surgeons. 2008. *Welcome to the STS national database.* http://www.sts.org/sections/stsnationaldatabase (accessed September 25, 2008).

Speier, C., J. S. Valacich, and I. Vessey. 1997. Task interruption and information presentation on individual decision making. Paper read at Conference on Information Systems, Atlanta, GA.

Steinbrook, R. 2002. The debate over residents' work hours. *New England Journal of Medicine* 347(16):1296-1302.

Tarnow-Mordi, W. O., C. Hau, A. Warden, and A. J. Shearer. 2000. Hospital mortality in relation to staff workload: A 4-year study in an adult intensive-care unit. *Lancet* 356(9225):185-189.

Tibby, S., J. Correa-West, A. Durward, L. Ferguson, and I. Murdoch. 2004. Adverse events in a paediatric intensive care unit: Relationship to workload, skill mix and staff supervision. *Intensive Care Medicine* 30(6):1160-1166.

Todd, B. A., A. Resnick, R. Stuhlemmer, J. B. Morris, and J. Mullen. 2004. Challenges of the 80-hour resident work rules: Collaboration between surgeons and nonphysician practitioners. *Surgical Clinics of North America* 84(6):1573-1586.

Tucker, A. L., and S. J. Spear. 2006. Operational failures and interruptions in hospital nursing. *Health Services Research* 41(3 Pt 1):643-662.

Vander Wood, M. A., K. L. O'Connell, and J. J. Pilcher. 2007. Teamwork on a cognitive task during a night of sleep deprivation and sustained operations. Paper read at Proceedings of the Human Factors and Ergonomics Society 51st Annual Meeting, Baltimore, MD.

Vaughn, D. M., C. L. Stout, B. L. McCampbell, J. R. Groves, A. I. Richardson, W. K. Thompson, M. L. Dalton, and D. K. Nakayama. 2008. Three-year results of mandated work hour restrictions: Attending and resident perspectives and effects in a community hospital. *American Surgeon* 74(6):542-546.

Vidyarthi, A. R., A. D. Auerbach, R. M. Wachter, and P. P. Katz. 2007. The impact of duty hours on resident self reports of errors. *Journal of General Internal Medicine* 22(2):205-209.

Vincent, C., S. Taylor-Adams, and N. Stanhope. 1998. Framework for analysing risk and safety in clinical medicine. *BMJ* 316(7138):1154-1157.

Viney, M. A. 2008. *Redesigning equipment to reduce time wasted and increase time spent at bedside: From the experts.* http://www.rwjf.org/pr/product.jsp?id=30632 (accessed July 22, 2008).

Vogeli, C., A. E. Shields, T. A. Lee, T. B. Gibson, W. D. Marder, K. B. Weiss, and D. Blumenthal. 2007. Multiple chronic conditions: Prevalence, health consequences, and implications for quality, care management, and costs. *Journal of General Internal Medicine* 22(Suppl 3):391-395.

Volpp, K. G. 2008. A delicate balance: Physician work hours, patient safety, and organizational efficiency. *Circulation* 117(20):2580-2582.

Volpp, K. G., and C. P. Landrigan. 2008. Building physician work hour regulations from first principles and best evidence. *JAMA* 300(10):1197-1199.

Wachter, R. M. 2006. The hospitalist movement 10 years later: Life as a Swiss army knife. *Medscape General Medicine* 8(3):30.

Wallach, S. L., K. Alam, N. Diaz, and D. Shine. 2006. How do internal medicine residency programs evaluate their resident float experiences? *Southern Medical Journal* 99(9):919-923.

Watson, J. C. 2005. Impact of the ACGME work hour requirements: A neurology resident and program director survey. *Neurology* 64(2):E11-E15.

Weinger, M. B., S. B. Reddy, and J. M. Slagle. 2004. Multiple measures of anesthesia workload during teaching and nonteaching cases. *Anesthesia and Analgesia* 98(5):1419-1425.

Whang, E. E., M. M. Mello, S. W. Ashley, and M. J. Zinner. 2003. Implementing resident work hour limitations: Lessons from the New York State experience. *Annals of Surgery* 237(4):449-455.

Wiener, E. L., R. E. Curry, and M. L. Faustina. 1984. Vigilance and task load: In search of the inverted U. *Human Factors* 26(2):215-222.

Wong, J. G., E. S. Holmboe, and S. J. Huot. 2004. Teaching and learning in an 80-hour work week: A novel day-float rotation for medical residents. *Journal of General Internal Medicine* 19(5 Pt 2):519-523.

Yoon, H. H. 2007. Adapting to duty-hour limits—Four years on. *New England Journal of Medicine* 356(26):2668-2670.

4

Improving the Resident Learning Environment

The primary goal of graduate medical training is for residents to achieve sufficient competence to deliver safe and effective patient care when they enter into practice. The inherent inexperience of residents as they train need not affect patient safety if they are adequately supervised by more experienced physicians guiding them toward gradual independence. Reduced work hours implemented in 2003, some believe, pose a risk to the acquisition of competencies and to the continuity of care from both an educational and a patient safety perspective.

The committee could not determine the full positive or negative effects of the 2003 limits on educational outcomes because sufficient data on those outcomes are not yet available. However, substantial evidence about how people learn stresses the importance of having a reasonable workload, sufficient time for reflection, and the need for sleep to consolidate learning. New educational designs (e.g., curriculum restructuring, competency-based training, simulation-based training) along with workload and scheduling redesigns should be promoted to incorporate these approaches into the resident environment to maximize learning within fewer duty hours.

Although residents are critically important to delivering direct patient care in teaching hospitals, the fundamental goal of residency training is education. It is through residency that physicians-in-training are transformed from novices into experienced professionals, providing society with competent and compassionate healers for the future. The Association of American Medical Colleges (AAMC, 2006) has recently reaffirmed that residents are "first and foremost learners" and that "a resident's educational needs should be the primary determinant of any assigned patient care services." Similarly, the Accreditation Council for Graduate Medical Education (ACGME) has established in its requirements for all residency programs that (1) the learning objectives of the program must not be compromised by excessive reliance on residents to fulfill service obligations, and (2) didactic and clinical education must have priority in the allotment of residents' time and energy (ACGME, 2008b). Throughout the history of residency training, hospitals have insisted that trainees perform an extraordinary range and amount of ancillary responsibilities that are often noneducational in nature (Ludmerer, 1999). However, while education may be the primary objective of residency training, the nature of residency training—participating in direct patient care—requires that patient safety never be separated from that education. Residency programs implicitly assume responsibility for protecting the patient during the educational experience, thus forming a "social contract" between patients and teaching care settings. In this setting, patients agree to have doctors in training at various milestones in their education, with variations in skills and competencies, provide their care in exchange for a social good—the production of future doctors.

In order to better understand graduate medical education, this chapter looks at key educational principles underlying residency training, the way in which the 2003 duty hour limits have affected them, and at how residency training can be informed by the research literature on the way people learn. It concludes with a look at what is known about educational outcomes in residency programs as they have adapted to the 2003 duty hour limits and presents illustrative innovative educational approaches that may facilitate adaptations to resident duty hours and scheduling.

EDUCATIONAL PRINCIPLES

Three cardinal educational principles underlie residency education: (1) the gradual or graded assumption of responsibility for patient care while under supervision, (2) adequate time to engage in reflective learning, and (3) sufficient continuity in the care of individual patients to understand the natural evolution of illness and to reinforce professionalism and its obligations. Educationally, what matters most in residency training is not the number of duty hours but whether an adequate learning environment exists to satisfy these three principles during those hours (Ludmerer, 1999). Instead of enhancing the learning environment for residents, implementation of the 2003 ACGME requirements is perceived by some educators to have weakened the educational environment in many programs (Charap, 2004; Fitzgibbons et al., 2006; Ludmerer and Johns, 2005; Ryan, 2005), pushing education away from key elements (e.g., adequate time for teaching and reflective learning) that would promote safety and better supervision. To change residency programs so that these positive elements can be enhanced instead of diminished, the interplay of many organizational factors must be supported and reinforced for effective training to result (Salas and Cannon-Bowers, 2000, 2001).

Graded Responsibility for Patient Care Under Supervision

An intrinsic challenge of graduate medical education (GME) has been to find a balance between the educational needs of residents, who require increasing independence as they learn, and the safety needs of patients, who may benefit from being cared for by more experienced physicians. The tension between these two aspects has become more obvious over the past few decades along with a growing attention to safety in medical care. Patients admitted to hospitals have been much sicker, and mistakes of omission and commission by any care provider may have more adverse consequences today than before (Ludmerer, 1999).

A defining characteristic of GME is the assumption of progressively greater patient care responsibility by residents. This type of training is necessary, lest the country face the predicament of future patients' being cared for by inadequately trained doctors (Kennedy et al., 2007). Residents can become effective independent physicians and assume full responsibility for patient care only after having acquired the competencies necessary to manage patients safely and well. To acquire this capacity, residents conduct initial evaluations of patients, make preliminary decisions about diagnosis and therapy, perform procedures, and administer treatments under the level of supervision appropriate for their developing competency—with the understanding that all residents are accountable to attending physicians. The tension that results from the need of the resident to have gradual responsibility under appropriate supervision and the desire to provide optimal and safe care is always present and must be managed carefully to protect patients.

Despite limited research on the use of on-the-job training (OJT) in health care, OJT has been widely used and validated in other fields as an effective training method (Barron et al., 1997; Becker, 1975; Mincer, 1962; Rothwell and Kazanas, 2004; Veum, 1999). In medicine, the validity of a graded responsibility model through in-hospital OJT has been grounded in its compelling inherent logic and rationale, and endorsed by generations of experienced teachers (Kennedy et al., 2005). However, it has not been evaluated systematically against an alternative education model. Aspects of the graded responsibility model are supported in the psychological literature, in particular a five-stage model of skill acquisition: novice, advanced beginner, competent, proficient, and expert (Batalden et al., 2002; Dreyfus and Dreyfus, 1986). In the context of medical residency, the intent of in-hospital training is to deepen existing competencies and teach new ones in a manner that moves residents further along the pathway from novice to expert (Jacobs, 2003; Rothwell and Kazanas, 2004). At times it can be difficult for attendings and faculty clinicians to assess the competency level of individual residents and determine the ideal degree of interaction that might suit them, but efforts to do so more effectively have been examined (Kennedy et al., 2007) and further development and learning of such methods may be useful in determining optimal supervision levels for individual residents.

Role of Supervision in Providing Graded Responsibility

Along the pathway of skill acquisition, supervision is the single most important element upon which this education model depends. In this context, supervision in medicine has been defined as (Kilminster and Jolly, 2000):

> The provision of monitoring, guidance and feedback on matters of personal, professional and educational development in the context of the doctor's care of patients. This would include the ability to anticipate a doctor's strengths and weaknesses in particular clinical situations in order to maximize patient safety.

Supervisory practices that enhance resident learning and performance are (1) the involvement of role models and mentors who demonstrate appropriate professional practice (Hough, 2008); (2) specific learning objectives communicated to learners in advance of their interactions with patients; (3) periodic assessment of how well learners have met those objectives (Jacobs, 2003; Rothwell and Kazanas, 2004; Salas and Cannon-Bowers, 2000); and (4) timely and actionable feedback to residents (Arco, 2008). This report raises concerns regarding the current application of supervisory practices in the context of both learning and patient safety.

Links Between Supervision and Patient Safety

Supervision was a key issue when patient safety and long duty hours were examined in 1987 by the Bell Commission, which originally recommended the 80-hour duty limit for residents, and it remains so today. Even prior to the Bell Commission's findings, the grand jury for the *Zion* case stated the following as part of its ruling (New York Supreme Court, 1986):

> A hospital is not the place for recently graduated doctors to grow and develop in isolation; rather it is a place where the learning process should continue under strict supervision. Thus, medical decisions, whether in an emergency room or on a hospital floor should not be made by inexperi-

enced interns and junior residents without in-person consultations with more senior physicians.

Dr. Bell himself has subsequently written repeatedly that better supervision, not only regulation of hours, is the key to improving the quality of patient care (Bell, 1993, 2003, 2007). And since the time of the *Zion* case, the increasing complexity of patients' illnesses and advancement of medical tools has strengthened the need for good supervision.

After the 1984 *Libby Zion* case brought attention to the issue of resident duty hours and fatigue on patient safety, several reports were published that examined the link between medical errors and resident supervision. A review of the effects of supervision by Kilminster and Jolly (2000) found that "supervision has a positive effect on patient outcomes and that lack of supervision is harmful for patients." The authors view supervision as a distinct intervention with variable outcomes depending on the work and learning environment and its orientation toward teaching.

A number of studies have found that closer resident supervision can lead to fewer errors and improved quality of care (Fallon et al., 1993; Gennis and Gennis, 1993; Singh et al., 2007; Sox et al., 1998). An attending physician's review of a resident's report on a patient case is more likely to result in a change in patient management when the attending sees the patient directly (Gennis and Gennis, 1993), and the impact of better supervision is likely to be more marked among less experienced residents (Fallon et al., 1993). Studies report higher death rates when residents are under poor supervision in surgery, anesthesia, emergency medicine, obstetrics, and pediatrics (McKee and Black, 1992), and report decreased complications and mortality rates when surgical residents are supported by the presence of attendings (Fallon et al., 1993). Residents' compliance with care guidelines has been found to be greater under direct supervision (Sox et al., 1998). Direct supervision of residents can also help them acquire skills more quickly and increase their comfort level in performing invasive procedures (Huang et al., 2006; Osborn et al., 1993; Smith et al., 2004). Finally, residents tend to use more resources, such as test ordering, when they are less supervised (Griffith et al., 1996).

Supervision in Practice

Since the time of the Bell Commission, requirements for supervision have been strengthened in Medicare reimbursement policies and ACGME guidance. ACGME requires "sound supervision" policies from institutions and program directors (ACGME, 2007, 2008b). Under these principles, however, there is latitude in the way each program outlines how graduated responsibility and supervision will interact, and how supervision is implemented in practice (ACGME, 2008b). For example, in an intensive care unit (ICU) the supervising attending might be onsite 24 hours a day, or be expected to be readily available by phone (e.g., within 5 minutes) and able to be at the bedside within a reasonable period (e.g., 20 minutes to 1 hour). For insurance payment purposes, attending physicians are required by Medicare's 1996 Teaching Physician Presence Rules to include progress notes and documentation of their presence during operative procedures in a patient's medical record.[1]

While residents are required to consult with their supervising attending physician about their assessment of a patient, the proposed treatment plan, and any key decisions in the patient's course of treatment, residents perform many of their duties without "over-the-shoulder" supervision. The degree of direct supervision varies by specialty, rotation, the tasks residents are undertaking, and the resident's year of training. An example of graded responsibility is illustrated by first-year surgical residents gaining exposure to what are considered more fundamental skills, such as performing basic suturing skills and placing central and arterial lines, but being expected to master such procedures by their second year of residency. Likewise, a second-year surgical resident might be restricted to performing a laparascopic cholecystectomy from the left side of the operating table (where visibility of the operation is greater and access to the organ easier), but by their fourth year in training that resident would expected to know how to perform the procedure from both sides of the table (Brody School of Medicine, 2008). A supervisor is generally present or accessible in each of these instances, but the degree of supervision may depend on the competence level individual residents demonstrate for each acquired skill; with some residents requiring more hands-on guidance than others.

Good Supervisory Practices

Especially important in the supervisory relationship are the following: continuity in mentoring over time, the supervisor's skill at providing oversight and promoting intellectual autonomy among trainees, and the opportunity for both trainee and supervisor to reflect on their work (Kilminster and Jolly, 2000). Of course, resident supervisors need to be clinically competent themselves as well as informed regarding effective learning processes. In particular, the way in which they communicate their knowledge is what matters to resident training. Trainees need clear feedback about their judgments; corrections must be conveyed unambiguously so that trainees are aware of potential mistakes and any weaknesses they may have (Kluger and

---

[1]CMS (Centers for Medicare and Medicaid Services). 2005. 42 CFR 4172(a) evolution of Medicare billing regulations. *Medicare Claims Processing Manual.*

DeNisi, 1996). Helpful supervisory behaviors include giving direct guidance on clinical work; discussing links between theory and practice; participating in joint problem solving; and offering feedback, reassurance, and role modeling (Kilminster and Jolly, 2000). Rigidity, intolerance, lack of empathy, failure to offer support, lack of concern with teaching, and overemphasis on the evaluative aspects of supervision can have negative impacts by generating defensive behaviors that interfere with learning (Kilminster and Jolly, 2000; Kluger and DeNisi, 1996).

There has been no formal requirement for attendings to be trained to perform their supervisory role. However, faculty can be taught to be better teachers and supervisors. Pioneering work by Skeff and other colleagues (Litzelman et al., 1998; Skeff, 1998) has been instrumental in raising awareness of the need for supervisors to be instructed in their roles, as has work by other investigators (Bishop, 1998; Cote and Leclere, 2000; Kilminster and Jolly, 2000; Meyers et al., 2007; Williams and Webb, 1994). For better supervision to flourish, medical faculties need to place a higher priority on their educational mission. This entails greater institutional willingness to develop and promote clinician educators, the creation of "academies of medical educators," mission-based budgeting, and related strategies to fund clinical teaching and supervision (Ludmerer, 2004). Return from investing in proper supervision can have a profound and long reach: the role modeling that residents witness forms the basis for the effective supervision of future physicians and the potential for improved patient outcomes for years to come.

Impact of 2003 Duty Hour Rules on Faculty Availability

A major concern stemming from the 2003 duty hour regulations is the effect they have had on the availability of faculty and senior residents for supervision and teaching with additional workload shifting to them (Arora et al., 2008; Coverdill et al., 2006a,b; Hutter et al., 2006). Some program responses to the 2003 duty hour limits indicate that the new regulations may have exacerbated preexisting shortcomings in the time for supervision and added new ones. Examples include reports of how reduced resident duty hours have shifted the workload to attendings and more senior residents, leaving them less time for listening to resident presentations, asking them questions, providing advice, or allowing residents to make the primary diagnosis (Barden et al., 2002; Harrison and Allen, 2006; Shojania et al., 2006). Additionally, supervision has generally been less at night and during extended shifts when junior residents (and their patients) would benefit from more supervision, not less, since the risks for poor patient outcomes are known to be greater at these times (Huang et al., 2006; Kilminster and Jolly, 2000; Landrigan et al., 2004; Shojania et al., 2006; Shulkin, 2008).

In one study, the clinical internal medicine faculty reported their belief that they now spend more time on patient care than teaching and supervising residents because of shifting workloads (see Figure 4-1). Almost 75 percent of key clinical faculty believed the duty hour regulations limited opportunities for both didactic and bedside teaching. The researchers noted the potential of the regulations for adverse consequences on faculty recruitment and retention due to potential increases in clinical responsibility (Reed et al., 2007). Another survey of attending physicians came to similar conclusions: less time for teaching, less satisfaction with professional growth and development, and decreased educational stimulation from work. Attending physicians reported a decline in the amount of time dedicated to didactic teaching, and residents missing educational conferences more often because more time was consumed by rounds (Arora and Meltzer, 2008).

Removing Barriers to Communication

In addition to lack of time, other barriers to good communication and supervision include lack of agreement on circumstances for consultation and institutional cultures that discourage communication. What needs to be supervised and when are often not clearly defined for most residencies, but this dialogue should occur. Farnan and colleagues (Farnan et al., 2007) examined the preferences of both internal medicine residents and their supervisors across four types of clinical scenarios involving specific critical decision making on the part of residents. Residents and attendings agreed that immediate contact was necessary and should be required when there was a transfer of an existing patient into the ICU, when cardiac arrest occurred, and when a resident performed an invasive procedure. Attendings desired notification more often than residents wanted to contact them for transfers from the ICU ($p = .0009$), transfers from an outside facility ($p = .001$), patients' receiving vasoactive medications for the first time ($p = .02$), or initiation of intravenous antibiotics. Clarification of expectations for consultations with supervisors in all programs would be beneficial.

In some situations, teaching physicians humiliate residents who provide them with insufficient patient information or consider residents "weak," insecure, and lacking in knowledge, skill, and judgment if they ask for help, thus suppressing needed discussion or calls for help even when residents know they are over their heads (Hoff et al., 2006; Kilminster and Jolly, 2000; Shojania et al., 2006). Residents do not want to be seen as unable to make their own judgments without support or to be viewed as a nuisance by the attending, nor do they always want to admit to gaps in their knowledge and skills or give up their autonomy (Farnan et al., 2008). This avoidance of discussion with teachers undermines a critical role of supervision—to help residents become aware of their cognitive biases and

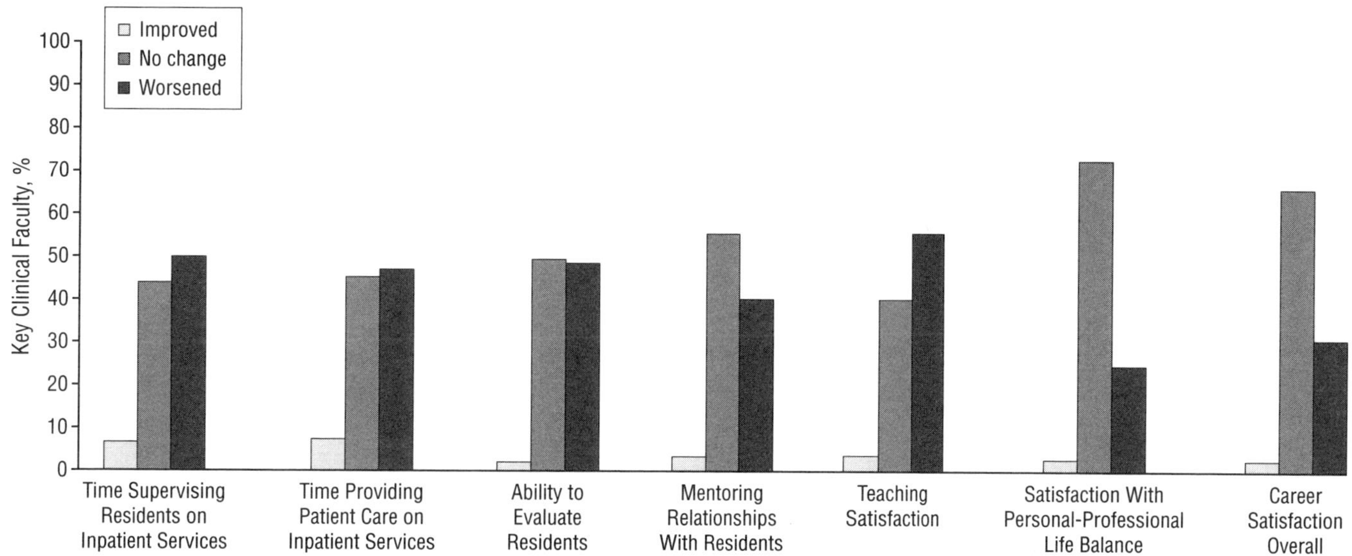

FIGURE 4-1 View of 111 key clinical faculty on the effect of duty hour regulations on faculty workload satisfaction.
NOTE: Sign test, $p < .001$ for all outcomes.
SOURCE: Reed et al., 2007. Copyright © 2007, American Medical Association. All rights reserved.

to develop effective ways for gathering and interpreting patient information (Groopman, 2008). Furthermore, residents intimidated about asking questions or requesting help can present a risk to patient safety (e.g., by taking undue time to reach decisions on courses of care).

To promote patient safety, medical teaching environments must support learners and the entire supervision and learning process. Teaching physicians must be ready to coach, back up, and aid a resident in providing quality patient care. It is the responsibility of house officers to call for help when they are unsure about what to do, but it is an attending's responsibility to foster conditions in which necessary consultations can take place. Supervisory behavior should include demonstrating how to act constructively upon recognizing a mistake. An effective teacher provides opportunity and sufficient time for learners to reflect on their own experiences (Langer, 1990). Furthermore, an effective residency program develops, rewards, and supports those physician supervisors who behave as appropriate role models for residents. Training for supervisors may need to be provided to help instill a greater sense of supervisory leadership among them and develop skills that will help residents learn more effectively.

The committee believes in the primacy of education in residency training, the value of supervision to guide residents to gradual independence and ensure patient safety, and the importance of having well-trained faculty for that role. The committee agrees that support for teaching time and recognition of its importance in assessing professional development of faculty should be encouraged. In conjunction with the evidence on error and patient safety in Chapter 6, the committee recommends the following:

Recommendation 4-1: To increase patient safety and enhance education for residents, the ACGME should ensure that programs provide adequate, direct, onsite supervision for residents. The ACGME should require

- **Residency Review Committees, in conjunction with teaching institutions and program directors, to establish measurable standards of supervision for each level of doctor in training, as appropriate to their specialty; and**
- **First-year residents not to be on duty without having immediate access to a residency program-approved supervisory physician in-house.**

There is no standard definition of whom or what level of "senior clinician" qualifies to act as a supervisor to residents, although any patient's attending physician is ultimately responsible for the care received. The committee recognized that this definition can depend on the specialty being pursued, the task being performed or taught, the competency level of individual residents, and the complexity of patient cases being cared for. While an attending-level supervisor is the ideal for all residents to be taught and guided under, for practical purposes and to avoid exacerbating the limited supply of supervising staff, the committee concluded that a senior resident (equivalent to a PGY-3) is an acceptable minimal level of experience to serve a supervisory role to more junior residents (PGY-1 and PGY-2 residents). For residents in their third year or higher of training, more senior clinicians (i.e., attendings, faculty, fellows) should provide supervision. In many cases, particularly in overseeing surgical procedures or dealing with highly complex cases such as interventional angiographic and intracranial procedures, an attending-level supervisor should always be required. Hospitalists and other senior-level staff can also serve to provide in-hospital resident supervision when needed.

The committee suggests that in-house supervisors be readily available to first-year residents, to help with any aspect of patient care duties, including on nights and weekends. It is hoped that these supervisors are not so overburdened with other clinical responsibilities such that their ability to supervise is compromised. Furthermore, supervisors need not necessarily be a member of the same team or service as the first year resident. It is expected, however, that they be a senior resident or higher level physician in the same specialty training program as the first-year resident (i.e., internal medicine first-years should have an internal resident senior with whom they can consult, pediatrics with pediatrics, surgery with surgery, obstetrics-gynecology with obstetrics-gynecology, etc.). The committee believes that residency programs and specialties would benefit from creating their own supervisory guidelines to ensure adequate supervision is provided for all resident levels at all times.

The committee also stresses the importance of enhancing supervisory leadership, by encouraging that supervisors at all levels (e.g., attendings and PGY-3s and above) be pro-active in their role: making conscientious efforts to contact their residents on a regular basis; providing feedback and constructive instruction (regarding diagnoses, treatment plans, professional behavior, or other attributes); and consistently helping residents identify areas to improve patient safety and their own patient care. Communication should not be left solely to the discretion of residents to contact their supervisors to address concerns or clarify questions they may have. Supervisor-initiated contact, regardless of resident competency level, can serve to catch problems with treatment plans or handle unexpected events sooner than waiting for interns or residents to contact them, ultimately helping to prevent patient harm. As previously mentioned, faculty and other supervisors may need to be trained in this type of interaction to introduce it on a broad scale throughout residency programs and help make it a part of the training and safety culture.

Ensuring a Workload That Allows Adequate Time for Reflective Learning

A second cardinal educational principle of graduate medical training is having time for reflective learning. This manifests in the regular pacing of mindful reflection in a physician's practice, during diagnosis, in the process of treatment, and upon case completion. From the beginning of the modern residency, medical educators have emphasized the importance to learning of allowing residents sufficient time to reflect on their work—although the reality of practice today does not always exhibit this ideal. Mindful reflection involves openness to new information and implicit awareness of multiple perspectives and possibilities (Langer, 1990). It is far better for the intellectual growth of residents, educators have argued, to have house officers study fewer patients in depth rather than many patients superficially. The pioneering medical educator Abraham Flexner spoke to this point: Medicine is best learned through the "intensive and thorough study of relatively few patients" (Flexner, 1925, p. 270). To allow residents the opportunity to reflect, medical educators strove to ensure that the "caseload" of house officers was not too burdensome, leaving them more time to read, contemplate, attend conferences and rounds, and monitor their patients carefully (Ludmerer, 1999).

Impact of 2003 Duty Hour Rules on Workload and Learning

For most of the 20th century, this cardinal educational principle that residents should have time for reflection was honored. However, in 1984, along with the implementation of prospective payment for hospitals and the need to control costs, the already decreasing average length of hospital stay continued to fall by one-third of what it had been before (7.3 days in 1980 to 4.8 in 2004) (Kozak et al., 2006). As more cases shifted to outpatient care, the remaining inpatients had a greater severity and complexity of health care needs. Although residents have managed an escalating number of admissions and discharges for the past two decades, thereby allowing hospital "throughput" to be maintained at a high level, the educational costs to their training have been significant (Ludmerer, 1999).

The regulations implemented in 2003 limiting resident duty hours have had the unintended consequence of worsening the situation in many programs (Hutter et al., 2006; Vidyarthi et al., 2006). By not decreasing residents' workload along with their work hours (Charap, 2004; Fitzgibbons et al., 2006; Ludmerer and Johns, 2005; Ryan, 2005), the already hectic pace at which residents worked has become faster than ever. One resident described the "frantic mentality" that engulfed the wards, with residents and interns "rushing from task to task and then out of the hospital" (Ryan, 2005, p. 82). He also described "the marginalization of learning" that resulted from this intense pace, "Success in the medical wards, moving patients along, and getting things done often require efficiency above all else. . . . [This] makes education about the diagnosis and management of disease feel like a hindrance, a drag on the steady progress through the day" (Ryan, 2005, p. 83). The patient load confronting residents during the era of "throughput" has raised concerns for patients' safety and the quality of their care.

In organizational research generally, extremely difficult performance demands yield dysfunctional consequences unless there is substantial support to make the demands manageable. Consequences include extreme stress, pressures on personal time, burnout, and in some cases, inappropriate or unethical behavior (Sejits and Latham, 2005). Teaching hospitals need to address the question: Have residents been given the means to succeed at maximizing learning while providing quality patient care?

Cognitive Load Theory

To fully understand how workload affects resident learning and performance, it is important to appreciate the implications of cognitive load theory. This theory deals with the amount of cognitive information a person is able to absorb, process, and retain from any given task. Some forms of cognitive load are useful in achieving goals while others waste mental resources. In learning, the goal is to minimize the inefficient or wasteful forms of cognitive load and maximize its useful forms. Cognitive load theory represents a universal set of learning principles demonstrated to result in efficient educational environments when designed with human cognitive learning processes in mind (Clark et al., 2006). Controlled experimental research studies are the basis for these principles (e.g., Mayer et al., 1996; Sweller et al., 1990).

Three types of cognitive information (or load) are relevant to training: intrinsic, germane, and extraneous. Intrinsic load is the mental work imposed by the complexity of the learning content (e.g., units of knowledge to be acquired). Germane or relevant load is mental work imposed by instructional activities that benefit learning (e.g., constructing a report to express what the learner has understood after completing task). Extraneous or irrelevant load is mental work that is irrelevant to the learning goal (e.g., "scut" work), usually under the control of the managers of the learning environment. Therefore, cognitive load is the mental work that a task or situation imposes. Humans are known to have limited cognitive capacity, which is why efficient instruction substitutes for the novice's lack of sophisticated knowledge—by segmenting and sequencing content in ways that reduce the amount of new information novices must process at one time (e.g., by controlling the complexity of tasks novices perform; by guiding their attention to critical information).

The traditional educational principle, that house officers should study problems in depth, has received substantial theoretical and empirical support from education and psychology research (Langer, 1990; Pollock et al., 2002). Experts in these disciplines have coined the terms "reflective learning" and "mindfulness" to describe the most important requirements for learning, which align with the principles of cognitive load theory. *Reflection* means deliberate recall and review of an event, typically an event in which the learner is actively involved. *Mindfulness* is systematic, careful attention, a heightened awareness. (The opposite of mindfulness is the automatic processing of information in routine habit-based ways.) Both reflective learning and mindfulness require that learning be paced, giving the learner time to engage in these critical intellectual activities. This goes hand-in-hand with segmenting cognitive load in order to optimize the information a learner retains.

The supervisor or instructor's role in this process is to focus residents' attention on appropriate content. For example, when talking about a case, there is a reflective discussion of possible alternatives for patient treatment. By questioning and modeling how he or she thinks and showing how the time is structured to do this, the instructor creates an active information processing on the part of the resident, showing him or her how to consider alternatives systematically and weigh what is most important (Clark et al., 2006; Langer, 1990; Richardson, 2005; Smith et al., 2004). This may occur more easily in inpatient settings, where patient visits can be longer and allow more time for reflection than in outpatient settings, where visits are typically 15 minutes long. However, since a residents' service is to be educational for them, time should be made to discuss cases and allow residents to reflect on decisions made and actions taken whether in the inpatient or outpatient setting. Integration of what has just happened with prior knowledge is important and should occur as close in time to the event as possible (Clark et al., 2006; Linn et al., 2006). This process of reflection permits residents to consolidate pieces of information into a bigger, richer understanding they can recall and apply (Chi, 2000; Cooper et al., 2001), and the importance of the role a supervisor plays in this cannot be overstated.

Therefore, the design of the learning environment itself—largely dictated by the structure of a hospital's system—affects how residents learn. Learning takes place best in the context of activity—that is, learning by doing. Resident learning is enhanced by ongoing collaborative social interaction and ready access to knowledgeable colleagues (Brown et al., 1989), reasons why onsite training and supervision are important for residents.

If faculty have modeled appropriate ways of learning by doing, then residents can learn to generate alternative courses of action for consideration based on prior modeling demonstrated by the attending (Clark et al., 2006; Groopman, 2008).

Training programs that mingle activities unrelated to learning place additional burdens on the working memory of would-be learners (Clark et al., 2006). The result is slower, inefficient, and potentially reduced or inappropriate learning (e.g., short cuts, poorly executed procedures). Learning is enhanced when the tasks that people perform (those that go into their working memory) focus on content that is relevant and germane to what they are trying to learn. For example, doing patient intakes and performing surgical procedures are relevant to learning about patient care. Patient transport, scheduling, or certain kinds of paperwork are not. The notion of cognitive load theory is to build a learning environment that helps active processing and avoids placing tasks in the environment that might hinder it (Clark et al., 2006; Sweller et al., 1990). The more relevant the content is, the more efficient is the learning, so that less time is needed to learn the same amount of material (Clark et al., 2006).

Optimizing Workload to Improve Learning

In addition to feedback from instructors, time for reflection, and relevant content, behavioral and brain research has demonstrated that workload is related to performance as well. For any given task, an optimal level of workload exists that yields the highest level of performance. On a graph, this relationship between learning on the *y*-axis and census and workload on the *x*-axis would appear as an inverted U function. A departure in either direction off the plateau of the inverted U is expected to result in lower performance (Chewning and Harrell, 1990; Choo, 1995; Wiener et al., 1984). Too little work stimulates less learning, and work overload undermines a resident's ability to absorb new information. Thought processes become fragmented, and judgment deteriorates. One can be the most responsible, knowledgeable, and thoughtful resident, but if the system gives a resident responsibility for too many patients, stress develops and learning does not take place, placing safe care at risk. The optimal level of workload varies with individual expertise, so a novice can absorb less workload than an expert. Haney and colleagues (2006) have explored this well-known principle for its applicability to medical residency. They find that for new learners (i.e., interns) the peak of this curve is sensitive to the total number of patients in their census. In contrast, for more experienced learners, the peak is sensitive to case variety and the severity of illness in new admissions. Translated to residency education, these concepts validate the idea that residents' caseloads should be of manageable size and variety so that residents may learn most effectively (Epstein, 1999; Ericsson, 2002; Ericsson and Charness, 1994; Ericsson and Krampe, 1993; Langer, 1990; Plack and Greenberg, 2005).

Less attention has been paid to workload than regulation of hours. Workload is affected not only by patient demands but also by enabling societal, educational, and hospital structures. Compressing the volume of work that residents must do into less time after the 2003 ACGME duty hour limits might have been done in the hopes that the level of educational outcomes would not decline. However, cognitive thresholds exist beyond which additional workload becomes counterproductive to learning and performance. Hence, the committee recommends in Chapter 3 that noneducational work be minimized, that Residency Review Committees develop workload guidance that is specialty-specific by year of residency, and that residents be given enough time to conduct thorough evaluations of patients and to engage in reflective learning based on their clinical experiences.

Additionally, sleep is necessary to consolidate memory to help people retrieve what they have learned (Gais et al., 2000; Huber et al., 2004; Plihal and Born, 1997; Stickgold et al., 2000). Learning is much less effective without adequate sleep, substantially decreasing the ability to retrieve information from one's long-term memory.

Finding the Balance Between Continuity of Care and Educational Opportunities

Continuity of care is the third principle of quality resident education. Residents require continuity in the care of their patients to understand the normal course of illness and to act in the best interest of patients. By following patients from admission to discharge, residents see the results of their treatments, learn how to respond to complications or complexities that arise from treatment, and better understand how to treat the "whole" patient. Decades of experience in preparing residents for independent practice have convinced medical educators that residents must have multiple opportunities, under supervision, to participate directly in the care of "their" patients from the inception of an illness through the entire course of diagnosis, treatment, and recovery.

In underscoring the importance of *continuity of care* in resident education, a clear distinction must be made between continuity of care in the hours or days of an acute illness, and continuity over the weeks, months, and years of a chronic disease or disability. The acute care circumstance is particularly an issue in the context of resident duty hours as certain specialties advocate for retention of the 30-hour extended duty period for at least a portion of their trainees. The committee has taken this into account in recommending adjustments to duty hours in Chapter 7. Like fully trained physicians in practice, residents clearly cannot be on duty 24/7; handovers to competent colleagues at appropriate intervals are essential. Indeed, ensuring effective handover of patient care responsibility is another critical skill that residents must acquire and is addressed in Chapter 8.

The committee concludes that societal concern about the adverse effects of resident fatigue during extended duty periods can be addressed by providing protected time for sleep and preserving the long-term societal benefits that stem from residents having ample continuity of care experiences during their training.

IMPACT OF 2003 LIMITS ON EDUCATIONAL OUTCOMES

This section addresses how the 2003 ACGME duty hour limits have affected the quality of the resident learning experience as determined by resident and faculty perceptions of education being provided, measured exposure to procedural cases, time available for didactic sessions and study, and test performance. Limitations of many studies from which the evidence is drawn make a wider generalization of conclusions difficult. Nonetheless, this body of research calls attention to how resident education may be affected by changes caused by duty hour limits. Overall, studies tend to suggest that individual programs were able to adjust to the new duty hour rules while still upholding educational standards. Note that programs struggling with maintaining standards are unlikely to be represented in the literature.

Perceptions of Effects on Education

Surveys of resident education represent perceptions of residents, faculty members, and program directors regarding the impact of duty hour regulations on resident education. These surveys vary in their sample sizes, ranging from single institutions to multiple institutions in a particular specialty. In several surveys, residents did not report improved educational satisfaction as a result of duty hour changes, nor did they report significant decreases in the volume of key clinical experiences or in trainees' perceptions of "preparedness" (AAMC, 2008; Jagsi and Surender, 2004; Jagsi et al., 2006; Vidyarthi et al., 2006). These resident reports essentially convey the perception that duty hour regulations had little or no change in the quality of their education or their satisfaction with it. In one single institution survey of plastic surgeons, however, residents reported increased satisfaction with their educational experience, stating that they were more alert, more satisfied with their time for didactics and study, and that they believed patient care had improved since implementing the new duty hours (Basu et al., 2004).

Surveys of faculty perceptions tend to yield more negative reports. The prevailing belief of faculty in these studies from a few institutions and across various specialties is that resident education has been compromised (Cohen-Gadol et al., 2005; Espey et al., 2007; Immerman et al., 2007; Lieberman et al., 2005; Vaughn et al., 2008; Winslow et al., 2004). In contrast, authors of one study note that residency programs were indeed able to maintain their standards by reinventing their approach to presenting educational content (de Virgilio et al., 2006).

Effects on Procedure Volume and Case Mix

The volume of procedures performed is one quantitative measure of resident training, particularly for surgical specialties, and is often used as a quality measure for physicians in practice. Most of the studies evaluated the numbers of procedures performed in single institutions and are largely specific to one or two medical specialties.

Effects on Procedure Volume for Residents

A systematic review of 54 studies examined the effects of the duty hour regulations on residents in internal medicine, pediatrics, OB/GYN, surgery, and family medicine, psychiatry, and radiology. Of these studies, only OB/GYN and surgery assessed procedure volume. Results showed that numbers of procedures remained relatively unchanged for residents in 3 of the OB/GYN studies. In the 25 surgical studies, however, the effect of duty hour regulations on volume was unclear: some found operative volume to increase, whereas others found that volume decreased (Fletcher et al., 2005). Research that employed surveys or other methods for resident self-report of case volume more often found that trainees perceived no change or a decline in operating time since the implementation of duty hour limits (Barden et al., 2002; Kort et al., 2004; Zuckerman et al., 2005). One study reported both actual case counts and perceptions; this pediatric surgery unit maintained the number of procedures performed 1 year before implementing duty hour changes and 1 year after (47 and 44 procedures, respectively), with matching perceptions among residents of steady caseload and maintenance of patient management skills (Spencer and Teitelbaum, 2005). A majority of the studies examined that relied on available surgical case logs and databases also showed no significant change in resident case volume or time in the operating room, maintaining numbers and hours from previous years (Durkin et al., 2008; Ferguson et al., 2005; Shin et al., 2008). One such study, however, showed a significant increase in total procedural volume for graduating residents (de Virgilio et al., 2006).

Regarding the surgical exposure of senior residents, one study found an increase in case volume specifically for PGY-5 residents (a 51-case increase post duty hour implementation: 339 vs. 390, $p = .05$) while volume remained stable for more junior residents (Ferguson et al., 2005). Another survey specific to chief residents also reported an increase in the total number of yearly operations they participated in, despite the finding of no volume change for more junior residents (Barden et al., 2002). Other studies examining the surgical volume for senior residents affirm no significant change in the number of cases performed as a result of duty hour regulations (Malangoni et al., 2005; Mendoza and Britt, 2005; Spencer and Teitelbaum, 2005). The authors in one of the studies concluded that no relationship existed between duty hours and procedure volume (Mendoza and Britt, 2005).

Effects on Case Mix for Residents

Case mix exposure has also been a concern of specialty boards (ABMS, 2008a). The more often routine cases are performed, the less opportunity residents have of expanding their educational base to learn new procedures or treatment steps. The teaching value of operative cases lies in the complexity or uniqueness of cases in addition to case volume.

Impact of duty hour limits on case mix varies across assessments. In a national survey of neurosurgical training programs, 41 percent of residents and 74 percent of program directors believed that chief residents operated on less complex cases post-implementation (Cohen-Gadol et al., 2005). Regarding the case mix of surgical procedures assessed in the studies above, senior residents in one report were found to perform more endoscopic and vascular access procedures than before (Spencer and Teitelbaum, 2005), while senior residents in the other study reported that the types of cases they performed remained essentially the same (Malangoni et al., 2005). Along with volume and complexity of operative cases, important lessons are also learned from postoperative patient care by following treatment outcomes and learning how to treat potential complications. Less perioperative exposure for residents in general has been observed, although this has not been well quantified.

Overall, the focus of the literature has been more on procedural volume and less on appropriate case mix, perioperative time and nonoperative didactic opportunities, quality of instruction, or success of competencies attained, which several educators have claimed may have equal or greater impact on a resident's learning capacity.

Effects on Time for Didactic Sessions and Study

The research addressing the impact of duty hour regulation on didactic education suggests that resident attendance at formal didactic sessions is down, either because they are not available when sessions are offered or because workload precludes their attendance (Arora and Meltzer, 2008; Reed et al., 2007). Responses by chief residents to a survey conducted by Horwitz and colleagues (2006) indicated that despite no change in the number of hours scheduled for educational activities there appeared to be "a decrease in intern attendance at conferences, and many reduced third-year elective time." Some reports have found that time for independent study, reading, and case preparation has increased (Basu et al., 2004; Vaughn et al., 2008; Zuckerman et al., 2005), perhaps replacing time lost for conference attendance (Lin et al., 2006). Although residents may compensate during separate time now available for independent reading and study, this is not universally observed.

Licensing Exam, Board Certification, and In-Training Exam Results

Residency training can take 3 to 7 years to complete depending on the degree of specialization or subspecialization pursued. The first cohort of 3-year residents trained entirely under 2003 ACGME limits finished in June 2006 so it is not surprising that educational outcomes data are just beginning to emerge. Examination pass rates provide a readily available objective measure of such outcomes, and data from these measures are presented below.

The U.S. Medical Licensing Exam Step 3, generally taken at the end of the first year of residency, assesses whether a medical doctor has sufficient knowledge and clinical decision-making skills to deliver medical care independently. Scores on this test have remained at fairly consistent levels for U.S. graduates of allopathic medical schools before and after the 2003 ACGME duty hour rules. First-time takers between 1999 and 2003 had a 94-95 percent pass rate, whereas 96 percent of first-time takers from these schools passed in 2004-2006 (n = 16,395 in 2006) (USMLE, 2008).

After completing accredited residencies and fellowships, graduate medical trainees take additional exams if they choose to become board certified (ABMS, 2008b). Data on board certification rates are not publicly available for each specialty. Only non-surgical specialties have had a resident cohort complete all of its training since the 2003 work limits were put in place. First-time taker pass rates for the American Board of Internal Medicine certifying exam were 92 percent for 2003, 92 percent for 2004, 92 percent for 2005, 91 percent for 2006, and 94 percent for 2007 (ABIM, 2008). Previous years' scores were 86 percent for 2000, 88 percent for 2001, and 87 percent for 2002.[2] Changes in board pass rates are susceptible to various factors, such as the number of U.S. medical school graduates entering a specific specialty, making it difficult to conclude whether resident duty hour

---

[2]Personal communication, L. J. Grosso, Director of Psychometric Operations, American Board of Internal Medicine, April 29, 2008.

adjustments had a direct impact. Testimony to the committee reported that there are declines in pediatric board pass rates and thoracic and orthopedic surgery, but there have been no detailed analyses of the factors that might have contributed to these declines (ABMS, 2008a).

A few institution-specific studies of surgical programs (e.g., trauma surgeons) report on their success in maintaining procedural volume and their scores on the American Board of Surgery In-Training Examination (ABSITE). To assess resident ABSITE scores before and after the implementation of duty hours, researchers surveyed general surgical residents in a New York program and found that scores increased on average for all 29 respondents to the survey. This increase was statistically significant for junior but not senior residents (Barden et al., 2002). In response to duty hour regulations, a trauma care program of 46 residents employed different schedules, made curriculum changes, and increased the number of residents in the program. Anticipating that residents would not do more reading on their own time, the program also added new meetings, tests, weekly assignments, and mock oral exams geared toward helping residents achieve high ABSITE scores. While ABSITE scores and pass rates did not change significantly between the two time periods, substantial financial and faculty resources were required to support these program changes (de Virgilio, 2008; de Virgilio et al., 2006). One study that reviewed resident ABSITE scores and operative logs before and after duty hour limits showed no change in case volume and a significant increase in overall ABSITE performance largely due to improved scores on its basic science component (Durkin et al., 2008).

The American Board of Medical Specialties (ABMS) anticipates training repercussions under the current 80-hour limit for certain types of training programs (e.g., emergency, trauma) because of less exposure to patients with unusual complexity of illnesses, long surgical cases, or management of postoperative complications. The ABMS notes that further reduction of total weekly duty hours may result in the need to lengthen the training period of various specialties (ABMS, 2008a). However, with the recommendations of this report, the committee has attempted to avoid having to extend training duration for medical and surgical residents by maintaining the 80-hour limit to provide flexibility for programs that need those hours. Redesign of curriculums, schedules, medical teams, caseload, and staffing are all factors to consider when attempting to reorganize training programs in a manner that will meet residents' educational needs within duty hour limits.

REDESIGNING EDUCATION AND EDUCATIONAL INNOVATIONS

Educators in the medical field have acknowledged that changes in duty hour rules provide an opportune time to redefine educational practices and improve overall resident learning environments (Skeff et al., 2004; Wong et al., 2004). Indeed, regardless of changes in duty hours, educational redesign may be the only reasonable response to statements made by other leaders in the field suggesting that aspects of current training systems for residents fail to reflect the constant changes taking place in medical practice (Arora et al., 2005). Redesign presents an opportunity to take the 20th-century model of GME training and transform it into a 21st-century model suited to the increasingly demanding needs of patients and the healthcare system. The new era of GME can continue to make greater use of new educational technologies, which have driven so much of the change in practice over the past few decades, innovative approaches to curricula, and greater training in systems strategies. Several interventions have been mentioned already (e.g., adjusting schedules and workload, reformatting curricula, increasing practice tests, reorganizing staff). This section illustrates other interventions that could be employed in response to duty hours and changing practice. ACGME itself has encouraged innovative new practices that would serve this purpose, supporting them through its Committee on Innovation (ACGME, 2008a).

Educational Redesign Interventions

The following educational redesign interventions demonstrate innovation in targeting the content of what residents learn and incorporating techniques that enhance learning (e.g., contextual learning, supervisory feedback) discussed earlier in this chapter. Using techniques to help residents learn more efficiently became increasingly important when programs adapted to the 2003 duty hour limits and had less time in which to teach residents.

Competency-Based Design

As a way to maintain educational outcomes given the reduced hour limits, some residency programs are focusing on a competency-based approach to their medical training. Competency-based teaching is a concept that proposes to replace the current time-based educational model of residency, which requires residents to complete their learning in a fixed number of years, with a model in which the completion of training is determined by the demonstration of required competencies. In this way, residents can progress and acquire knowledge and skills in a manner that is more attuned to their individual abilities at any given time. To help enhance the effectiveness of this approach, educators should be encouraged to utilize better tools for measuring skill acquisition (Satish and Streufert, 2002; Satish et al., 2001) and advance trainees according to their individual progress.

A competency-based training curriculum was implemented in one institution's neurosurgical department. Evaluation of its effectiveness yielded promising results. Not only did the residents succeed in mastering the prescribed neurosurgical skills, but the time taken to master the procedures was reduced by several months compared to traditional time requirements (Long, 2000). Similar results were achieved using competency-based teaching interventions for invasive skills to a small number of general surgery interns. The authors of this study acknowledged that residents in their program often lacked basic invasive skills until asked to perform on patients. To better prepare them to perform such procedures the interns were taken to the cadaver laboratory where faculty instructed them directly, providing hands-on direction and feedback, as they conducted endotracheal tube insertions, chest tube insertions, and venous cutdowns. The goal was for interns to correctly perform these procedures within 120 seconds. Prior to instruction, seven of the eight interns failed the tasks of endotracheal tube insertion and venous cutdown, and five out of eight failed the chest tube insertion (Martin et al., 1998). After instruction, there were no failures, no complications, and the time taken to complete each task was significantly reduced. These skills were transferable to the clinical setting where these interns performed these procedures multiple times with minimal complications (Martin et al., 1998). Different simulated settings have been shown to improve resident practice in other specialties as well, such as anesthesiology and radiology (Sica et al., 1999; Wong, 2004).

Long (2000) acknowledged that such training may be easier to determine in procedural fields, since outcomes can be more clearly defined, but widely used skills such as accurate history taking, physical examination, interpreting diagnostic data, and sound patient management were also included. Shifting attention to milestones of learning, rather than time in place, can promote the integration of proper learning experiences. Furthermore, reduction in actual time taken to acquire competencies is particularly beneficial in light of reduced duty hours, indicating that more time for training may not be necessary for residents to learn their needed clinical skills. Further evaluations in other specialties would be helpful in assessing the extent to which such training is practical or beneficial across different residency programs and specialties, recognizing that competency-based training "is just one of many potentially useful approaches that may have a role at various stages of the educational process" (Leung, 2002).

Simulation-Based Training to Support Educational Designs

Simulation, both high and low tech, has been increasingly studied to train residents more efficiently and has been gaining acceptance as a method through which to teach and measure competency-based education. With the duty hour constraints limiting their schedules, these techniques provide additional opportunities for residents to gain applicable clinical knowledge without risk to patient safety. Simulation-based training enables trainees to learn the necessary competencies (i.e., knowledge, skills, attitudes) (Salas et al., 2005) and has been shown to improve performance in clinical skills, such as procedural training (Lindquist et al., 2008; Medina et al., 2000; Sica et al., 1999; Wong, 2004), and in nonclinical skills, including communication, cooperation, leadership, and decision making (Medina et al., 2000; Østergaard, 2004; Sica et al., 1999).

Simulation-based training is an effective training strategy when utilized properly (Salas et al., 2008). Practice must be guided (through crafted scenarios and timely, diagnostic feedback) to keep residents focused on learning key competencies (Salas and Burke, 2002). Allowing skills to be "practiced, assessed, diagnosed, remedied, and reinforced" all at once can create effective learning environments that require less time than real-life settings (Salas and Burke, 2002, p. 120). Both medical students and physicians have identified simulation-based training as a valuable tool for educational purposes (Bond and Spillane, 2002; Bond et al., 2001; Gordon, 2000; Gordon et al., 2001; Halamek et al., 2000). Several electronic tools have also been shown to provide residents with learning opportunities in the absence of available faculty, helping them to learn more efficiently on their own (Cook et al., 2008).

Assessing effectiveness of specific simulation courses or methods is beyond the scope of this study. Numerous types and levels of simulation exist, each for different intents, purposes, and costs. The committee encourages examination and evaluation of the various simulation tools and methods that might serve to support educational redesign solutions, innovational training, and student evaluation.

Long-Block Design

In an effort to move away from service-oriented inpatient training toward education-oriented training, an Educational Innovations Project sponsored by ACGME was piloted by Warm and colleagues (2006) for ambulatory care training in internal medicine. A continuous year-long ambulatory group practice, called a "long-block," was created (beginning in a resident's 17th month of training) and separated from traditional inpatient responsibilities. This long-block practice replaced sporadic ambulatory training rotations previously completed over 3 years and was scheduled to comply with duty hour limits. Results showed positive outcomes in multiple areas including increased resident and patient satisfaction and improvements in quality processes, outcome measures, and care continuity (Warm et al., 2008). Residents reported more time for learning and increased ability to focus in clinic with fewer interruptions in the long-block setting. They also reported better patient relationships and increased ownership of patient care (Warm et al., 2008).

Quality Improvement Techniques

Other redesign efforts, such as those based on teaching quality improvement (QI) techniques to residents, have also been effective. For example, one internal medicine residency program sought to achieve one of ACGME's new six core competencies—practice-based learning and improvement—which addresses "the need to teach and evaluate residents' ability to apply quality improvement in their medical practice." The faculty of this program chose to teach this competency by modifying the curriculum readings to focus on quality of care (which other programs have done), but also by having residents reflect on their work with faculty and evaluate their practice performances (Holmboe et al., 2005). The outcomes showed that residents involved in this education intervention were more likely to perform quality of care measures for their diabetic patients, which resulted in more positive patient outcomes and improved resident satisfaction with their education (Holmboe et al., 2005).

A more recent study also aimed to teach internal medicine residents QI concepts and assessment techniques. Positive outcomes resulted after redesigning ambulatory block rotations and introducing a new curriculum specifically geared to achieve QI and assessment goals (Oyler et al., 2008). The authors note that teaching these skills can be difficult with limited staff availability or familiarity with the topics, but that using the American Board of Internal Medicine's practice improvement module for preventive services was useful in overcoming these challenges. The new curriculum improved resident confidence with assessing QI and learning how to apply QI practices in their continuity clinics (Oyler et al., 2008).

The educational redesign approaches presented above each has its strengths for teaching residents more effectively and efficiently. Additional innovations exist (Wong, 2006) and more are encouraged. It will be important to keep in mind the different needs of individual specialties, programs, and institutions when considering how best to redesign the educational content while complying with duty hour limits.

CONCLUSION

The committee concludes that the full effects of implementing the 2003 ACGME duty hour regulations on resident education remain unclear. The lack of published studies in most disciplines make assessments of educational outcomes difficult. There seems to be a general impression from residents that their educational quality has remained relatively unchanged since the implementation of duty hour restrictions, while supervisors and faculty perceive that education has deteriorated in some instances. Looking at quantitative measures of educational outcomes, different programs and specialties have reported varying degrees of maintaining procedural volume or resident test scores since the 2003 duty hour limits. However, many of the more rigorous studies reported programs that managed either to sustain or improve these outcomes. With demanding workloads and less time in which to teach or learn, a new approach for graduate medical education—befitting the evolving medical landscape of the 21st century—is necessary.

The committee's approach has been to focus on the aspects of current resident work within the given duty hour limits that can have positive effects on resident learning. Among those factors are redesigning residency program schedules so that they provide time needed for rest and recovery to consolidate learning, establishing appropriate workloads that allow time for reflection to enhance learning, strengthening supervision, and encouraging approaches to curricula and training that improve overall learning environments.

REFERENCES

AAMC (Association of American Medical Colleges). 2006. *Compact between postdoctoral appointees and their mentors December 2006.* http://www.aamc.org/research/postdoc compact/postdoccompact.pdf (accessed August 12, 2008).

———. 2008. Resident duty schedules and resident education. Presentation by Debra Weinstein to the Committee on Optimizing Graduate Medical Trainee (Resident) Hours and Work Schedules to Improve Patient Safety, May 8, 2008, Washington, DC.

ABIM (American Board of Internal Medicine). 2008. *Residency program pass rates 2005-2007.* Philadelphia, PA: American Board of Internal Medicine.

ABMS (American Board of Medical Specialties). 2008a. Presentation by Kevin Weiss to the Committee on Optimizing Graduate Medical Trainee (Resident) Hours and Work Schedules to Improve Patient Safety, May 8, 2008, Washington, DC.

———. 2008b. *What board certification means.* http://www.abms.org/About_Board_Certification/means.aspx (accessed March 17, 2008).

ACGME (Accreditation Council for Graduate Medical Education). 2007. *ACGME institutional requirements: Checklist, effective: July 1, 2007.* http://www.acgme.org/acWebsite/irc/IRCheckList-07.pdf (accessed 2/28/2008).

———. 2008a. *CI–Committee on Innovation.* http://www.acgme.org/acWebsite/ci/ci_welcome.asp (accessed October 22, 2008).

———. 2008b. *Program director guide to the common program requirements.* http://www.acgme.org/acWebsite/navPages/nav_commonpr.asp (accessed August 12, 2008).

Arco, L. 2008. Feedback for improving staff training and performance in behavioral treatment programs. *Behavioral Interventions* 23(1):39-64.

Arora, V., and D. Meltzer. 2008. Effect of ACGME duty hours on attending physician teaching and satisfaction. *Archives of Internal Medicine* 168(11):1226-1228.

Arora, V., S. Guardiano, D. Donaldson, I. Storch, and P. Hemstreet. 2005. Closing the gap between internal medicine training and practice: Recommendations from recent graduates. *American Journal of Medicine* 118(6):680-685.

Arora, V. M., E. Georgitis, J. N. Woodruff, H. J. Humphrey, and D. O. Meltzer. 2008. Improving sleep hygiene—reply. *Archives of Internal Medicine* 168(11):1230.

Barden, C. B., M. C. Specht, M. D. McCarter, J. M. Daly, and T. J. Fahey III. 2002. Effects of limited work hours on surgical training. *Journal of the American College of Surgeons* 195(4):531-538.

Barron, J. M., M. C. Berger, and D. A. Black. 1997. How well do we measure training? *Journal of Labor Economics* 15(3):507-528.

Basu, C. B., L. M. Chen, L. H. Hollier, Jr., and S. M. Shenaq. 2004. The effect of the Accreditation Council for Graduate Medical Education duty hours policy on plastic surgery resident education and patient care: An outcomes study. *Plastic & Reconstructive Surgery* 114(7):1878-1886.

Batalden, P., D. Leach, S. Swing, H. Dreyfus, and S. Dreyfus. 2002. General competencies and accreditation in graduate medical education. *Health Affairs* 21(5):103-111.

Becker, G. (1975). *Human capital*, 2nd ed. New York: Columbia University Press.

Bell, B. M. 1993. Supervision, not regulation of hours, is the key to improving the quality of patient care. *JAMA* 269(3):403-404.

———. 2003. Reconsideration of the New York State laws rationalizing the supervision and the working conditions of residents. *Einstein Journal of Biological Medicine* 20(1):36-40.

———. 2007. Resident duty hour reform and mortality in hospitalized patients. *JAMA* 298(24):2865-2866.

Bishop, V. 1998. *Clinical supervision in practice. Some questions, answers and guidelines.* Basingstoke: Macmillan Press.

Bond, W. F., and L. Spillane. 2002. The use of simulation for emergency medicine resident assessment. *Academic Emergency Medicine* 9(11):1295-1299.

Bond, W. F., M. Kostenbader, and J. F. McCarthy. 2001. Prehospital and hospital-based health care providers' experience with a human patient simulator. *Prehospital Emergency Care* 5(3):284-287.

Brody School of Medicine. 2008. *Department of surgery: Residency curriculum.* http://www.ecu.edu/cs-dhs/surgery/curriculum.cfm (accessed November 7, 2008).

Brown, J. S., A. Collins, and P. Duguid. 1989. Situated cognition and the culture of learning. *Educational Researcher* 18(1):32-42.

Charap, M. M. D. 2004. Reducing resident work hours: Unproven assumptions and unforeseen outcomes. *Annals of Internal Medicine* 140(10):814-815.

Chewning, E. G., and A. M. Harrell. 1990. The effect of information load on decision makers' cue utilization levels and decision quality in a financial distress decision task. *Accounting, Organizations and Society* 15(6):527-542.

Chi, M. T. H. 2000. Self-explaining expository texts: The dual processes of generating inferences and repairing mental models. In *Advances in instructional psychology educational design and cognitive science*, Vol. 5, edited by R. Glaser. Mahwah, NJ: Lawrence Erlbaum Associates.

Choo, F. 1995. Auditors' judgment performance under stress: A test of the predicted relationship by three theoretical models. *Journal of Accounting, Auditing, and Finance* 10(3):611-641.

Clark, R., F. Nguyen, and J. Sweller. 2006. *Efficiency in learning: Evidence-based guidelines to manage cognitive load.* New York: John Wiley & Sons, Inc.

Cohen-Gadol, A. A., D. G. Piepgras, S. Krishnamurthy, and R. D. Fessler. 2005. Resident duty hours reform: Results of a national survey of the program directors and residents in neurosurgery training programs. *Neurosurgery* 56(2):398-403.

Cook, D., T. Beckman, K. Thomas, and W. Thompson. 2008. Adapting web-based instruction to residents' knowledge improves learning efficiency. *Journal of General Internal Medicine* 23(7):985-990.

Cooper, G., S. Tindall-Ford, P. Chandler, and J. Sweller. 2001. Learning by imagining. *Journal of Experimental Psychology: Applied* 7(1):68-82.

Cote, L., and H. Leclere. 2000. How clinical teachers perceive the doctor-patient relationship and themselves as role models. *Academic Medicine* 75(11):1117-1124.

Coverdill, J. E., G. L. Adrales, W. Finlay, J. D. Mellinger, K. D. Anderson, B. W. Bonnell, J. B. Cofer, D. B. Dorner, C. Haisch, K. L. Harold, P. M. Termuhlen, and A. L. Webb. 2006a. How surgical faculty and residents assess the first year of the Accreditation Council for Graduate Medical Education duty-hour restrictions: Results of a multi-institutional study. *American Journal of Surgery* 191(1):11-16.

Coverdill, J. E., W. Finlay, G. L. Adrales, J. D. Mellinger, K. D. Anderson, B. W. Bonnell, J. B. Cofer, D. B. Dorner, C. Haisch, K. L. Harold, P. M. Termuhlen, and A. L. B. Webb. 2006b. Duty-hour restrictions and the work of surgical faculty: Results of a multi-institutional study. *Academic Medicine* 81(1):50-56.

de Virgilio, C. 2008. The 80-hour resident workweek: A perspective from a university-affiliated Los Angeles County surgical training program. Presentation to the Committee on Optimizing Graduate Medical Trainee (Resident) Hours and Work Schedules to Improve Patient Safety, May 8, 2008, Irvine, CA.

de Virgilio, C., A. Yaghoubian, R. J. Lewis, B. E. Stabile, and B. A. Putnam. 2006. The 80-hour resident workweek does not adversely affect patient outcomes or resident education. *Journal of Surgical Education* 63(6):435-439.

Dreyfus, H. L., and S. E. Dreyfus. 1986. Five steps from novice to expert, edited by Mind over Machine. New York: The Free Press. Pp. 16-51.

Durkin, E. T., R. McDonald, A. Munoz, and D. Mahvi. 2008. The impact of work hour restrictions on surgical resident education. *Journal of Surgical Education* 65(1): 54-60.

Epstein, R. M. 1999. Mindful practice. *JAMA* 282(9):833-839.

Ericsson, K. A. 2002. Attaining excellence through deliberate practice: Insights from the study of expert performance. In *The pursuit of excellence through education*, edited by M. E. Ferrari. Mahwah, NJ: Lawrence Erlbaum Associates. Pp. 21-56.

Ericsson, K. A., and N. Charness. 1994. Expert performance: Its structure and acquisition. *American Psychologist* 49(8):725-747.

Ericsson, K. A., and R. T. Krampe. 1993. The role of deliberate practice in the acquisition of expert performance. *Psychological Review* 100(3):363-406.

Espey, E., T. Ogburn, and E. Puscheck. 2007. Impact of duty hour limitations on resident and student education in obstetrics and gynecology. *Journal of Reproductive Medicine for the Obstetrician and Gynecologist* 52(5):345-348.

Fallon, W. F., Jr., R. L. Wears, and J. J. Tepas III. 1993. Resident supervision in the operating room: Does this impact on outcome? *Journal of Trauma-Injury Infection & Critical Care* 35(4):556-560.

Farnan, J. M., D. O. Meltzer, H. J. Humphrey, and V. Arora. 2007. Assessing supervision in internal medicine residency training. Paper read at Society of General Internal Medicine National Meeting, April 28, Toronto.

Farnan, J. M., J. K. Johnson, D. O. Meltzer, H. J. Humphrey, and V. M. Arora. 2008. Resident uncertainty in clinical decision making and impact on patient care: A qualitative study. *Quality and Safety in Health Care* 17(2):122-126.

Ferguson, C. M., K. C. Kellogg, M. M. Hutter, and A. L. Warshaw. 2005. Effect of work-hour reforms on operative case volume of surgical residents. *Current Surgery* 62(5):535-538.

Fitzgibbons, J. P., D. R. Bordley, L. R. Berkowitz, B. W. Miller, and M. C. Henderson. 2006. Redesigning residency education in internal medicine: A position paper from the Association of Program Directors in Internal Medicine. *Annals of Internal Medicine* 144(12):920-926.

Fletcher, K. E., W. Underwood III, S. Q. Davis, R. S. Mangrulkar, L. F. McMahon, Jr., and S. Saint. 2005. Effects of work hour reduction on residents' lives: A systematic review. *JAMA* 294(9):1088-1100.

Flexner, A. 1925. *Medical education: A comparative study*. New York: Macmillan Co.

Gais, S., W. Plihal, U. Wagner, and J. Born. 2000. Early sleep triggers memory for early visual discrimination skills. *Nature Neuroscience* 3(12):1335-1339.

Gennis, V. M., and M. A. Gennis. 1993. Supervision in the outpatient clinic: Effects on teaching and patient care. *Journal of General Internal Medicine* 8(7):378-380.

Gordon, J. A. 2000. The human patient simulator: Acceptance and efficacy as a teaching tool for students. The medical readiness trainer team. *Academic Medicine* 75(5):522.

Gordon, J. A., W. M. Wilkerson, D. W. Shaffer, and E. G. Armstrong. 2001. "Practicing" medicine without risk: Students' and educators' responses to high-fidelity patient simulation. *Academic Medicine* 76(5):469-472.

Griffith III, C. H., N. S. Desai, J. F. Wilson, E. A. Griffith, K. J. Powell, and E. C. Rich. 1996. Housestaff experience, workload, and test ordering in a neonatal intensive care unit. *Academic Medicine* 71(10):1106-1108.

Groopman, J. 2008. *How doctors think*. New York: Mariner Books.

Halamek, L. P., D. M. Kaegi, D. M. Gaba, Y. A. Sowb, B. C. Smith, B. E. Smith, and S. K. Howard. 2000. Time for a new paradigm in pediatric medical education: Teaching neonatal resuscitation in a simulated delivery room environment. *Pediatrics* 106(4):E45.

Haney, E. M., C. Nicolaidis, A. Hunter, B. K. Chan, T. G. Cooney, and J. L. Bowen. 2006. Relationship between resident workload and self-perceived learning on inpatient medicine wards: A longitudinal study. *BMC Medical Education* 6:35.

Harrison, R., and E. Allen. 2006. Teaching internal medicine residents in the new era. Inpatient attending with duty-hour regulations. *Journal of General Internal Medicine* 21(5):447-452.

Hoff, T. J., H. Pohl, and J. Bartfield. 2006. Teaching but not learning: How medical residency programs handle errors. *Journal of Organizational Behavior* 27(7):869-896.

Holmboe, E. S., L. M. Prince, and M. Green. 2005. Teaching and improving quality of care in a primary care internal medicine residency clinic. *Academic Medicine* 80(6):571-577.

Horwitz, L. I., H. M. Krumholz, S. J. Huot, and M. L. Green. 2006. Internal medicine residents' clinical and didactic experiences after work hour regulation: A survey of chief residents. *Journal of General Internal Medicine* 21(9):961-965.

Hough, M. C. 2008. Learning, decisions and transformation in critical care nursing practice. *Nursing Ethics* 15(3):322-331.

Huang, G. C., C. C. Smith, C. E. Gordon, D. J. Feller-Kopman, R. B. Davis, R. S. Phillips, and S. N. Weingart. 2006. Beyond the comfort zone: Residents assess their comfort performing inpatient medical procedures. *American Journal of Medicine* 119(1):71.e17-71.e24.

Huber, R., M. F. Ghilardi, M. Massimini, and G. Tononi. 2004. Local sleep and learning. *Nature* 430(6995):78-81.

Hutter, M. M., K. C. Kellogg, C. M. Ferguson, W. M. Abbott, and A. L. Warshaw. 2006. The impact of the 80-hour resident workweek on surgical residents and attending surgeons. *Annals of Surgery* 243(6):864-871.

Immerman, I., E. N. Kubiak, and J. D. Zuckerman. 2007. Resident work-hour rules: A survey of residents' and program directors' opinions and attitudes. *American Journal of Orthopedics* 36(12):E172-E179.

Jacobs, R. L. 2003. *Structured on-the-job training: Unleashing employee expertise in the workplace*, 2nd ed. San Francisco, CA: Berrett-Koehler Publishers, Inc.

Jagsi, R., and R. Surender. 2004. Regulation of junior doctors' work hours: An analysis of British and American doctors' experiences and attitudes. *Social Science & Medicine* 58(11):2181-2191.

Jagsi, R., J. Shapiro, J. S. Weissman, D. J. Dorer, and D. F. Weinstein. 2006. The educational impact of ACGME limits on resident and fellow duty hours: A pre-post survey study. *Academic Medicine* 81(12):1059-1068.

Kennedy, T. J., G. Regehr, G. R. Baker, and L. A. Lingard. 2005. Progressive independence in clinical training: A tradition worth defending? *Academic Medicine* 80(10 Suppl): S106-S111.

Kennedy, T. J. T., L. Lingard, G. R. Baker, L. Kitchen, and G. Regehr. 2007. Clinical oversight: Conceptualizing the relationship between supervision and safety. *Journal of General Internal Medicine* 22(8):1080-1085.

Kilminster, S. M., and B. C. Jolly. 2000. Effective supervision in clinical practice settings: A literature review. *Medical Education* 34(10):827-840.

Kluger, A. N., and A. DeNisi. 1996. The effects of feedback interventions on performance: Historical review, a meta-analysis and a preliminary feedback intervention theory. *Psychological Bulletin* 119:254-284.

Kort, K. C., L. A. Pavone, E. Jensen, E. Haque, N. Newman, and D. Kittur. 2004. Resident perceptions of the impact of work-hour restrictions on health care delivery and surgical education: Time for transformational change. *Surgery* 136(4):861-871.

Kozak, L. J., C. J. DeFrances, and M. J. Hall. 2006. *National Hospital Discharge Survey: 2004 annual summary with detailed diagnosis and procedure data.* Hyattsville, MD: National Center for Health Statistics.

Landrigan, C. P., J. M. Rothschild, J. W. Cronin, R. Kaushal, E. Burdick, J. T. Katz, C. M. Lilly, P. H. Stone, S. W. Lockley, D. W. Bates, and C. A. Czeisler. 2004. Effect of reducing interns' work hours on serious medical errors in intensive care units. *New England Journal of Medicine* 351(18):1838-1848.

Langer, E. J. 1990. *The power of mindful learning.* Cambridge, MA: Da Capo Press.

Leung, W. C. 2002. Competency based medical training: Review. *BMJ* 325(7366):693-696.

Lieberman, J. D., J. A. Olenwine, W. Finley, and G. G. Nicholas. 2005. Residency reform: Anticipated effects of ACGME guidelines on general surgery and internal medicine residency programs. *Current Surgery* 62(2):231-236.

Lin, G. A., D. C. Beck, and J. M. Garbutt. 2006. Residents' perceptions of the effects of work hour limitations at a large teaching hospital. *Academic Medicine* 81(1):63-67.

Lindquist, L., K. Gleason, M. McDaniel, A. Doeksen, and D. Liss. 2008. Teaching medication reconciliation through simulation: A patient safety initiative for second year medical students. *Journal of General Internal Medicine* 23(7):998-1001.

Linn, M. C., H. S. Lee, R. Tinker, F. Husic, and J. L. Chiu. 2006. Inquiry learning: Teaching and assessing knowledge integration in science. *Science* 313(5790):1049-1050.

Litzelman, D. K., G. A. Stratos, D. J. Marriott, and K. M. Skeff. 1998. Factorial validation of a widely disseminated educational framework for evaluating clinical teachers. *Academic Medicine* 73(6):688-695.

Long, D. M. 2000. Competency-based residency training: The next advance in graduate medical education. *Academic Medicine* 75(12):1178-1183.

Ludmerer, K. M. 1999. *Time to heal: American medical education from the turn of the century to the era of managed care.* New York: Oxford University Press.

———. 2004. Learner-centered medical education. *New England Journal of Medicine* 351(12): 1163-1164.

Ludmerer, K. M., and M. M. Johns. 2005. Reforming graduate medical education. *JAMA* 294(9):1083-1087.

Malangoni, M. A., J. J. Como, C. Mancuso, and C. J. Yowler. 2005. Life after 80 hours: The impact of resident work hours mandates on trauma and emergency experience and work effort for senior residents and faculty. *Journal of Trauma-Injury Infection & Critical Care* 58(4):758-761.

Martin, M., B. Vashisht, E. Frezza, T. Ferone, B. Lopez, M. Pahuja, R. K. Spence, F. Y. Bhora, and A. H. Harken. 1998. Competency-based instruction in critical invasive skills improves both resident performance and patient safety. *Surgery* 124(2):313-317.

Mayer, R. E., W. Bove, A. Bryman, R. Mars, and L. Tapangco. 1996. When less is more: Meaningful learning from visual and verbal summaries of science textbook lessons. *Journal of Educational Psychology* 88(1):64-73.

McKee, M., and N. Black. 1992. Does the current use of junior doctors in the United Kingdom affect the quality of medical care? *Social Science & Medicine* 34(5):549-558.

Medina, L. S., J. M. Racadio, and H. A. Schwid. 2000. Computers in radiology. The sedation, analgesia, and contrast media computerized simulator: A new approach to train and evaluate radiologists' responses to critical incidents. *Pediatric Radiology* 30(5):299-305.

Mendoza, K. A., and L. D. Britt. 2005. Resident operative experience during the transition to work-hour reform. *Archives of Surgery* 140(2):137-145.

Meyers, F. J., S. E. Weinberger, J. P. Fitzgibbons, J. Glassroth, F. D. Duffy, and C. P. Clayton. 2007. Redesigning residency training in internal medicine: The consensus report of the Alliance for Academic Internal Medicine Education Redesign Task Force. *Academic Medicine* 82(12):1211-1219.

Mincer, J. 1962. On-the-job training: Costs, returns, and some implications. *Journal of Political Economy* 70(5):50-79.

New York Supreme Court. 1986. *Report of the fourth grand jury for the April/May term of 1986 concerning the care and treatment of a patient and the supervision of interns and junior residents at a hospital in New York county.* New York: New York Supreme Court.

Osborn, L. M., J. R. Sargent, and S. D. Williams. 1993. Effects of time-in-clinic, clinic setting, and faculty supervision on the continuity clinic experience. *Pediatrics* 91(6):1089-1093.

Østergaard, D. 2004. National medical simulation training program in Denmark. *Critical Care Medicine* 32(2 Suppl):S58-S60.

Oyler, J., L. Vinci, V. Arora, and J. Johnson. 2008. Teaching internal medicine residents quality improvement techniques using the ABIM's practice improvement modules. *Journal of General Internal Medicine* 23.

Plack, M. M., and L. Greenberg. 2005. The reflective practitioner: Reaching for excellence in practice. *Pediatrics* 116(6):1546-1552.

Plihal, W., and J. Born. 1997. Effects of early and late nocturnal sleep on declarative and procedural memory. *Journal of Cognitive Neuroscience* 9(4):534-547.

Pollock, E., P. Chandler, and J. Sweller. 2002. Assimilating complex information. *Learning and Instruction* 12(1):61-86.

Reed, D. A., R. B. Levine, R. G. Miller, B. H. Ashar, E. B. Bass, T. N. Rice, and J. Cofrancesco, Jr. 2007. Effect of residency duty-hour limits: Views of key clinical faculty. *Archives of Internal Medicine* 167(14):1487-1492.

Richardson, W. S. 2005. Teaching evidence-based practice on foot. *Evidence-Based Medicine* 8:100-103.

Rothwell, W. J., and H. C. Kazanas. 2004. *Improving on-the-job training: How to establish and operate a comprehensive OJT program.* 2nd ed. New York: John Wiley and Sons.

Ryan, J. 2005. Unintended consequences: The Accreditation Council for Graduate Medical Education work-hour rules in practice. *Annals of Internal Medicine* 143(1):82-83.

Salas, E., and C. S. Burke. 2002. Simulation for training is effective when. . . . *Quality and Safety in Health Care* 11(2):119-120.

Salas, E., and J. A. Cannon-Bowers. 2000. Designing training systems systematically. In *The Blackwell handbook of principles of organizational behavior*, edited by E. A. Locke. Malden, MA: Blackwell Publisher Ltd. Pp. 43-59.

———. 2001. The science of training: A decade of progress. *Annual Review of Psychology* 52:471-499.

Salas, E., K. A. Wilson, C. S. Burke, and H. A. Priest. 2005. Using simulation-based training to improve patient safety: What does it take? *Joint Commission Journal on Quality & Patient Safety* 31(7):363-371.

Salas, E., K. A. Wilson, E. H. Lazzara, H. B. King, J. S. Augenstein, D. W. Robinson, and D. J. Birnbach. 2008. Simulation-based training for patient safety: 10 principles that matter. *Journal of Patient Safety* 4(1):3-8.

Satish, U., and S. Streufert. 2002. Value of a cognitive simulation in medicine: Towards optimizing decision making performance of healthcare personnel. *Quality and Safety in Health Care* 11(2):163-167.

Satish, U., S. Streufert, R. Marshall, J. S. Smith, S. Powers, P. Gorman, and T. Krummel. 2001. Strategic management simulations is a novel way to measure resident competencies. *American Journal of Surgery* 181(6):557-561.

Sejits, G., and G. P. Latham. 2005. Learning versus performance goals: When should each be used? *Academy of Management Executive* 19:124-131.

Shin, S., R. Britt, and L. D. Britt. 2008. Effect of the 80-hour work week on resident case coverage: Corrected article. *Journal of the American College of Surgeons* 207(1):148-150.

Shojania, K. G., K. E. Fletcher, and S. Saint. 2006. Graduate medical education and patient safety: A busy—and occasionally hazardous—intersection. *Annals of Internal Medicine* 145(8):592-598.

Shulkin, D. J. 2008. Like night and day—Shedding light on off-hours care. *New England Journal of Medicine* 358(20):2091-2093.

Sica, G. T., D. M. Barron, R. Blum, T. H. Frenna, and D. B. Raemer. 1999. Computerized realistic simulation: A teaching module for crisis management in radiology. *American Journal of Roentgenology* 172(2):301-304.

Singh, H., E. J. Thomas, L. A. Petersen, and D. M. Studdert. 2007. Medical errors involving trainees: A study of closed malpractice claims from 5 insurers. *Archives of Internal Medicine* 167(19):2030-2036.

Skeff, K. M. 1998. An educational framework for the analysis of teaching. *Substance Abuse* 9(2):61-75.

Skeff, K. M., S. Ezeji-Okoye, P. Pompei, and S. Rockson. 2004. Benefits of resident work hours regulation. *Annals of Internal Medicine* 140(10):816-817.

Smith, C. A., A. B. Varkey, A. T. Evans, and B. M. Reilly. 2004. Evaluating the performance of inpatient attending physicians: A new instrument for today's teaching hospitals. *Journal of General Internal Medicine* 19(7):766-771.

Sox, C. M., H. R. Burstin, E. J. Orav, A. Conn, G. Setnik, D. W. Rucker, P. Dasse, and T. A. Brennan. 1998. The effect of supervision of residents on quality of care in five university-affiliated emergency departments. *Academic Medicine* 73(7):776-782.

Spencer, A. U., and D. H. Teitelbaum. 2005. Impact of work-hour restrictions on residents' operative volume on a subspecialty surgical service. *Journal of the American College of Surgeons* 200(5):670-676.

Stickgold, R., L. James, and J. A. Hobson. 2000. Visual discrimination learning requires sleep after training. *Nature Neuroscience* 3(12):1237-1238.

Sweller, J., P. Chandler, P. Tierney, and M. Cooper. 1990. Cognitive load and selective attention as factors in the structuring of technical matter. *Journal of Experimental Psychology: General* 119:176-192.

USMLE (United States Medical Licensing Examination®). 2008. USLME performance data. http://www.usmle.org/Scores_Transcrtips/performance.html (accessed September 10, 2008).

Vaughn, D. M., C. L. Stout, B. L. McCampbell, J. R. Groves, A. I. Richardson, W. K. Thompson, M. L. Dalton, and D. K. Nakayama. 2008. Three-year results of mandated work hour restrictions: Attending and resident perspectives and effects in a community hospital. *American Surgeon* 74(6):542-546.

Veum, J. R. 1999. Training, wages and the human capital model. *Southern Economic Journal* 65(3):526-538.

Vidyarthi, A. R., P. P. Katz, S. D. Wall, R. M. Wachter, and A. D. Auerbach. 2006. Impact of reduced duty hours on residents' educational satisfaction at the University of California, San Francisco. *Academic Medicine* 81(1):76-81.

Warm, E., D. Schauer, T. Diers, B. Mathis, Y. Neirouz, J. Boex, and G. Rouan. 2008. The ambulatory long-block: An Accreditation Council for Graduate Medical Education (ACGME) Educational Innovations Project (EIP). *Journal of General Internal Medicine* 23(7):921-926.

Wiener, E. L., R. E. Curry, and M. L. Faustina. 1984. Vigilance and task load: In search of the inverted U. *Human Factors* 26(2):215-222.

Williams, P. L., and C. Webb. 1994. Clinical supervision skills: A Delphi and critical incident technique study. *Medical Teacher* 16(2):139-157.

Winslow, E. R., L. Berger, and M. E. Klingensmith. 2004. Has the 80-hour work week increased faculty hours? *Current Surgery* 61(6):602-608.

Wong, A. K. 2004. Full scale computer simulators in anesthesia training and evaluation. *Canadian Journal of Anesthesia* 51(5):455-464.

———. 2006. Curriculum development in anesthesia: Basic theoretical principles. *Canadian Journal of Anesthesia* 53(9):950-960.

Wong, J. G., E. S. Holmboe, and S. J. Huot. 2004. Teaching and learning in an 80-hour work week: A novel day-float rotation for medical residents. *Journal of General Internal Medicine* 19(5 Pt 2):519-523.

Zuckerman, J. D., E. N. Kubiak, I. Immerman, and P. DiCesare. 2005. The early effects of Code 405 work rules on attitudes of orthopaedic residents and attending surgeons. *Journal of Bone and Joint Surgery—Series A* 87(4):903-908.

5

Impact of Duty Hours on Resident Well-Being

As residents acquire needed skills during their educational training, the degree of fatigue and workload they experience places them at risk for workplace injury, driving incidents, decreased physical and mental health, and weakened professional and personal relationships. This chapter looks at the risks associated with each of these consequences because of working long hours and how they affect residents' general well-being. The committee recommends that transportation alternatives and adjustments to work hours and schedules be put in place to prevent the harm that may be caused to residents by the current work environment.

Workers' schedules and lengthy work hours can affect their safety and psychological, social, and physical well-being. Residents are no exception. A review by Caruso assessing the impact of long work hours on the general U.S. worker population revealed that working 50 hours or more a week can have detrimental effects on workers, placing them at risk for sleep deprivation or fatigue, declines in alertness or concentration, depression, poorer general health (including weight gain, cardiovascular decline, and muscular pain), and injuries (Caruso, 2006). Resident physicians, who typically work well over 50 hours a week, may therefore be at risk for these negative effects on their health and well-being, although there may be some counterbalancing effect in pursuing their desired career goal and working in a collegial environment. Residents may thrive on and enjoy the extensive and intensive training paramount to acquiring the necessary skills to become a physician, but the time and workload demands this places on them can impact their health and safety, and potentially affect their personal and professional relationships (Cohen, 2002; Papp et al., 2006).

Since the design of duty hours can affect all these aspects of a resident's life, this chapter presents available evidence to guide the development of recommendations that promote resident well-being. First, resident physical safety is examined with respect to increased work-related injuries and driving incidents due to fatigue, followed by an examination of resident burnout, depression, and physical health. The impact of fatigue on personal and professional relationships and overall quality of life is also addressed.

RESIDENT SAFETY

Although safety concerns for residents stem from activities that take place within a hospital, their demanding duty hours can create safety risks when they leave the hospital as well. The empirical literature highlights two main sources of resident physical injury: injuries experienced while delivering care, such as accidental needlesticks and exposure to blood-borne pathogens, and motor vehicle accidents.

Resident Work Injuries

Most on-the-job injuries of residents are accidental injuries, especially percutaneous ones (those that penetrate the skin). Several well-described multi-institutional survey-based studies have substantiated that injuries are more likely when residents are fatigued.

A prospective cohort study surveyed 2,737 interns (first-year residents) nationwide in a number of medical specialties in 2002-2003 before Accreditation Council for Graduate Medical Education (ACGME) duty hour reform (Ayas et al., 2006).[1] Results of the survey show that first-year residents reported a higher rate of exposure to injury when fatigued. Responding to monthly web surveys, residents reported 1,551 instances in which they were exposed to contaminated bodily fluids, 498 of which occurred through percutaneous injuries. First-year residents reported more than twice as many percutaneous injuries at night than during the day (1.48 per 1,000 opportunities vs. 0.70 per 1,000 opportunities; odds ratio [OR] = 2.04, confidence interval [CI] = 1.98-2.11) and sustaining such injuries nearly twice as often while working extended shifts (i.e., working 24 consecutive hours or more) compared to working a day shift only (1.31 per 1,000 opportunities vs. 0.76 per 1,000 opportunities; OR = 1.61, CI = 1.46-1.78). Lack of concentration and fatigue were cited as major reasons for these injuries (64 percent and 31 percent, respectively), with fatigue more frequently cited as a contributing factor when residents worked at night

[1]Medical specialties included internal medicine, surgery, obstetrics-gynecology, pathology, family medicine, psychology, pediatrics, and emergency care.

and when they worked extended shifts ($p < .001$). These results differed by specialty, with obstetrics-gynecology (OB/GYN), pathology, and surgery residents citing more injuries than others (Ayas et al., 2006).

Self-reported accounts of fatigue were also positively associated with risk of injury involving sharp medical instruments and devices in a study of 109 medical trainees (e.g., medical students, residents, fellows) in five academic medical centers in the United States and Canada. Conducted between 2000 and 2004, the study found that trainees were at three times greater risk of fatigue-related injury than other healthcare workers (including attendings and nurses) (relative risk = 2.03, CI = 1.41-2.94). Injury among trainees was associated with less sleep before an injury and longer work hours per week. The week prior to the injury, medical trainees slept a median 6 hours per night compared to nontrainees' 6.75 hours ($p < .001$). Medical trainees worked on average 70 hours per week compared to other healthcare workers' 40 hours per week, and they had also been at work on average 1.5 hours longer than other healthcare workers when injuries occurred (Fisman et al., 2007). Although the study included 1 year of data gathering post-ACGME duty hour reform, no attempt was made to determine whether the risk of a fatigue-related injury decreased during 2003-2004.

A major risk of percutaneous injury is exposure to blood-borne pathogens (e.g., HIV and hepatitis B and C). A retrospective review to assess whether resident exposure to blood-borne pathogens varied during a given 24-hour period found that residents ($n = 782$) were exposed more often at night (Parks et al., 2000). Exposures resulted from needle punctures (75 percent of incidents), cuts (13 percent), and splashes of infected body fluids (12 percent). Over a 5-year period (November 1993-July 1998), the overall relative risk of accidental exposure to these pathogens was 1.5 times higher during nighttime hours (6 p.m.-6 a.m.) than during the day (6 a.m.-6 p.m.); the highest rate tended to occur from midnight to 1 a.m., and the lowest from 6 a.m. to 7 a.m. Exposures were concentrated in five specialties: anesthesiology (30 percent), internal medicine (20 percent), surgery (16 percent), OB/GYN (11 percent), and pediatrics (5 percent) and rarely occurred in outpatient clinics. First- and second-year residents were the most likely to be exposed to blood-borne pathogens (56 percent of total, 75 percent of resident exposures) (Parks et al., 2000), and anecdotal accounts indicate that this occurs because they perform activities such as blood-drawing more commonly than senior residents or attendings. A more recent study by Landrigan and colleagues attempted to assess incidence rates of occupational exposure to blood and other bodily fluids pre-post ACGME limits, and found that reported rates of exposure for 2003 and 2004 were nearly the same (21.6 percent), which the authors attributed to a minimal change in actual hours worked (Landrigan et al., 2008).

Resident injuries are often affected by fatigue, sleep loss, and lower concentration levels (and not necessarily by skill level). Preventing and mitigating fatigue and sleep loss whenever possible may help sustain improved concentration levels and thus reduce the occurrence of injuries among residents.

Driving Incidents

Concerns regarding resident safety extend beyond the workplace. Driving home after an extended duty period on call can also be hazardous to residents' well-being. As the following studies indicate, residents are more likely to be involved in a car crash or to receive a citation when driving after working long duty periods than after working shorter ones. Fatigued and sleepy residents on the road potentially affect not only themselves but the public as well, raising further concerns for public safety.

In 1996, a survey of pediatric residents (n = 62) and faculty (n = 72) at one institution showed that, on average, residents managed to sleep 2.7 hours when on call and 7.2 hours when not on call, while faculty recalled sleeping undisturbed for an average of 6.5 hours each night. Responses revealed that residents fell asleep more frequently at red lights (40 percent vs. 12.5 percent) and while driving (23 percent vs. 11 percent) than did faculty and were involved in more motor vehicle crashes (20 vs. 11) (Marcus and Loughlin, 1996). In addition, residents who fell asleep behind the wheel did so most frequently after being on duty (90 percent of incidents occurred after approximately a 33-hour shift). These results indicate that the hours of rest one receives each night and the duration of duty periods may seriously impact one's driving capabilities.

More recently, a national sample of 682 interns who completed 12 monthly surveys reported being involved in 133 crashes during the year, 131 of which occurred upon leaving work (Barger et al., 2005). Interns were 2.3 times more likely to be involved in a crash after working extended shifts (their duty periods averaged 32 hours, during which they averaged less than 3 hours sleep) than those not working extended duty periods. These first-year residents were 5.9 times more likely to experience near-miss crashes after extended duty periods than after non-extended shifts. After five extended duty periods in a month, the risk of falling asleep while driving or stopped in traffic significantly increased (while driving: OR = 2.39, CI = 2.31-2.46; stopped: OR = 3.69, CI = 3.60-3.77) (Barger et al., 2005). Similarly, an earlier survey conducted by Steele and colleagues showed that emergency medical residents were at greater risk of being involved in near-miss or collision incidents after working a night shift, and that the prevalence of incidents was positively correlated with the number of night shifts a resident worked per month (Steele et al., 1999). The Barger et al.

study was conducted prior to the 2003 duty hour limits, but residents are still allowed to work periods of 30 consecutive hours more than five times a month.

In another study, resident performance after working 4 weeks of heavy call (defined as working on average 90 hours per week and being on call every fourth or fifth night) was found to be comparable to resident performance with blood alcohol levels of 0.04-0.05 g per 100 mL of blood.[2] This study of 34 pediatric residents also found that residents on heavy call for 4 weeks (sleeping on average slightly more than 6 hours per night) were less alert and sleepier than those on light call (defined as working only 44 hours per week on average), who averaged about 7.5 hours of sleep per night as measured by wrist actigraphy. Reaction times were also slower for residents on heavy call than those on light call (242.5 milliseconds [ms] vs. 225.9 ms, $p < .001$). In addition, residents on the heavy call schedule performed more poorly in the driving simulator than those on light call (lane variability: 7.0 feet vs. 5.5 feet, $p < .001$; speed variability 4.1 miles per hour [mph] vs. 2.4 mph, $p < .001$) (Arnedt et al., 2005).

Two separate population-based case-control studies conducted to determine the greatest risk factors for sleepy drivers also support the results of the above studies on residents. The first study of North Carolina drivers involved in a sleep-related crash showed they were more likely to work multiple jobs, night shifts, or other unusual schedules and averaged fewer hours of sleep per night than drivers who were not involved in a recent crash (Stutts et al., 2003). The second study determined that injuries from sleep-related crashes occurred more often among drivers who had slept less than 5 hours in the previous 24 hours (Connor et al., 2002). These studies clearly demonstrate that sleepiness and fatigue are serious risks for driving incidents, which is why mitigating these factors for residents will be important to their safety.

Although residents are at high risk for fatigue-related car crashes, they, like many other healthy but sleep-deprived adults, often fail to recognize their degree of impairment (Arnedt et al., 2005; Van Dongen et al., 2003; Woodrow et al., 2008). If a resident does not recognize this risk or is not aware of his or her level of impairment and is involved in a collision when driving after a shift in the hospital, responsibility for the resulting injuries has been known to fall on the resident in the past. In one case, the hospital at which a resident worked was found not liable for impaired driving incidents caused by their residents, as a court ruling in Illinois established: "There is no liability imputed to health care providers for injuries to third

---

[2]It is considered a crime to drive with a blood alcohol level of 0.08 g per 100 mL of blood throughout the United States (Insurance Institute for Highway Safety, 2008) and with a level of 0.04 g for commercial drivers (FMCSA, 2008).

parties who are not patients in the hospital" (IPRO, 2007). Therefore, while the committee recommends that hospitals institute transportation services to help prevent these incidents, residents should be aware of all risks associated with deciding to drive after working extended hours and should know that responsibility for their actions ultimately rests with them.

Improving Resident Safety

To reduce physical harm to residents, the committee believes that it is important to address the level of acute and chronic sleep deprivation and fatigue they experience. Although needlesticks or other sharps injuries to residents will not be eliminated altogether, strategies to increase sleep should help reduce these events. Recommendations for duty hours and work schedules that incorporate ways to protect residents against acute and chronic sleep loss and fatigue can be found in Chapter 7.

Regarding driving incidents, the committee found only one study that measured incidents involving residents after the 2003 rules were adopted, and it showed no significant change in motor vehicle accidents or near-miss motor vehicle incidents compared to before implementation for pediatric residents at 3 institutions (Landrigan et al., 2008). Extended duration shifts of 30 hours are still permissible, and the allowable frequency of long call duty periods per month (seven to nine per month depending on averaging and the ability to remain under 80 hours per week) is associated with a greater likelihood of falling asleep at the wheel (Barger et al., 2005). Since fatigued residents are often unable to accurately evaluate their ability to remain alert during their drive home after an extended duty period, to help prevent driving incidents due to fatigue or sleepiness the committee recommends that medical training institutions take some responsibility by implementing the following:

> **Recommendation 5-1: The committee recommends that sponsoring institutions immediately begin to provide safe transportation options (e.g., taxi or public transportation vouchers) for any resident who for any reason is too fatigued to drive home safely.**

This recommendation will be particularly important until further adjustments to resident work schedules are made as recommended by the committee in Chapter 7, which incorporate time for sleep after being on extended duty for more than 16 hours. The committee recognizes that for such practices to become widely instituted, a culture will need to develop among residents and other staff that is more attuned to the risks of fatigue or sleep deprivation. Because sleeping is a voluntary and local behavior, the committee believes that residents should own the responsibility of one's own fatigue levels. Thus, they should behave in a manner that reflects accountability both on a personal and professional level when making decisions to drive after being on extended duty. Institutions should include education about the risks associated with fatigue and sleep deprivation in the basic curriculum of medical students and promote greater awareness of the topic among residents and all medical staff (ACGME, 2007; Jha et al., 2005). Such education would help residents to be more cognizant of their risks. However, because residents and others are not always self-aware when fatigued, one option that the committee suggests is to have institutions provide transportation, both to and from the hospital, as the default scenario for residents on the days they are scheduled to be on duty for more than 16 hours. This would then not be dependent on someone making a fatigue assessment of residents; instead it would be based on hours worked. The committee also supports evaluating alternatives, such as hospitals providing onsite space to allow residents to sleep before driving home after these long shifts without this counting toward duty hour limits when transportation services are unavailable. Evidence suggests that naps are often effective in dispelling drowsiness sufficiently to be able to drive (Philip et al., 2006). However, residents indicated anecdotally that they would prefer to go home to have longer periods of uninterrupted sleep. Alternatives should be assessed to ensure that residents would not opt out of using services provided and continue unsafe driving.

RESIDENT WELL-BEING AND QUALITY OF LIFE

Residents' well-being refers to their state of overall mental and physical health and how these factors, among others, can affect their general quality of life. This section discusses aspects of mental health such as levels of resident burnout and depression, concerns regarding their physical fitness, satisfaction with their personal and professional lives, and how these aspects have been impacted by ACGME's duty hour regulations or fatigue.

Before discussing burnout and depression, definitions may clarify the differences between these two similar symptoms experienced by residents. Originally coined by Freudenberger in 1974, the term "burnout" described a state of exhaustion or extreme fatigue resulting from an excessive demand of energy, strength, or resources, in turn causing individuals to become cynical about their work (Douglas Institute, 2008). Although considered a vague notion for several years, more complete definitions came to include physical and mental exhaustion observed by those in professions requiring continuous contact with others. Maslach and colleagues eventually identified three widely recognized core elements of burnout: emotional exhaustion—depleted energy from overwhelming work demands; depersonalization—personal detachment from one's job; and lack of personal accomplishment due to self-perceptions of inefficiency (Maslach et al., 1997).

Depression, on the other hand, is characterized by "depressed mood, inability to derive pleasure from things, weight loss or gain, insomnia or hypersomnia, psychomotoric agitation or retardation, fatigue or loss of energy, feelings of insufficiency or guilt, indecisiveness or inability to concentrate, and thoughts about death and suicide" (Brenninkmeijer et al., 2001). Substantial evidence concerning the distinctions between burnout and depression can be found in a literature review by Glass and McKnight (1996) that empirically investigated the relationship between the two. The authors concluded that burnout and depression are not identical, yet they have symptoms in common, such as emotional exhaustion, that are positively related to both (Brenninkmeijer et al., 2001; Glass and McKnight, 1996).

Burnout

The empiric literature focuses on three main issues: the prevalence of burnout in residents, the factors associated with burnout, and the impact of changes in duty hours on resident burnout. Studies focused on the impact of duty hour regulations tended to be of small numbers of residents, single institutions, and specialty-specific. As discussed below, the data are mixed—residents do experience high levels of burnout, but burnout is not necessarily associated with the numbers of hours worked or slept. Instead, burnout among residents has been found to be more highly associated with managing a heavy workload or exposure to high work intensity (Thomas, 2004).

Prevalence of Burnout

Burnout is quite prevalent among residents, with rates varying from 41 to 76 percent (Fahrenkopf et al., 2008; Thomas, 2004). A study of 321 residents in one institution found that 50 percent reported experiencing burnout during their training as measured by the Maslach Burnout Inventory (MBI), a validated, widely used questionnaire. Although there were varying rates of burnout across specialties (27 to 75 percent), these differences were not statistically significant. The number of hours worked was also not associated with increased risk of burnout (i.e., residents working more than 80 hours per week were not more likely to experience burnout than those working 80 hours or less). However, first-year residents were more likely to report burnout than more senior residents (77.3 percent and 41.8 percent, respectively) (Martini et al., 2004). A longitudinal study of 47 internal medicine interns the year prior to ACGME limits found that the prevalence of burnout increased and empathy decreased during their first year of residency. Only 4.3 percent of residents reported high levels of burnout at the beginning of the year compared to 55.3 percent at the end of the year ($p < .0001$) (Rosen et al., 2006). Although increased sleep deprivation was not associated with increased burnout, it was associated with higher rates of depression.

Factors Associated with Burnout

Several factors can contribute to the dimensions of burnout. A literature review assessing 15 studies of resident burnout published between 1983 and 2004 found that burnout was associated less with sleep deprivation than with work intensity and work interference with home life (Thomas, 2004). Work intensity according to residents was often related to feelings of being overwhelmed by work demands or workload and having insufficient time to plan or manage them (Biaggi et al., 2003; Nyssen et al., 2003). Observations of this sort can be related to a perceived lack of control over one's job (Nyssen et al., 2003). An additional study points to stress over financial strains or debt that many residents experience and how this may play a role in producing emotional exhaustion (Collier et al., 2002). Although sleep deprivation and lack of leisure time are still commonly cited by residents as reasons for burnout (Thomas and Brennan, 2000), specialty-specific studies ($n < 130$) have shown that despite these claims by residents, no statistically significant correlation was found between hours slept, hours worked, or sleep deprivation and burnout (Fahrenkopf et al., 2008; Rosen et al., 2006). These findings underscore that duty hours are merely one factor affecting resident performance and that modifying other factors as well—for example, moderating workload—can help improve overall training experiences.

Impact of Duty Hour Regulations on Burnout

Evidence of whether the 2003 ACGME duty hour limits reduced burnout is mixed, but no studies have shown that duty hour reductions or limits have increased its prevalence. Duty hour regulations did not decrease symptoms of burnout in a study of 33 surgical residents in six institutions (Gelfand et al., 2004). Another study of internal medicine residents from one institution surveyed in May 2003 ($n = 121$) and May 2004 ($n = 106$) found that a reduction in duty hours (from 74.6 hours per week to 67.1 hours per week) was associated with decreased emotional exhaustion (42 percent vs. 29 percent). There were however, no significant changes in depersonalization as measured by the MBI or perceptions of personal achievement (Gopal et al., 2005). A third study, comparing survey responses of 115 internal medicine residents in 2001 and 118 internal medicine residents in 2004, also found that although the number of residents reporting emotional exhaustion as measured by the MBI decreased significantly from 53 to 40 percent after the implementation of duty hour regulations, there was no significant change in the percentage of residents with total scores meeting the burnout criteria (Goitein et al., 2005). In contrast, a study comparing the scores of 220 pediatric residents from three large programs found a statistically significant decrease in the burnout rates before and after the 2003 duty hour limitations (75.4 percent versus 57.0 percent) (Landrigan et al., 2008).

It is important to note here that the committee's proposed changes in duty hours without appropriate adjustments of workload could possibly have an unintended consequence of leading to more stress or burnout. For example, one method of moderating resident workload is to reduce or limit the number of patient cases that a resident can handle per duty period. However, if all less complex patient cases are taken over by physician extenders and only more complex patients are concentrated on resident teams (as a way to increase the educational value of time spent on duty), the new level of work intensity could cause some degree of burnout unless caseload is adjusted for patient severity. Because of this, burnout should be an outcome that is studied with the proposed interventions.

Depression and Mood

Depression is a mood disorder that can affect job performance, personal and professional interactions, and health. Studies of depression in residents generally present data on prevalence of depression among residents and the impact of duty hour regulations on depression rates. Studies of the latter type tend to be small and specialty-specific. The study data tend to report depression based on screening instruments rather than diagnoses of clinical depression.

Prevalence of Depression

Statistics regarding the prevalence of depression among residents vary widely from 7 to 56 percent based on different validated tools used to screen for depression or detect clinical depression (Becker et al., 2006; Bellini et al., 2002; Fahrenkopf et al., 2008; Goitein et al., 2005; Gopal et al., 2005; Shanafelt et al., 2002). One study of 125 OB/GYN residents recruited from 23 randomly selected programs across the United States found that more than one-third of participants (34.2 percent) were depressed, according to the Center for Epidemiological Studies-Depression Scale (Becker et al., 2006). Just prior to duty hour regulations, Fahrenkopf et al. (2008) found that among 123 pediatric residents evaluated, 20 percent were at high risk for depression (determined through the Harvard National Depression Screening Day Scale, which measures depressive symptoms, not criteria for a diagnosis of depression). Ninety-six percent of these residents also met the criteria for burnout (measured through the MBI) and more often reported having poor health and having difficulty concentrating at work than their nondepressed colleagues (Fahrenkopf et al., 2008). Becker also noted high rates of burnout among residents who were depressed.

At least one study conducted prior to the 2003 regulations suggests that sleep deprivation may be associated with the development of moderate depression among interns (Rosen et al., 2006). In addition to finding that the prevalence of chronic sleep deprivation increased from 9 percent at the beginning of the year to 43 percent at the end of the year, Rosen and colleagues reported that the prevalence of moderate depression (as measured by the Beck Depression Inventory-Short Form) among residents also increased as the year progressed (4.3 percent to 29.8 percent; $p = .0002$) and was associated with chronic sleep deprivation (OR = 7; $p = .014$). In fact, chronically sleep-deprived interns had a seven times greater likelihood of developing depression during their first year of residency than colleagues who obtained more sleep (Rosen et al., 2006). Further research is needed to determine whether depression rates vary across specialties.

Impact of Duty Hour Limits on Depression

Only three studies have evaluated depression rates in residents after the institution of duty hour regulations. Two of the three studies were limited to a single institution and focused on a single specialty, internal medicine. Although Gopal and colleagues (2005) reported that fewer residents had a positive result on a depression screening instrument after the first year of duty hour regulations than before the regulations were implemented, the results were not statistically significant. Nor were there statistically significant differences in the increased percentage of internal medicine residents who screened positive on an unnamed depression screening questionnaire (Goitein et al., 2005). The third study, involving 220 residents from three large pediatric residency programs, found no change in the rates of depression before and after the institution of duty hour limitations (Landrigan et al., 2008). From these studies, it appears that the ACGME regulations had no significant impact on the prevalence of depression.

Only one single-institution study of pediatric residents assessed the mood and fatigue levels of residents who worked night float shifts and found that feelings of depression among night float residents can be more prevalent than among residents on day shifts (Cavallo et al., 2002).

Effects on Physical Health

In addition to affecting mood, at least one study suggests that the sleep deprivation experienced by residents may have other adverse effects on their health. Baldwin and Daugherty's (2004) survey of 3,604 randomly selected postgraduate year 1 (PGY-1) and PGY-2 residents during 1998-1999 revealed that residents who reported obtaining 5 hours of sleep or less per night were more likely to report increased use of alcohol (OR = 1.52), had "taken medications to stay awake" (OR = 1.91), and experienced a significant weight change (OR = 1.51). Almost one-quarter of the participants (22 percent) reported obtaining 5 hours or less of sleep on a regular basis, and two-thirds reported obtaining 6 hours or less of sleep on a regular basis throughout the year (Baldwin and Daugherty, 2004). A more recent web-based survey of 3,971 emergency medicine residents revealed that almost half of the participants (45 percent) were excessively sleepy (a score of >10 on the Epworth Sleepiness Scale), and that approximately one-third of the participants had used medications and/or alcohol to help them fall asleep at least four times in the past month (Handel et al., 2006).

The significant changes in weight reported by residents who regularly obtained 5 or fewer hours of sleep per night (Baldwin and Doughtery, 2004) is not surprising in light of recent findings related to sleep loss, weight gain, and changes in appetite regulation. In the past 7 years, at least 12 epidemiologic studies have documented a dose-dependent relationship between sleep duration and increased body mass index. Sample sizes ranged from 422 participants to more than 68,000 participants, with some studies focused on specific occupational groups (e.g., truck drivers [*n* = 4,878] or registered nurses [*n* = 68,183]). Despite being conducted in different areas of the world (Brazil, Canada, Europe, Japan, and the United States), using different methodologies, and including varying degrees of control for other related variables (e.g., parental weight, depression, shift work), the findings have been quite similar: short sleep durations are associated with greater risks of weight gain and obesity. Although the exact mechanisms linking sleep deprivation to weight gain are unknown, a number of well-controlled laboratory experiments suggest that sleep restriction alters the levels of leptin and other hormones involved in the regulation of appetite (Guilleminault et al., 2003; Spiegel et al., 2004a, 2005).

Other contributions to weight gain can arise from the simple fact that residents have limited time for leisure activities and often lack sufficient opportunities, or energy, to exercise. Anecdotal accounts suggest that residents do not take the advice they give their own patients to exercise regularly and eat healthy foods, admitting to a less healthy lifestyle during their training (Glines, 2004).

Additional health risks due to sleep restriction or sleep deprivation have been demonstrated, such as increased risk of developing various types of diabetes (Ayas et al., 2003; Spiegel et al., 2004b; Van Helder et al., 2003). Although the incidence of residents' being overweight or developing diabetes is unknown, the evidence from both epidemiological and laboratory studies implies that residents who routinely obtain limited amounts of sleep may be at higher risk for these health outcomes.

Regarding their physical and mental well-being, it appears that residents still experience stress and burnout, which can affect their health. The varying quality of the research conducted on these issues suggests that future research may benefit from using standardized measures of quality of life, depression, and well-being, in order to assess the impact of current regulations on health and quality of life. Research to determine the association between burnout, sleep deprivation, and depression would be useful as well.

Quality of Life

Residents are full-time caregivers at work and supportive family members and friends at home. As physicians interacting closely with their healthcare team and with patients, their health and attitude are vital to their success and necessarily have impacts on those around them. The committee thought it important to examine the effects of fatigue and duty hour adjustments on residents' roles outside the hospital, recognizing that success in their training must be understood in the context of their overall lives.

Effects of Duty Hour Regulations on Quality of Life

Most studies that examine resident quality of life are based on surveys of residents at single institutions or in a single geographic area. The term "quality of life" was often used ambiguously or not clearly defined in the studies, and many incorporated burnout, stress, or depression as part of their definition. Rather than using a standard, validated instrument to measure residents' quality of life, institutions developed their own surveys. Despite these methodological weaknesses, findings were similar: most residents believed that their quality of life improved as a result of duty hour regulations.

For example, 128 residents from four training programs adhering to ACGME duty hour regulations were surveyed for their impressions of how the rules would continue to affect future residents. The results indicated a strong agreement (by a Likert-type fixed response scale from "strongly agree" to "strongly disagree") that hour restrictions would have marked benefits on residents' personal lives in the future. The degree of improvement foreseen varied by specialty. Family medicine residents felt most positively about the regulations in terms of better quality of life, followed by internal medicine residents, and to a lesser extent, OB/GYN residents. Surgical residents were the least likely to agree that the regulations would have a positive effect on their quality of life (Zonia et al., 2005).

Yet two separate surveys of surgical residents (98 residents from four programs and 29 residents and 8 faculty from a single program), both administered after duty hour regulations were implemented, reported that these residents believed that those regulations had positive effects on their quality of life. They reported having more time to spend with family and friends, being able attend to important nonmedical responsibilities, and being happier and less tired (Barden et al., 2002; Kort et al., 2004). Another single, one-time survey of 12 plastic surgery residents administered 6 months after implementation of duty hour regulations found residents to be less fatigued as a result of decreased hours. These residents also saw improvements in quality of life and morale, as well as improvements in spousal, family, and other relationships (Basu et al., 2004).

A systematic review by Fletcher et al. (2005) examined how the quality of life in various medical specialties was affected by duty hour reductions. The measures of quality of life in this review encompassed several of the factors examined in this chapter, including mood factors, sleep, relationships, health, and education. The results were mixed for nearly all measures and across specialties, indicating "that there may not be uniform benefits for residents from these changes" (Fletcher et al., 2005, p. 1098).

Differences Between Junior and Senior Residents

Survey responses from 48 orthopedic residents indicated that junior residents felt that their quality of life was better because of duty hour regulations, while senior residents were more neutral. Responses from 39 orthopedic attendings also had improved perceptions of their quality of life. The difference between junior and senior residents' perceptions was attributed to situations in which senior residents had to do work they previously had done as junior residents, which would not have been necessary before implementation of the regulations. This may be valid only for senior residents who began their training before the implementation of regulations (Zuckerman et al., 2005). A different study that gathered 554 surveys from orthopedic surgical residents across the country showed that PGY-3 and more junior residents, who worked in excess of 80 hours per week more frequently than their senior peers, still had more positive attitudes toward duty hour regulations than the senior residents. Nonetheless, residents in this study overall (PGY-4, -5, and -6 residents made up 68 percent of 495 responses) reported an improved quality of life (Kusuma et al., 2007).

In general it seems that reduced hours improve residents' perception of their quality of life, and no study was reviewed that showed duty hour restrictions were associated with poorer quality of life.

Effects of Fatigue on Professional Relationships

Residents' perception of their quality of life can be affected by their professional relationships as much as their personal ones. Satisfaction at the workplace seems to play an important role in resident well-being and depends on factors such as relationships with colleagues and patients, personal performance, and work schedules.

Professionalism is also a key component of a resident's training and should typify the working relationships that residents forge. It is based on the concepts of patients as the primary focus, patient autonomy, and social justice (Project of the ABIM Foundation et al., 2002)—the same concepts on which patient-centered care is founded. Patient centeredness, as defined in the Institute of Medicine (IOM) Quality Chasm series, "encompasses qualities of compassion, empathy, and responsiveness to the needs, values, and expressed preferences of the individual patient" (IOM, 2001, p. 48).

Effect of Fatigue on Professionalism

Given the intensity of work that residents experience, and their susceptibility to personal and professional stress, it is not surprising that some facets of their work, namely efforts toward patient centeredness or professionalism, may at times be neglected. For example, the Committee of Interns and Residents provides reports of residents actively avoiding care conversations with a patient's family members out of fatigue. Other residents reported growing resentful toward their patients because of feeling too exhausted or depressed to provide adequate care (CIR/SEIU Healthcare, 2007). Relationships with coworkers are also affected. One survey study found that sleep-deprived residents (5 hours or less of sleep per night) were significantly more likely (between 1.41 and 1.87 times more) to be involved in serious conflicts with other residents, attendings, or nursing staff (Baldwin and Daugherty, 2004).

Impact of Reduced Duty Hours on Professionalism

Although professionalism is difficult to measure, a few methods exist that attempt to capture a physician's level of professionalism, including surveys of peer assessment, faculty assessments, and self-reflection, as well as objective clinical exams (Cohen, 2006; Swick, 2000). Professionalism is acquired both formally and informally. Formally, it is taught infrequently or incidentally through lectures and conferences. Informally, professionalism is modeled daily by medical colleagues and implicitly required through the appropriate expectations of patients and their loved ones. In a study of 169 internal medicine, neurology, and family practice residents in three hospitals, 45 percent of the residents studied believed that professionalism decreased after duty hours were reduced because of having less time to talk with patients and families, leading to fewer opportunities to participate in shared decision making. However, 32 percent of residents perceived no change and 19 percent believed professionalism improved due to reduced fatigue, allowing for increased reserves of empathy, compassion, and sensitivity to patients and colleagues (Ratanawongsa et al., 2006).

In a systematic review by Fletcher and colleagues, the perceived effect of reduced work hours on professionalism was mixed. Multiple studies of internal medicine residents found varied opinions regarding the effects of schedule interventions on a resident's sense of professionalism: some believed patient-physician relationships, patient care, and continuity of care had improved, while others felt it had decreased or stayed the same (Fletcher et al., 2005). However, a more recent study by Fletcher and her colleagues reported anecdotes from residents who feel they do not always participate in important patient care activities at times (e.g., family meetings) in order to comply with duty hour regulations (Fletcher et al., 2008).

CONCLUSION

Medical training exposes residents to real risks regarding their overall health and quality of life. Varied study methods and reports by residents on the impact of duty hour regulations on aspects of their mental health and professionalism make it difficult to clearly gauge the degree to which working reduced hours truly improves their outlook or satisfaction with life. From the literature, it appears that residents generally feel that reduced hours have positive effects on their well-being and personal life. Yet, several of these positive comments are accompanied by negative perceptions of the impact on their educational training (Fletcher et al., 2005; Gopal et al., 2005; Whang et al., 2003) or on patient safety (Shanafelt et al., 2002; West et al., 2006), which are discussed in Chapters 4 and 6, respectively.

These contrasting sentiments suggest that altering duty hours alone is not a comprehensive strategy to improve the resident experience. Furthermore, promoting resident well-being does more than simply help residents feel better. Protecting physicians' health fitness could help increase patient safety and care, as error rates by residents at high risk for depression have suggested (Fahrenkopf et al., 2008). The committee suggests that other changes, such as enhanced supervision and team support by other staff, may help counter feelings of being overwhelmed that can lead to burnout, depression, and decreased professionalism. Although adjusting resident duty hours can impact resident well-being and may help residents balance the many requirements of training, merely changing trainee schedules cannot substitute for a professional, supportive, and responsive learning environment to promote their success.

REFERENCES

ACGME (Accreditation Council for Graduate Medical Education). 2007. *Common program requirements.* http://acgme.org/acWebsite/dutyHours/dh_dutyhoursCommonPR07012007.pdf (accessed March 23, 2008).

Arnedt, J. T., J. Owens, M. Crouch, J. Stahl, and M. A. Carskadon. 2005. Neurobehavioral performance of residents after heavy night call vs after alcohol ingestion. *JAMA* 294(9):1025-1033.

Ayas, N. T., D. P. White, W. K. Al-Delaimy, J. E. Manson, M. J. Stampfer, F. E. Speizer, S. Patel, and F. B. Hu. 2003. A prospective study of self-reported sleep duration and incident diabetes in women. *Diabetes Care* 26(2):380-384.

Ayas, N. T., L. K. Barger, B. E. Cade, D. M. Hashimoto, B. Rosner, J. W. Cronin, F. E. Speizer, and C. A. Czeisler. 2006. Extended work duration and the risk of self-reported percutaneous injuries in interns. *JAMA* 296(9):1055-1062.

Baldwin, D. C., Jr., and S. R. Daugherty. 2004. Sleep deprivation and fatigue in residency training: Results of a national survey of first- and second-year residents. *Sleep* 27(2):217-223.

Barden, C. B., M. C. Specht, M. D. McCarter, J. M. Daly, and T. J. Fahey III. 2002. Effects of limited work hours on surgical training. *Journal of the American College of Surgeons* 195(4):531-538.

Barger, L. K., B. E. Cade, N. T. Ayas, J. W. Cronin, B. Rosner, F. E. Speizer, C. A. Czeisler, and Harvard Work Hours Health and Safety Group. 2005. Extended work shifts and the risk of motor vehicle crashes among interns. *New England Journal of Medicine* 352(2):125-134.

Basu, C. B., L. M. Chen, L. H. Hollier, Jr., and S. M. Shenaq. 2004. The effect of the Accreditation Council for Graduate Medical Education duty hours policy on plastic surgery resident education and patient care: An outcomes study. *Plastic & Reconstructive Surgery* 114(7):1878-1886.

Becker, J. L., M. P. Milad, and S. C. Klock. 2006. Burnout, depression, and career satisfaction: Cross-sectional study of obstetrics and gynecology residents. *American Journal of Obstetrics & Gynecology* 195(5):1444-1449.

Bellini, L. M., M. Baime, and J. A. Shea. 2002. Variation of mood and empathy during internship. *JAMA* 287(23):3143-3146.

Biaggi, P., S. Peter, and E. Ulich. 2003. Stressors, emotional exhaustion and aversion to patients in residents and chief residents—What can be done? *Swiss Medical Weekly* 133(23-24):339-346.

Brenninkmeijer, V., N. W. Van Yperen, and B. P. Buunk. 2001. Burnout and depression are not identical twins: Is superiority a distinguishing feature? *Personality and Individual Differences* 30:873-880.

Caruso, C. C. 2006. Possible broad impacts of long work hours. *Industrial Health* 44(4):531-536.

Cavallo, A., J. Jaskiewicz, and M. D. Ris. 2002. Impact of night-float rotation on sleep, mood, and alertness: The resident's perception. *Chronobiology International: The Journal of Biological & Medical Rhythm Research* 19(5):893.

CIR/SEIU Healthcare (Committee of Interns and Residents-Service Employees International Union). 2007. Presentation by L. Toni Lewis to the Committee on Optimizing Graduate Medical Trainee (Resident) Hours and Work Schedules to Improve Patient Safety December 3, 2007, Washington, DC.

Cohen, J. J. 2002. Heeding the plea to deal with resident stress. *Annals of Internal Medicine* 136(5):394-395.

———. 2006. Professionalism in medical education, an American perspective: From evidence to accountability. *Medical Education* 40(7):607-617.

Collier, V. U., J. D. McCue, A. Markus, and L. Smith. 2002. Stress in medical residency: Status quo after a decade of reform? *Annals of Internal Medicine* 136(5):384-390.

Connor, J., R. Norton, S. Ameratunga, E. Robinson, I. Civil, R. Dunn, J. Bailey, and R. Jackson. 2002. Driver sleepiness and risk of serious injury to car occupants: Population based case control study. *BMJ* 324(7346):1125-1128.

Douglas Institute. 2008. *What is burn-out and depression?* http://burnout.douglas.qc.ca/burnout/history.html (accessed July 10, 2008).

Fahrenkopf, A. M., T. C. Sectish, L. K. Barger, P. J. Sharek, D. Lewin, V. W. Chiang, S. Edwards, B. L. Wiedermann, and C. P. Landrigan. 2008. Rates of medication errors among depressed and burnt out residents: Prospective cohort study. *BMJ* 336(7642):488-491.

Fisman, D. N., A. D. Harris, M. Rubin, G. S. Sorock, and M. A. Mittleman. 2007. Fatigue increases the risk of injury from sharp devices in medical trainees: Results from a case-crossover study. *Infection Control and Hospital Epidemiology* 28(1):10-17.

Fletcher, K. E., W. Underwood III, S. Q. Davis, R. S. Mangrulkar, L. F. McMahon, Jr., and S. Saint. 2005. Effects of work hour reduction on residents' lives: A systematic review. *JAMA* 294(9):1088-1100.

Fletcher, K. E., V. Parekh, L. Halasyamani, S. R. Kaufman, M. Schapira, K. Ertl, and S. Saint. 2008. Work hour rules and contributors to patient care mistakes: A focus group study with internal medicine residents. *Journal of Hospital Medicine* 3(3):228-237.

FMCSA (Federal Motor Carrier Safety Administration). 2008. *Alcohol and drug rules.* http://www.fmcsa.dot.gov/safety-security/safety-initiatives/drugs/engtesting.htm (accessed December 3, 2008).

Gelfand, D. V., Y. D. Podnos, J. C. Carmichael, D. J. Saltzman, S. E. Wilson, and R. A. Williams. 2004. Effect of the 80-hour workweek on resident burnout. *Archives of Surgery* 139(9):933-938.

Glass, D. C., and J. D. McKnight. 1996. Perceived control, depressive symptomatology, and professional burnout: A review of the evidence. *Psychology & Health* 11(1):23-48.

Glines, M. E. 2004. The effect of work hour regulations on personal development during residency. *Annals of Internal Medicine* 140(10):818-819.

Goitein, L., T. D. Shanafelt, J. E. Wipf, C. G. Slatore, and A. L. Back. 2005. The effects of work-hour limitations on resident well-being, patient care, and education in an internal medicine residency program. *Archives of Internal Medicine* 165(22):2601-2606.

Gopal, R., J. J. Glasheen, T. J. Miyoshi, and A. V. Prochazka. 2005. Burnout and internal medicine resident work-hour restrictions. *Archives of Internal Medicine* 165(22):2595-2600.

Guilleminault, C., N. B. Powell, S. Martinez, C. Kushida, T. Raffray, L. Palombini, and P. Philip. 2003. Preliminary observations on the effects of sleep time in a sleep restriction paradigm. *Sleep Medicine* 4(3):177-184.

Handel, D., A. Raja, and C. Lindsell. 2006. The use of sleep aids among emergency medicine residents: A web based survey. *BMC Health Services Research* 6(1):136.

Insurance Institute for Highway Safety. 2008. *DUI/DWI laws.* http://www.iihs.org/laws/dui.aspx (accessed June 17, 2008).

IOM (Institute of Medicine). 2001. *Crossing the quality chasm: A new health system for the 21st century.* Washington, DC: National Academy Press.

IPRO. 2007. Legal watch: Hospital not liable for sleep-deprived intern's motor vehicle crash. *Resident Times* 1(1):Spring.

Jha, A. K., B. W. Duncan, and D. W. Bates. 2005. Fatigue—Sleepiness and medical errors. Chapter 46. *Making Health Care Safer: A Critical Analysis of Patient Safety Practices.* Rockville, MD: Agency for Healthcare Research and Quality.

Kort, K. C., L. A. Pavone, E. Jensen, E. Haque, N. Newman, and D. Kittur. 2004. Resident perceptions of the impact of work-hour restrictions on health care delivery and surgical education: Time for transformational change. *Surgery* 136(4):861-871.

Kusuma, S. K., S. Mehta, M. Sirkin, A. J. Yates, T. Miclau, K. J. Templeton, and G. E. Friedlaender. 2007. Measuring the attitudes and impact of the eighty-hour workweek rules on orthopaedic surgery residents. *Journal of Bone & Joint Surgery-American Volume* 89(3):679-685.

Landrigan, C. P., A. M. Fahrenkopf, D. Lewin, P. J. Sharek, L. K. Barger, M. Eisner, S. Edwards, V. W. Chiang, B. L. Wiedermann, and T. C. Sectish. 2008. Effects of the Accreditation Council for Graduate Medical Education duty hour limits on sleep, work hours, and safety. *Pediatrics* 122:250-258.

Marcus, C. L., and G. M. Loughlin. 1996. Effect of sleep deprivation on driving safety in housestaff. *Sleep* 19(10):763-766.

Martini, S., C. L. Arfken, A. Churchill, and R. Balon. 2004. Burnout comparison among residents in different medical specialties. *Academic Psychiatry* 28(3):240-242.

Maslach, C., S. E. Jackson, and M. P. Leiter. 1997. Maslach burnout inventory: Third edition. In *Evaluating stress: A book of resources*, edited by C. P. Zalaquett and R. J. Woor. Lanham, MD: Rowman & Littlefield Publishers, Inc. Pp. 191-218.

Nyssen, A. S., I. Hansez, P. Baele, M. Lamy, and V. De Keyser. 2003. Occupational stress and burnout in anaesthesia. *British Journal of Anaesthesia* 90(3):333-337.

Papp, K. K., C. M. Miller, and K. P. Strohl. 2006. Graduate medical training, learning, relationships, and sleep loss. *Sleep Medicine Reviews* 10(5):339-345.

Parks, D. K., R. J. Yetman, M. C. McNeese, K. Burau, and M. H. Smolensky. 2000. Day-night pattern in accidental exposures to blood-borne pathogens among medical students and residents. *Chronobiology International* 17(1):61-70.

Philip, P., J. Taillard, N. Moore, S. Delord, C. Valtat, P. Sagaspe, and B. Bioulac. 2006. The effects of coffee and napping on nighttime highway driving: A randomized trial. *Annals of Internal Medicine* 144(11):785-791.

Project of the ABIM Foundation, ACP-ASIM Foundation, and European Federation of Internal Medicine. 2002. Medical professionalism in the new millennium: A physician charter. *Annals of Internal Medicine* 136(3):243-246.

Ratanawongsa, N., S. Bolen, E. E. Howell, D. E. Kern, S. D. Sisson, and D. Larriviere. 2006. Residents' perceptions of professionalism in training and practice: Barriers, promoters, and duty hour requirements. *Journal of General Internal Medicine* 21(7):758-763.

Rosen, I. M., P. A. Gimotty, J. A. Shea, and L. M. Bellini. 2006. Evolution of sleep quantity, sleep deprivation, mood disturbances, empathy, and burnout among interns. *Academic Medicine* 81(1):82-85.

Shanafelt, T. D., K. A. Bradley, J. E. Wipf, and A. L. Back. 2002. Burnout and self-reported patient care in an internal medicine residency program. *Annals of Internal Medicine* 136(5):358-367.

Spiegel, K., R. Leproult, M. L'Hermite-Baleriaux, G. Copinschi, P. D. Penev, and E. Van Cauter. 2004a. Leptin levels are dependent on sleep duration: Relationships with sympathovagal balance, carbohydrate regulation, cortisol, and thyrotropin. *Journal of Clinical Endocrinology and Metabolism* 89(11):5762-5771.

Spiegel, K., E. Tasali, P. D. Penev, and E. Van Cauter. 2004b. Sleep curtailment in healthy young men is associated with decreased leptin levels, elevated ghrelin levels, and increased hunger and appetite. *Archives of Internal Medicine* 141:846-850.

Spiegel, K., R. Leproult, and E. Van Cauter. 2005. Metabolic and endocrine changes. In *Sleep deprivation: Basic science, physiology, and behavior*, edited by C. A. Kushida. New York: Marcel Dekker. Pp. 293-318.

Steele, M. T., O. J. Ma, W. A. Watson, H. A. Thomas, and R. L. Muelleman. 1999. The occupational risk of motor vehicle collisions for emergency medicine residents. *Academic Emergency Medicine* 6(10):1050-1053.

Stutts, J. C., J. W. Wilkins, J. S. Osberg, and B. V. Vaughn. 2003. Driver risk factors for sleep-related crashes. *Accident Analysis and Prevention* 35(3):321-331.

Swick, H. M. 2000. Toward a normative definition of medical professionalism. *Academic Medicine* 75(6):612-616.

Thomas, E. J., and T. A. Brennan. 2000. Incidence and types of preventable adverse events in elderly patients: Population based review of medical records. *BMJ* 320(7237):741-744.

Thomas, N. K. 2004. Resident burnout. *JAMA* 292(23):2880-2889.

Van Dongen, H. P. A., G. Maislin, J. M. Mullington, and D. F. Dinges. 2003. The cumulative cost of additional wakefulness: Dose-response effects on neurobehavioral functions and sleep physiology from chronic sleep restriction and total sleep deprivation. *Sleep* 26(2):117-126.

Van Helder, T., J. D. Symons, and M. W. Radomski. 2003. Effects of sleep deprivation and exercise on glucose tolerance. *Aviation, Space, & Environmental Medicine* 74(4):487-492.

West, C. P., M. M. Huschka, P. J. Novotny, J. A. Sloan, J. C. Kolars, T. M. Habermann, and T. D. Shanafelt. 2006. Association of perceived medical errors with resident distress and empathy: A prospective longitudinal study. *JAMA* 296(9):1071-1078.

Whang, E. E., M. M. Mello, S. W. Ashley, and M. J. Zinner. 2003. Implementing resident work hour limitations: Lessons from the New York State experience. *Annals of Surgery* 237(4):449-455.

Woodrow, S. I., J. Park, B. J. Murray, C. Wang, M. Bernstein, R. K. Reznick, and S. J. Hamstra. 2008. Differences in the perceived impact of sleep deprivation among surgical and non-surgical residents. *Medical Education* 42(5):459-467.

Zonia, S. C., R. J. LaBaere II, M. Stommel, and D. D. Tomaszewski. 2005. Resident attitudes regarding the impact of the 80-duty-hours work standards. *Journal of the American Osteopathic Association* 105(7):307-313.

Zuckerman, J. D., E. N. Kubiak, I. Immerman, and P. DiCesare. 2005. The early effects of code 405 work rules on attitudes of orthopaedic residents and attending surgeons. *Journal of Bone and Joint Surgery—Series A* 87(4):903-908.

6

Contributors to Error in the Training Environment

Residents can make errors, but the proportion of errors they make relative to those of other healthcare workers is unknown. Inexperience, fatigue, inadequate supervision, workload intensity, and other work system factors (poor handover practices, inadequate medication labeling) contribute to errors by residents as they may for all health care workers. Data are insufficient to determine the relative contribution of each of these factors. Because residents are in supervised training programs and work within teams, many mistakes can be intercepted before they can harm patients.

Uncertainty surrounds the impact of the 2003 reduction of resident duty hours on patient safety (adverse patient outcomes) and whether further adjustments to duty hours might diminish unsafe conditions (e.g., sleep deprivation) and reduce errors. The few national studies that have attempted to capture the impact of duty hour reform show no evidence of harm as measured by mortality rates. A well-designed randomized trial in two intensive care units of a single institution found a reduction in rates of serious medical error committed by first-year residents when their extended duty periods (up to 30 hours) were reduced to 16 hours, total weekly work hours were also reduced, and they obtained more sleep. The study found no statistically significant difference in unit-wide preventable adverse events or patient mortality between the reduced duty hour and standard hours. Nor was it able to isolate the effect of the shorter shift from reduced total workweek hours, increased sleep, having an additional intern, or increased handovers. A larger-scale, multicenter trial with sufficient statistical power would be necessary to confirm the positive findings in other settings and for residents in other training years.

This chapter examines what is known about the relationship between resident duty hours and patient safety. By definition the performance of trainees is imperfect as they learn, and they, just as other healthcare professionals, will make errors. The response of the system to those errors and its actions to prevent future errors determine the safety of patients. First, this chapter discusses what is known about the overall frequency of medical errors in hospitals by all staff and the resulting patient harm. Then it examines what evidence is available on the relative contribution of residents to the overall patient safety burden in teaching hospitals, and examines whether the degree to which resident fatigue contributes to the occurrence of error can be ascertained. The chapter continues with a discussion of the results of two natural experiments (the 1989 New York State and the 2003 Accreditation Council for Graduate Medical Education [ACGME] national duty hour reforms). Then a detailed review follows of the effects of an interventional study in which both total duty hours and the 30-hour duty period were further constrained from the limits allowable under the 2003 ACGME duty hour rules. Finally, literature on how other factors contribute to hospital errors, including the influence of poorly designed work systems on individual performance is considered.

The discussion that follows presents research that helps answer five broad questions:

1. Do residents make errors that contribute to patient harm?
2. Is resident fatigue from long duty hours among the most significant risks to patient safety?
3. Did the 2003 reduction in resident duty hours affect patient safety?
4. Would further reductions in resident duty hours improve patient safety?
5. What factors in the resident work and learning environment contribute to error?

The committee's answers to these questions will be drawn together in this chapter in a final section of conclusions. The next chapter (Chapter 7) looks to the human performance and sleep literature on how adults perform under scheduling practices that contribute to sleep deprivation, and contains the committee's recommendations on adjustments to duty hours.

MEASURING HOSPITAL-BASED ERROR RATES AND RESIDENT INVOLVEMENT

This Institute of Medicine (IOM) study grew out of questions about how significant a part residents play within the universe of hospital errors that affect inpatients and to what degree the long duty hours and associated fatigue contribute to making errors (Dingell et al., 2007). The purpose here is to determine what is known scientifically about resident-associated errors and the degree to which fatigue and sleep deprivation of residents affect patient safety. Lessons learned from resident errors may reveal approaches for improving overall patient safety. Evidence on the subject is limited to a few studies.

Measuring Patient Safety

Before beginning, it is important to understand basic terms and approaches used in discussing and measuring patient safety.

Defining Medical Errors

A spectrum of medical errors may occur during the treatment and care of hospital patients. If it is a very serious error, death, injury, or other preventable harm (e.g., delays in treatment, extended days in hospital, complications) could result if an error is not intercepted and corrected. Other errors may have no or very little impact on a patient's condition or may be intercepted before they reach the patient and cause harm. The 2000 IOM report *To Err Is Human: Building a Safer Health System* presents an extensive analysis of safety and errors, based in large part on the research of James Reason and Charles Perrow. The framework, terms, and definitions used here are from that report (see Box 6-1).

Measuring Medical Errors

The measurement of patient safety is neither easy nor cost-free, and the ideal method for system-level surveillance has not been established. There are several types of measures commonly found in the literature that are used to assess patient safety (freedom from accidental injury). These include measuring the following:

- The occurrence of errors,
- The occurrence of adverse events (AEs) and preventable adverse events (PAEs), and
- Patient outcomes such as injury or death or length of stay in the hospital.

Errors with the potential to harm patients tend to be classified in studies according to their seriousness and category (e.g., medication, diagnostic, procedural, or other errors). Different approaches to collecting data both

BOX 6-1
Taxonomy of Errors

Error: "... failure of a planned action to be completed as intended (i.e., error of execution) or the use of a wrong plan to achieve an aim (i.e., error of planning)" (p. 28). An error of execution could be an error of omission of an essential step, a critical piece of data, etc.; could be caused by a poorly designed system requiring staff to "work around" the design fault or miscommunications; an error of planning could result from a misdiagnosis or lack of knowledge about the patient's medical problem. Some errors are caught and corrected before they harm the patient.

Harm or adverse event: An unintended physical injury resulting from or contributed to by medical care rather than the underlying condition of the patient, that requires additional monitoring, treatment, or hospitalization or results in death. Not all adverse events are caused by errors.

Preventable adverse event (PAE): "An adverse event attributable to error ..." (p. 28).

Sentinel event: An unexpected occurrence (which may or may not result from an error) in a hospital patient's case, including actual or risk of death or serious physical or psychological injury (Joint Commission, 2007).

Negligent adverse event: A subset of preventable adverse events that satisfy a legal standard of negligence (i.e., the care provided did not meet the standard of care reasonably expected of an average physician qualified to care for the patient) (p. 28).

Safety: "... freedom from accidental injury" (p. 58).

SOURCE: IOM, 2000.

for internal hospital quality improvement efforts and for research purposes capture different pieces of data but not a whole picture of patient safety or the universe of error. Data sources include (1) voluntary reporting by patients and families; (2) mandatory or voluntary but facilitated reporting systems for healthcare workers; (3) direct, prospective observation of work being done in the hospital; (4) retrospective review of medical records using formal criteria or a "trigger tool" approach (i.e., clues in data that help predict adverse events) (Classen et al., 2008; Griffin and Classen, 2008); (5) use of administrative data on average length of stay, complication rates, readmission rates, and mortality; and (6) hybrid approaches that combine two or more of these methods.

No one method of data collection is ideal. The method used to identify medical errors and assess the preventability of a patient's death in the studies that produced the early IOM estimates used trained physicians conducting a structured implicit review of medical records. This method has been shown to have a low interrater reliability and other limitations (Hayward and Hofer, 2001), although other studies have found similar rates of preventable deaths. In recognition of this fact, institutions and researchers are increasingly employing a combination of different methods for collecting data on errors and analyzing them (Bates et al., 1995; Rothschild et al., 2005). In fact, one study that observed staff in a medical care unit and a coronary intensive care unit (ICU) reported that 62 percent of identified incidents were found through direct observation, 49 percent through chart review, 15 percent through solicited staff reporting, 7 percent through pharmacy reports including adverse drug event monitoring, and 4 percent through formal incident reporting (Rothschild et al., 2005). Only 23 percent of these events were identified by more than one approach.

The common feature of these methods is the reliance on frontline provider knowledge and description of the patient's treatment and condition to inform voluntary or mandatory reporting systems, or to record direct or indirect observations of care (e.g., medical records, non-participant observers). The reproducibility and precision of measurements of AEs and PAEs are limited (Classen et al., 2008; Hayward and Hofer, 2001). In particular, the determination of preventability is subjective and can change based on the state of medical knowledge available at the time of assessment.

Error-Reporting Systems

While national data on errors and PAEs are nearly nonexistent, more information exists at the hospital level since most now have voluntary error-reporting systems. The Joint Commission requires hospitals seeking accreditation to implement a voluntary reporting system for sentinel events, to conduct a root-cause analysis of reported events, and to prepare a corrective action plan to avoid similar incidents in the future (Joint Commission, 2007). These error-reporting systems can provide useful data, but they do not define the universe of errors, only those events recognized as problematic and reported by an observer or participant. Underreporting appears to be a common problem; such systems may detect fewer than 10 percent of adverse events (Classen et al., 2008; Rothschild et al., 2005), but the data provided nonetheless can have important uses to the reporting facility when they are embedded in a vigorous error elimination program. Such voluntary systems focus on the circumstances surrounding the adverse event and the systems involved, rather than identifying the individuals involved. Hence, even well-supported reporting systems do not typically note whether a resident was involved with the patient's care. Also, because of the complexity associated with some adverse events, it may be difficult to attribute the event to a specific individual or even to know exactly when it was committed.

Compliance with voluntary reporting systems by physicians and other clinicians depends in part on the importance given to safety issues by the organization's leadership, whether such data (when gathered) are actually used in respected improvement efforts, and importantly whether workers feel safe to discuss errors without fear of punishment, retribution, or other negative consequences (Garbutt et al., 2008; Kaldjian et al., 2008). These issues are discussed in Chapter 8. Although voluntary reporting systems cannot be used to define the frequency of harmful and other medical errors, they can be an important source of information to hospital leaders for identifying vulnerabilities in their systems that should be considered for corrective action. Along with risk management reports, patient complaints, error reports, quality assurance audits, and quality improvement reports, such systems can indicate areas for more detailed retrospective review, which can identify many more adverse events (Griffin and Classen, 2008). Error-reporting systems can provide data to assist in priority setting for quality improvement projects. The committee believes strongly that they can also be of educational value to doctors in training and should become an integral part of residency programs, as discussed in Chapter 8.

Determining the Universe of Errors and PAEs with Limited Data

As background for the committee's study of the impact of residents' duty hours on patient safety, it would be useful to follow a chain of inquiry and quantify, in order, the universe of medical errors, medical errors made in hospitals, medical errors made by residents, and medical errors made by residents in which fatigue is a contributing factor. The *universe* of medical errors affecting patient safety would encompass PAEs as defined earlier, including both fatal preventable errors and the larger number of nonfatal preventable errors. The data to determine the universe of errors and the subelements in the above-mentioned hierarchy are not available to present a full picture. This lack inhibits the ability of the medical community to track and guide progress on patient safety. It has constrained the ability of the committee to answer fully some of the important questions put forth by the sponsors of this inquiry. Nonetheless, this section of the chapter gathers available data to paint a partial picture of the relationship between residents, errors in hospitals, and patient safety.

Estimates of PAEs

U.S. short stay, non-federal hospitals treated and discharged 35 million inpatients in 2006 (DeFrances et al., 2008) and can produce miraculous cures, but an estimated 44,000-98,000 patients die from preventable errors (IOM, 2000). The broad range of that estimate reflects, in part, the methodological challenges mentioned above. The estimate of deaths was based on studies in which researchers examined hospital medical records from large samples of admissions in New York, Colorado, and Utah to determine whether the patients had experienced AEs as a consequence of medical errors (Brennan et al., 1991; Leape et al., 1991). A later study determined that 2.9 percent of admissions in Utah and Colorado and 3.7 percent of admissions in New York State experienced an AE; that 53 percent of Utah and Colorado events and 58 percent of the events in New York were attributable to errors and therefore were PAEs (Thomas et al., 1999). Another study by Thomas and colleagues determined that the AE rates in Utah and Colorado varied by teaching status: 4.0 percent in major teaching hospitals, 3.9 percent in minor teaching hospitals, and 2.5 percent in non-teaching and private hospitals. The study did not focus on case mix differences among individual hospitals or categories of hospitals. The researchers did not present sufficient data to explain the variation based on their available data (Thomas et al., 2000a). The estimated number of deaths resulting from PAEs was extrapolated from 1992 data by applying the death rates due to errors in the three states noted to the total of national hospital admissions in 1997. The committee uses the Thomas study (1999) as the basis for cost estimates of PAEs discussed in Chapter 9.

Experts believe that the rate of preventable deaths has not improved substantially since the report *To Err Is Human* brought these issues to the public's attention in 2000 (Leape and Berwick, 2005). A significant and unsatisfactory level of errors is also indicated by several smaller studies of medical errors in a single hospital or hospital service since that time (AHRQ, 2002; Forster et al., 2003; Hayward and Hofer, 2001; IOM, 2006; Leape and Berwick, 2005; Rothschild et al., 2005). No recent estimate of the universe of errors nationwide exists, and because studies use different definitions of errors and PAEs and a variety of inconsistent methodologies for identifying PAEs and calculating error rates, their results cannot be aggregated.

Assessing Patient Safety and Quality

In the absence of a national error-reporting system, several commercial organizations as well as the Centers for Medicare and Medicaid Services (CMS), the Agency for Healthcare Research and Quality (AHRQ), and the

Commonwealth Fund have developed alternative methods for assessing quality and safety using existing data sources. CMS posts provider-level quality measures, including indicators for hospitals, nursing homes, home health providers, and dialysis facilities to help consumers make more informed choices (HHS, 2008). AHRQ created national estimates of hospital quality from existing data sources for its annual National Healthcare Quality Report, which includes some indicators of safety, but not errors. For example, a composite indicator of selected generally avoidable postoperative complications shows that such adverse events occurred in 6.55 percent of cases in 2005, and that nearly one-quarter of surgical patients did not receive appropriately timed antibiotics (AHRQ, 2007). The improvements in quality according to a variety of ambulatory and hospital indicators used in AHRQ's National Healthcare Quality Reports amounted to only 1.5 percent per year between 2000 and 2005 (Brady et al., 2008). The Commonwealth Fund uses a safety indicator for U.S. hospitals—a construction of unexpected mortality, calculated by Jarman—that it tracks over time (Commonwealth Fund, 2008). The U.S. rate shows an improvement of 19 percent in the 2004-2006 period compared to 2000-2002. Nonetheless, both of these quality reports indicate the persistence of significant hospital mortality and injury related to conditions that generally should be avoidable or should be caught and treated before the patient dies, indicating the continuing need for improvement in patient care.

Errors and PAEs Involving Residents

The above "classic" studies involving statewide hospital AEs do not report errors or PAEs that were related specifically to residents' care although there would appear to be higher AE rates in teaching hospitals based on these data alone (Brennan et al., 1991; Leape et al., 1991; Thomas et al., 1999, 2000b). A more recent set of papers by Rothschild, Landrigan, Lockley, and colleagues examined resident error through a randomized trial in two critical care units at a single institution (Landrigan et al., 2004; Lockley et al., 2004; Rothschild et al., 2005). This section discusses the studies with a focus on the baseline incidence of errors while a later section of this chapter examines the effect of a scheduling intervention on error and PAE rates. Malpractice negligence claims provide another source of data (Gandhi et al., 2006; Regenbogen et al., 2007; Singh et al., 2007).

Incidence of Error and PAEs in ICUs

Rothschild (2005) and colleagues conducted a prospective observational study of two critical care units at a major urban teaching hospital. This study focused on errors made by all caregivers when first-year residents were following a traditional duty hour schedule. The authors found that 20.2 percent of patients suffered at least one AE and 45 percent of those AEs were found to be preventable (Rothschild et al., 2005). The authors note that their definition of an AE is more inclusive than the earlier study by Brennan et al. (1991) cited above and that the ICU setting of their trial would be expected to have higher medical error rates than other areas (Beckmann et al., 2003). The unit-wide error rates per 1,000 patient-days were 80.5 for all AEs, 36.2 for PAEs, and 149.7 for serious errors. Serious errors did not always result in harm to patients "either because the patient had sufficient reserve to buffer an error (nonintercepted serious error) or because the error was caught before reaching the patient or before harm developed" (Rothschild et al., 2005, p. 1697). The Rothschild data along with the national reports from AHRQ, CMS, and the Commonwealth Fund support the committee's conclusion that 8 years after publication of the IOM report *To Err Is Human* (2000), patient safety remains a serious issue in the United States (AHRQ, 2007; Commonwealth Fund, 2008; HHS, 2008).

The complementary article by Landrigan et al. (2004) reporting on data collected in the same setting but for a slightly shorter period describes differences in error rates unit-wide and for first-year residents. It found the rates per 1,000 patient-days involving all staff unit-wide were 38.6 for PAEs and 193.2 for serious errors. Incidents involving first-year residents working a schedule with overnight call every third night appear to make up a substantial portion of the reported errors, including 20.9 per 1,000 patient-days for PAEs and 136.0 per 1,000 patient-days for serious errors (Landrigan et al., 2004). Rothschild notes that compared to the unit-wide data, the "data on interns were somewhat more comprehensive because of the presence of the observers" who kept the interns under direct continuous observation, but that the unit-wide results were within the range identified by other studies (Rothschild et al., 2005, p. 1695). Thus, the error rates for other workers may have been underestimated relative to the error rates of first-year residents.

Errors and PAEs in Malpractice Claims

Another study that identified errors associated specifically with doctors in training (both residents and fellows) is based on 1,452 closed malpractice claims from five liability insurers in different parts of the country (Singh et al., 2007). Malpractice claims represent only a small proportion of errors and AEs—the more serious AEs for which negligence is assessed. It is unclear in what other ways these data might differ from the universe of PAEs. Singh identified 889 cases that reviewers determined to have included both an error and an adverse outcome; 240 (27 percent) involved trainees.

Residents were involved with 87 percent of the 240 cases involving trainees, and fellows were involved with 13 percent of those cases. Multiple trainees could have been involved in a single case, with interns involved in 13 percent of the 240 cases. The study's physician reviewers considered these doctors in training to have had at least a moderately important contributory role in those cases with a PAE.

A study of 307 diagnosis-related ambulatory care malpractice claims closed between 1984 and 2004 found that 181 such claims involved diagnostic errors that led to adverse outcomes (Gandhi et al., 2006). Of the 181 cases, trainees (intern, resident, or fellow) were identified as involved in 20 percent of them by trained reviewers. The study also identified several causes of breakdowns in the diagnostic process and concluded that multiple factors were involved. Researchers in a different study examined surgical malpractice claims, selecting a random sample of 444 cases for closer study. Among the 52 percent (n = 133) that included technical errors, the researchers determined that 9 percent involved poorly supervised residents (Regenbogen et al., 2007).

Conclusion About Whether Residents Make Errors

These studies provide enough evidence to answer the question: *Do residents make errors that contribute to patient harm?* Common sense and these studies lead to the conclusion that the answer is, Yes, they do. Additional information from resident surveys confirms this as well (Jagsi et al., 2005, 2008; Wu et al., 2003). Without more quantitative data, it is impossible to determine what proportion of all errors or what proportion of PAEs involve residents. Consequently, the magnitude of the impact of residents on patient safety is unknown.

FATIGUE AS A CONTRIBUTOR TO ERROR

A principal aim of this study is to determine the degree to which resident fatigue from long duty hours poses a significant risk to patient safety and whether there are interventions that might reduce that risk. As Howard and colleagues have observed, "continuous operational demands [of providing access to health care in hospitals 24 hours a day] present unique physiologic challenges to the humans who are called on to provide safe operations within these systems" (Howard et al., 2002b, p. 1281). While long work hours and fatigue appear to play a role, other systemic factors also contribute. Resident reports give some insight into how great a factor they believe fatigue to be. In a survey of two large teaching institutions just before the required 2003 ACGME duty hour limits were in force, medical and surgical specialty and subspecialty residents were asked what the contributing factors were for mistakes related to AEs. They reported that long work hours were a contributing factor in 19 percent of the mistakes observed, but they also noted that lack of supervision (20 percent), faulty handovers (15 percent), large patient caseloads (12 percent), and cross-covering too many patients (5 percent) were important factors (Jagsi et al., 2005). Working more than 80 hours in the past week was a significant predictor of caring for a patient with an AE in the last week (odds ratio 1.8) (Jagsi et al., 2005). Chapter 7 details the evidence base that establishes the link between fatiguing aspects of resident work-rest schedules and what is known about how fatigue affects human performance and the propensity for error.

Assessing Incidence of AEs Involving Fatigue

This section examines data from the U.S. Department of Veterans Affairs (VA) and from malpractice claims to evaluate the contribution of fatigue as a factor. The VA offers residency training through approximately 8,800 residency positions in its facilities (9 percent of U.S. total), and because residents from other facilities rotate through the VA, this training reaches about one-third of residents in training in any single year (Chang, 2007). The VA has a heavy emphasis on patient safety and has trained its staff in the value of reporting both AEs and close calls. The system has accumulated more than 10,000 root-cause analyses (RCAs) of individual serious incidents or groups of events since its inception in 1999. The analyses tend to look beyond the individuals involved with an AE to the underlying systemic causes. The database is not designed to identify the specific involvement of residents. It does, however, include fatigue as a "cause" choice on its structured data collection tool. Fewer than 4.5 percent of the VA RCA reports included fatigue as an associated factor and 0.7 percent included a more extensive discussion of fatigue-related causation. A review of a random sample of 4,742 reports drawn from approximately 180,000 reports from the same time period concerning less serious safety incidents showed that 1.0 to 3.3 percent included fatigue-associated causes.[1,2] It is unknown what percent of those cases associated with fatigue included fatigued residents because the VA does not routinely track residency status of the involved parties.

Fatigue related to medical errors is recorded in some cases in the Singh study of malpractice claims discussed above: 5 percent (n = 12) of the trainee

[1]Personal communication, J. P. Bagian, Director, VA National Center for Patient Safety, Department of Veterans Affairs, February 11, 2008.

[2]Personal communication, J. P. Bagian, Director, VA National Center for Patient Safety, Department of Veterans Affairs, February 14, 2008.

cases (less than 2 percent of PAE claims studied, $n = 889$) and 1 percent ($n = 6$) of the nontrainee PAE cases included fatigue as a factor. Since the fatigue of the provider is not routinely noted in medical records and legal case notes, it is not possible to know how frequently it was a factor but not noted as such; thus, the actual percentage of negligence cases in which the trainee was affected by fatigue is unknown. It is also possible that fatigue would be noted more frequently if healthcare workers were more aware of the role of fatigue and how to assess its role in creating unsafe conditions, and if workers were informed about the importance of adequate sleep. Such courses have been developed (e.g., those by NTSB [2008]) in response to fatigue-related incidents in other industries (Rosekind et al., 1994).

Better Conditions for Patient Safety Through Reducing Fatigue

One survey-based study and one prospective observational study of ICUs suggest that shorter work hours may lead to less fatigue and, as a result, to better patient safety (Jagsi et al., 2008; Landrigan et al., 2004). The survey incorporated questions on the relationship of duty hours and fatigue to the quality of care delivered, patient safety, and AEs. Responses of residents in 76 specialty and subspecialty programs at two institutions were obtained before and after the 2003 ACGME duty hour reforms. Residents in programs that reduced their workweek by at least 5 hours were found to be less likely to violate the 80-hour limit than prior to 2003 (16.6 percent vs. 44.0 percent) and less likely to have worked more than 30 continuous hours in the past week (11.4 percent vs. 40.8 percent). Days of significant fatigue in the past 4 weeks remained but were less (6.5 vs. 8.7). Fewer residents reported that "fatigue frequently or always affected the quality of care they provided" (14.6 percent vs. 9.2 percent) and that "fatigue frequently or always impacted the safety of patients that they cared for" (7.0 percent vs. 2.9 percent) and these differences are significant when compared to programs that did not reduce work hours (Jagsi et al., 2008, p. 496).

The ICU environment examined in the Rothschild study was the subject of a change in the resident duty schedule that resulted in fewer work hours per week (a mean 19.5 hours less per week) and a shorter consecutive duty period (no shifts over 16 hours). This allowed more hours of sleep (mean 5.8 hours per week) and presumably more rested interns (Landrigan et al., 2004; Lockley et al., 2004). Interns working on the intervention work schedule made 36 percent fewer serious medical errors ($p = .001$), but the difference in rates of PAEs was not statistically significant in comparing the two groups (Landrigan et al., 2004). Interns on the intervention schedule also had fewer attentional failures as measured by slow rolling eye movements (Lockley et al., 2004). Rothschild et al. (2005) found that 53 percent of the performance errors were slips (unintended acts) or lapses (omitted acts) rather than knowledge-based or rule-based errors (e.g., not following a protocol). Because sleep and fatigue were not the only factors that changed during the intervention, it is not possible to attribute all of the reduction in errors to reduced fatigue, but this study provides a substantial contrast to the VA and Singh data, which had found relatively low rates of fatigue noted in relation to PAEs.

Conclusion About Whether Fatigue Is a Significant Factor in Error

Clearly fatigue is a factor in some of the errors by medical workers in general and by residents in particular given their work-rest schedules. It is unresolved exactly what percentage of all errors that fatigue-based errors compose, with these sources reviewed suggesting a wide range from 5 to 36 percent. The potential impact of fatigue in the ICU studies (Landrigan et al., 2004; Lockley et al., 2004; Rothschild et al., 2005) was substantial, whereas the VA and the malpractice studies noted relatively little mention of fatigue as a factor in error reports. Overall, the committee concludes that the existing data are insufficient to determine if the current duty hours of residents and the fatigue resulting from them are the most significant causal factors for errors committed by residents or if resident errors occur more frequently than errors committed by other health workers.

Assessing Fatigue and Performance After Extended Duty Periods

A number of studies have noted poorer performance by residents post-call, but others find no difference. Friedman and colleagues' classic study showed that interns made almost twice as many errors when reading electrocardiograms after being up for 24 hours than when they had a night of sleep (Friedman et al., 1971). Additional studies also point out increased technical errors in simulated laparoscopic surgical skills after being up all night (Eastridge et al., 2003; Grantcharov et al., 2001), decrease in cognitive skills (Jacques et al., 1990; Robbins and Gottlieb, 1990) as well as in memory attention and coordination in surgical residents post-extended duty period (Kahol et al., 2008), and reduced psychological well-being and problems with alertness and coordination after an extended shift (Leonard et al., 1998). When Jacques et al. (1990) examined the effects of sleep loss on resident performance on the American Board of Family Practice in-training examinations, they found that the difference in test scores after a night without sleep was equivalent to the difference between third-year and first-year residents' performance.

Further, researchers have found that even with a call frequency no more often than every fourth night, which is typical of call schedules under the

ACGME 80-hour limit, residents do not fully recover between nights of overnight call (Saxena and George, 2005). This may imply that residents are not sufficiently recovering their lost sleep time between extended duty periods (Saxena and George, 2005). Howard et al. (2002a) have confirmed that residents prior to the 2003 duty hour limit were as sleepy before and after extended duty shifts and that their level of sleepiness matched that of persons with clinical sleep disorders. Obtaining sufficient sleep returned the residents to normal sleep levels (i.e., 2 hours more sleep per day over 4 days). Saxena et al. (2005) suggest that it is unlikely in an emergency that resident judgment would be impaired even in a sleep-deprived state, but that more routine tasks (e.g., medication reorders) might be missed, potentially leading to more serious consequences later. Indeed, a number of studies have found that medication errors are among the most common errors that residents make (Jagsi et al., 2005; Landrigan et al., 2004; Rothschild et al., 2005).

Other studies report no deficit in resident performance after being up all night (Ellman et al., 2005; Howard et al., 2002b; Jakubowicz et al., 2005). These results are not consistent with the extensive literature on human performance and acute sleep deprivation presented in Chapter 7. Several review articles note that research examining the effects of fatigue on the performance of healthcare personnel, not just residents, do not always come to the same conclusion, and these articles ascribe this to definitional and methodological differences (Howard et al., 2002a; Veasey et al., 2002; Weinger and Ancoli-Israel, 2002). Differences in reported results may have to do with the way sleep deprivation is defined, the degree of chronic sleep deprivation present both pre- and post-call (i.e., extended duty periods), and the presence of compensating factors that may have helped mitigate the performance of sleep-deprived residents (e.g., presence of rested and experienced team members). For example, in a retrospective review of 10 years of cases, Ellman et al. (2005) concluded that thoracic residents in an acute sleep-deprived state had patient outcomes (morbidity or mortality) and operative efficiency comparable to those who had not been on call the previous evening. This matches their previous findings for attending physicians (Ellman et al., 2004). In these retrospective studies, the definition of acute sleep deprivation was based on whether a resident or attending had started or ended an operation the previous night; there was no determination of the chronic sleep deficit for either population although it was assumed that in a university-based teaching program the attendings would not have a chronic sleep deficit. Just 3 percent of resident cases were performed by residents who met the acute sleep deprivation criteria. The report does not indicate whether the resident was assisted by an attending, which may have buffered the effects of sleep deprivation for the resident or to what degree resident errors were intercepted by an attending (Ellman et al., 2005). One could assume that a rested attending surgeon supervising these procedures would compensate for suboptimal resident performance and help prevent errors or surgical inefficiency.

Veasey and colleagues (2002) have commented on how residents were chronically sleep deprived in their baseline state prior to the 2003 duty hour reform. Kiernan et al. (2006) have suggested that some problems, such as declines in mood, previously associated with acute sleep deprivation may have been ameliorated by the 2003 duty hour limits. Their rationale is that the limits may have helped to reduce chronic sleep deprivation, giving residents sufficient reserve to better tolerate a night without sleep. Yet others, such as Saxena and colleagues, suggest that even on an every fourth night overnight call schedule, residents do not sufficiently recover from their sleep loss (Kiernan et al., 2006; Saxena and George, 2005). As Chapter 7 discusses, the buildup of sleep loss contributes more heavily to impaired states of alertness, cognition, and performance.

IMPACT OF REDUCED DUTY HOURS ON ERROR RATES AND PATIENT SAFETY

This section addresses two central questions: (1) Did the 2003 reduction in resident duty hours improve patient safety? (2) Would further reductions in resident duty hours improve patient safety? As noted above, attempting to isolate the effect of reducing resident duty hours on patient outcomes is difficult. For studies on a national level, drawing a direct link between hours worked by residents and patient outcomes is problematic, given available data, and for studies at individual institutions, it is difficult to obtain a sample size sufficiently powered to find statistically significant differences in mortality or other patient safety measures. The relationship between duty hours and patient safety has been a central element in the debate over reduced hours. Expectations were raised in 2003 that reducing duty hours would improve patient outcomes and safety, but others predicted that reducing hours could negatively affect patients in the short and long term because it required more frequent handover of patients from one resident to another. Such handovers, or "handoffs," are considered potentially risky for loss of information and continuity of care, and could also lessen the overall learning experience of residents (Fischer, 2004; Petersen et al., 1994, 1998).

Natural Experiments

A number of studies look at the effect of implementation of duty hour reforms throughout a single state or nationwide without controlling for the specifics of a scheduling intervention. As shown in Chapter 3, there are many different approaches to schedule resident hours, and there can be variations even within a single resident team that suit the specialty, patient characteristics, size of the resident team, system supports, and other factors. Thus, these studies examine outcomes of practices that have naturally evolved within and across teaching institutions. On the positive side, the datasets in these studies are sufficiently large to detect changes in mortality. Patient mortality became the center of attention in *To Err Is Human* and subsequent patient safety campaigns (e.g., the Institute for Healthcare Improvement's [IHI's] 100,000 Lives and 5 Million Lives campaigns) with projections of the number of lives potentially saved if there were focused attention on healthcare quality improvement (IHI, 2008; IOM, 2000; McCannon et al., 2007). These studies do not provide information on how many hours per week residents actually work or sleep.

New York State Studies

When New York State instituted an 80-hour workweek and limited extended duty periods to 24 + 3 hours in 1989, it provided a laboratory for implementation and evaluation of such changes and led the way for eventual adoption of an 80-hour week nationwide (Howard et al., 2004; Laine et al., 1993). In their systematic review of the literature available before implementation of the 2003 ACGME rules, Fletcher and colleagues found few duty hour and patient safety studies that adequately addressed their two criteria: (1) examined a system change to address work hours, fatigue, or sleep deprivation and (2) included an outcome directly related to patient safety (e.g., mortality, morbidity, error) (Fletcher et al., 2004). Three well-designed, but not randomized, studies found that duty hour reduction led to (1) more complications and test delays but neither increased nor decreased mortality (Laine et al., 1993), (2) a decrease in mortality in teaching hospitals that was equally apparent in non-teaching hospitals (Howard et al., 2004), and (3) more PAEs attributed specifically to handovers and cross-coverage of unfamiliar patients (Petersen et al., 1994). The first two studies of New York State are discussed below; the Petersen study is described within the context of continuity of care in Chapter 8.

Summaries of the two New York studies looking at patient outcomes before and after the 1989 New York State duty hour limitations follow. Laine et al. (1993) examined patient outcomes in a teaching hospital with residents on a general medical service by comparing all admissions during October of 1988 (pre) and October of 1989 (post) (Laine et al., 1993). They found an increased number of patients with at least one complication (35 percent vs. 22 percent; $p = .002$) and delays in residents ordering diagnostic tests (17 percent vs. 2 percent; $p < .001$) after the reduction in duty hours. However, these decreases in quality-of-care metrics did not result in more serious outcomes for patients. No difference was found for transfers of patients to intensive care, length of stay, or disposition at discharge; the study did not have sufficient power to detect a statistically significant change in mortality. Howard and colleagues (2004) found that mortality declined for congestive heart failure, acute myocardial infarction, and pneumonia after New York State's duty hour change, but this could not be attributed with confidence to duty hour reduction alone because mortality rates declined at both teaching and non-teaching hospitals in the state between 1988 and 1991; moreover, the study assigned teaching status to the hospital as a whole rather than to specific patients cared for by residents (Howard et al., 2004).

Another large-scale study examining surgical mortality in New York also found no change over time (Poulose et al., 2005). Poulose and colleagues examined changes in surgical outcomes based on five patient safety indicators and found no substantive change for New York State teaching hospitals compared to two control groups: non-teaching hospitals in the same state and teaching hospitals in another state that had not yet implemented an 80-hour limit (Poulose et al., 2005). The authors also suggest that this can best be interpreted as examining the system's global response to duty hour limits (e.g., schedule and supervision changes, substitution by others for residents' time) not resident work hours alone. The data examined were from the Healthcare Cost and Utilization Project (HCUP) Nationwide Inpatient Sample (NIS) from 1995 to 2001; the choice of surgical patient safety indicators was deliberate because of a higher level of agreement between administrative data and actual occurrences in surgery than medicine (73-81 percent vs. 32-70 percent).

Chapter 2 has noted the poor state of adherence to duty hour rules in New York State in the early years of implementation (DeBuono and Osten, 1998; Kennedy, 1998). Low levels of adherence suggest that the results of these otherwise well-designed studies comparing pre- and post-duty hours reform should be examined with caution since there may indeed have been little actual change in hours worked by residents. Additionally, New York State hospitals had very long lengths of stay and low managed care penetration, giving other reasons why the findings might not be generalizable outside of New York.

Nationwide Studies

Now this section turns to studies of the impact of the ACGME 2003 duty hour reforms. Three recent studies by Shetty and Bhattacharya (2007) and Volpp et al. (2007a,b) examining national mortality trends showed some improvement in mortality for medical patients but not surgical patients after duty hour regulations (Shetty and Bhattacharya, 2007; Volpp et al., 2007a,b). A fourth study by Prasad (2008) finds no change in ICU mortality after the duty hour reforms. Summaries of these studies follow. None of these studies were able to document actual hours worked by residents in the facilities studied.

Shetty and Bhattacharya (2007) compared changes, if any, in patient outcomes at teaching hospitals that should have been affected by the duty hour rule changes versus non-teaching hospitals. They examined a representative national dataset, the HCUP NIS, which is of sufficient size to have enough statistical power to detect changes in mortality. The researchers used non-teaching hospital services as a control (963,916 non-teaching patients and 548,029 teaching patients). To ensure that the teaching cases were indeed on teaching services, a cross-match was made to the presence of specific residency programs for each type of patient examined (e.g., internal medicine, orthopedics). After the 2003 duty hour limitations, there was a small but statistically significant improvement for medical but not surgical cases on teaching services, specifically a "0.25% decrease in the absolute risk for death ($p = .043$), which corresponded to a 3.75% decrease in relative risk in medical patients per hospitalization" (p. 76). The oldest patients, those more than 80 years old, and those with infectious diseases were most likely to benefit in the period after duty hour changes. Mortality decreased as the number of residents in a facility increased. The authors offer several possible explanations of why there were no observed changes for surgical patients in spite of the fact that the 2003 duty hour reforms reduced surgical resident hours most substantially. Their reasons included a smaller set of surgical cases, which may have limited the statistical power to detect change for this type of patient and the possibility that work conditions at least in the operating room may not have changed. One critique of the study is the nature of the dataset since the HCUP NIS looks at different hospitals each year and the dataset does not allow one to distinguish between single and multiple admissions for the same condition (Volpp et al., 2007b).

Since mortality may occur not long after discharge from a hospital, Volpp and colleagues scrutinized both in-hospital and post-discharge rates. They looked at mortality for Medicare beneficiaries and VA patients in acute care hospitals in the first 2 years after implementation of the 2003 limits (Volpp et al., 2007a,b). The main outcome measure is all-location mortality within 30 days of a first hospital admission for acute myocardial infarction, stroke, gastrointestinal bleeding, congestive heart failure, general surgery, orthopedic surgery, or vascular surgery. These are the AHRQ Quality Indicators that use mortality as an outcome measure. Each hospital is compared with itself over time. For Medicare beneficiaries, no significant change was found in the odds of risk-adjusted mortality in either the first or the second year of duty hour reforms based on data from the Medicare Provider Analysis and Review File (MEDPAR). There was a small increase in the relative mortality for stroke in more intensive teaching hospitals, but it appeared to be part of an ongoing trend that is divergent from non-teaching hospitals and a trend that started before reform.

In more teaching-intensive VA hospitals, there was a significant improvement for AMI on its own and for the other medical conditions combined (with or without AMI) in the second year of duty hour reform but no improvement for surgery. The VA health system is the largest single provider of residency training in the country. The authors suggest that a reason for the difference between the VA and Medicare patient outcomes may be a dose-response effect due to the markedly higher resident-to-bed ratios in VA hospitals than non-VA teaching hospitals, as well as differences in staffing models. They suggest that this may create a different balance between the consequences of resident fatigue and discontinuity in patient care in VA facilities versus others so that the VA experience is not generalizable to other facilities. Also, there may be other nonmeasured factors that contribute to the difference; for example, VA hospitals have electronic medical records, which may help diminish communication problems in transitions of care. Although not stated explicitly, this study like the one by Shetty and Bhattacharya documents the advantage of being a patient in a teaching hospital; risk-adjusted mortality rates were generally lower for hospitals with higher resident-to-bed ratios.

Prasad (2008) examined the pattern of adult in-hospital mortality in ICUs from July 1, 2001, through June 30, 2005, using the APACHE IV database, a voluntary multicenter ICU clinical registry, to determine whether there was an effect of the 2003 duty hour reforms on ICU mortality. They found a significant decline in risk-adjusted mortality in both teaching and non-teaching hospitals from before the implementation of duty hour rules to after. They eliminated patients from the sample whose care straddled the start date of the rules. The difference between the two settings over time was not significant: the adjusted odds ratio for mortality after the regulations was 0.89 (95 percent CI 0.87, 0.92; $p < .001$) overall, 0.88 (95 percent CI 0.85, 0.92; $p < .001$) in teaching hospitals, and 0.91 (95 percent CI 0.87, 0.95; $p < .001$) in non-teaching hospitals. The authors conclude that duty hour reforms did not have a positive or negative effect on major patient outcomes and it is possible that the positive and negative effects of reform may have offset each other, that mortality in the ICU environment may not have been sensitive to resident staffing patterns, or that ICUs may have made other compensating changes to maintain and improve patient outcomes (Prasad, 2008).

Institution-Specific Studies

Few studies to date have an adequate control group to isolate the specific effects of duty hours (e.g., Horwitz et al., 2007; Howard et al.,

2004; Landrigan et al., 2004; Poulose et al., 2005). Studies comparing data both before and after the 2003 ACGME limits without control groups may falsely read improvement in error rates or patient outcomes as being related to duty hour reform when in fact they may have nothing to do with resident work hours but reflect national trends toward improved quality of care (Horwitz et al., 2007). Additionally, smaller institution-specific studies often have insufficient statistical power to detect changes in mortality (e.g., Landrigan et al., 2004).

To control for temporal trends in practice and patient outcomes, Horowitz et al. (2007) carried out a retrospective single-center cohort study comparing outcomes for medical patients on a resident-hospitalist teaching service (n = 708) to a non-teaching service run by hospitalists (n = 2,954) (Horwitz et al., 2007). No adverse consequences for patients occurred under their new scheduling plan in which residents have no overnight call. The teaching service had a significant improvement relative to the hospitalist service from 2002-2003 to 2003-2004 on three measures of mean net adjusted change: ICU utilization decreased by 2 percent, discharge to home or rehabilitation versus elsewhere increased 5 percent, and pharmacist interventions to prevent error decreased by 1.9 interventions per 100 patient-days. Readmission rates, length of stay, and medication interactions were not found to be significantly different. There was also insufficient statistical power to detect changes in mortality. The remaining variables (length of hospital stay, 30-day readmission rate, and drug-drug interactions) were consistent across both services.

Bhavsar and colleagues performed a retrospective analysis of patient outcomes at a single facility for those with acute coronary syndrome, before and after duty hour regulation (Bhavsar et al., 2007). They assert that their program maintained—if not enhanced—the level of care because they had more well-rested residents to handle discharge planning as a result of their scheduling response to duty hour limits (an incremental increase in residents available at discharge through use of day float). No significant difference was detected for in-hospital patient mortality (4.2 percent before vs. 2.8 percent after, p = .23), but 6-month mortality (8.0 percent vs. 3.8 percent, p = .007) and risk-adjusted 6-month mortality (OR 0.53; 95 percent CI 0.28, 0.99, p = .05) improved. At the same time, there was increased adherence to quality prescribing practices for cardiac care at discharge and mean length of stay was reduced. There was no control group in this study and the cardiac care quality improvement program instituted at this facility may be at least part of the reason for the improvements rather than duty hours.

Surgical programs traditionally have had much longer duty hours than medical programs, so the adjustment to 80 hours was expected to be more difficult for these programs. The choice of 80 hours was seen by some stakeholders as arbitrary and not responsive to the special demands of surgery (Fischer, 2004). Despite these concerns, reports from several surgical programs found no change in mortality and other patient indicators or increased errors with the reduced work hours (de Virgilio et al., 2006; Kaafarani et al., 2005; Vaughn et al., 2008). DeVirgilio et al. (2006) examined mortality and morbidity for trauma patients at one institution before (July 1998-June 2003) and after (July 2003-June 2005) duty hour changes; adjusting to the reduced hours required an increase in their resident complement and hiring others to do some of the tasks previously done by residents (de Virgilio et al., 2006). They conducted a pre-post study without a comparison group and found no significant difference in patient mortality during the periods before and after the implementation of duty hour rules, despite a larger volume of patients, a higher injury severity score (7.9 to 9.6, $p < .0001$), and a greater portion of penetrating trauma (14.85 to 17.6 percent, $p < .0001$) among patients. Morbidity and raw mortality data come from their Trauma and Emergency Medicine Information System (TEMIS). They also observed no decline in operative experience for the residents or in their success rate in passing the General Surgery Board Examination. Thus, reassurance is given that there were no overt downturns in patient outcomes in this surgical program despite the reduction to an 80-hour week.

Kaarafani and colleagues (2005) similarly found no worsening in mortality and morbidity in either vascular or general surgery at a single institution based on surgical outcome data from the VA National Surgical Quality Improvement Program. Pre-intervention hours were longer (87-92 hours without a consistent 24 hours off every week) than the post-intervention period (80-87 hours from October 1, 2002, until January 1, 2003, and after that 80-hour weeks until September 30, 2003) (Kaafarani et al., 2005). The number of cases with an attending present increased. In the same way, another redesigned surgery program using a combination of apprenticeship, small-team, and night-float models was able to increase operative volume, improve ABSITE scores for PGY-1s and PGY-2s, and maintain previous patient mortality levels. In addition to remodeling its schedule, the program added 0.2 FTE (full-time equivalent) of physician assistant and nurse positions per resident (Schneider et al., 2007).

Conclusion About Patient Outcomes After Implementation of 80-Hour Duty Week

Smaller institution-specific studies allow easier identification of the actual duty hours worked by residents, how fatigued they may be, and the multiple programmatic changes made that help balance the reduction in resident hours (e.g., hire additional staff, remodel their education program, increase attending presence). These studies illustrate the complexity of teasing out not only the impact of duty hours alone, but also the impact on patient outcomes of different staffing configurations and scheduling practices whether in medical or surgical settings. Few studies to date have any type of concurrent control group (Horwitz et al., 2007; Howard et al., 2004). The national studies of mortality show that there is no evidence of widespread harm occurring after implementation of the limits (i.e., duty hour restrictions did not lead to an increase in mortality rates for the common conditions studied) and there may be modest improvements for medical if not surgical patients.

Interventional Study—Reducing Intern Duty Hours in the ICU Setting

The most rigorous scientific data on the direct impact of duty hours on patient safety available to the committee comes from three publications that describe overlapping aspects of the same prospective 1-year randomized trial in 2002 and 2003 (Landrigan et al., 2004; Lockley et al., 2004; Rothschild et al., 2005).[3] This trial compared "the rates of serious medical errors made by interns while they were working according to a traditional schedule with extended (24 hours or more) work shifts every other shift (an 'every third night' call schedule) and while they were working according to an intervention schedule that eliminated extended work shifts and reduced the number of hours worked per week" (Landrigan et al., 2004, p. 1838). The "intervention" schedule had shifts with a maximum of 16 consecutive hours. The study followed a sample of 20 interns who were randomly assigned to work 3-week rotations on both schedules in two ICUs—essentially a crossover experimental design. In contrast to many of the retrospective studies cited earlier, the authors carefully observed medical errors in real time, monitored and recorded actual hours worked, recorded hours slept, and measured intern fatigue. This was a well-designed and well-executed randomized controlled trial—although the randomization was only partial and the evaluations of medical errors could not be fully blinded. Medical error detection was by multiple means: primarily trained physician observers, with voluntary staff reporting, chart review, and computerized event detection monitors.

In this chapter, the focus is on the error reduction and patient safety aspects of this trial (Landrigan et al., 2004). Chapter 7 contains discussion of the associated sleep and fatigue data (Lockley et al., 2004). The study found that the intervention schedule with its shorter shifts resulted in more intern sleep time, decreased intern fatigue, and significantly fewer serious

[3] In addition to its close reading of the three published reports, the committee benefited from testimony by some of the study's principal investigators and from follow-up written and oral communications with Dr. Landrigan, lead author on the Landrigan et al. (2004) paper.

errors by interns. While there was a reduction in intern-related PAEs—a measure of harm that reached the patient—this outcome was not statistically significant. Among the interns participating, serious medical errors decreased by 36 percent (136 vs. 100 per 1,000 patient-days, $p < .001$), while intern PAEs declined by 27 percent (20.9 to 16.5 per 1,000 patient-days, $p = .21$). The authors did not report the proportion of patients in the two arms of the trial who suffered PAEs. Improvements occurred across the various categories of medical errors observed for interns working on the intervention schedule of 16-hour shifts; they made statistically significantly fewer serious diagnostic and medication errors but not fewer serious procedural errors. The committee also noted that intern-related diagnostic errors showed the greatest reduction, from 18.6 to 3.3 per 1,000 patient-days (Landrigan et al., 2004).

In addition to collecting data on errors made by interns, the study reported overall unit-wide error rates (serious errors 193.2 per 1,000 patient-days on the traditional schedule versus 158.4 on the intervention schedule, $p < .001$), but the intense real-time error monitoring processes were applied only to interns so the overall error data may be less complete. Patient populations were similar in volume, severity, and complexity across the two schedules, but the study did not detect effects on patient mortality and unit-wide PAEs (not just those by interns), which remained the same (38.6 vs. 38.5 per 1,000 patient-days). The authors suggest that larger-scale, multicenter trials would be needed to gain sufficient power to confirm their findings.

The committee concludes that, because of careful experimental design, the reported 36 percent reduction in the intern's rate of serious medical errors and other performance improvements appear to be largely a result of the intervention, rather than a result of confounding influences. The committee also notes that the schedule intervention actually incorporated five changes, each of which may have contributed to error reduction:

1. Total duty hours per week were reduced from about 80 to about 60 hours.[4]
2. The duration of long duty periods was reduced from about 30 to about 16 hours.
3. Sleep was significantly increased by an average of 5.8 hours per week.
4. Workload per intern was reduced under the intervention because the bed census, severity and complexity of patients, and number of ad-

[4]There are slight differences in the hours of work reported in the Landrigan and Lockley papers, from 79 to 63 hours or 84.9 to 65.4 hours, respectively.

missions were similar across the two schedules, but the intervention schedule used four rather than three interns to handle the work.

5. The number of handovers during the intervention increased, but during the intervention schedule there also was a designed increase in the overlap time between tours in order to perform handovers.

In the complementary paper by Rothschild et al. (2005), about 53 percent of errors were judged to be slips (unintended acts) and lapses (omitted acts) rather than rule-based errors (e.g., not following a protocol). Since the frequency of such errors tends to be increased by sleep loss, sleep deprivation may be a more important factor than hours worked. As might be expected, the duty hour reduction in this study did not provide an equivalent increase in sleep time: 19 additional minutes of sleep occurred per hour of duty hour reduction for a total increase of 5.8 hours per week, while the mean decrease in work was 19.5 hours (Lockley et al., 2004). Further, Lockley et al. (2004) state that the 16-hour shift schedule was still "long enough to indeed produce significant decrements in neurobehavioral performance owing to sleep deprivation" and required interns to "rise between 4 a.m. and 6 a.m., the time of maximal sleep propensity and efficiency in this age group, to review their patients' progress before morning rounds" (p. 1836). Still the shorter intern schedule was associated with less fatigue, more sleep overall, increased numbers of shifts where the intern had more sleep in the preceding 24 hours, and fewer electro-oculography (EOG)-defined attentional failures.

During the intervention, the interns may have committed fewer errors for several reasons. In addition to providing more sleep overall (5.8 hours per week), the intervention had an increased number of shifts in which the intern had had more than 4 hours of sleep in the previous 24 hours and had less fatigue as measured by fewer EOG-defined attentional failures ("defined as intrusion of slow-rolling eye movements into polysomnographically confirmed episodes of wakefulness during work hours").

The scientific rigor of the study results and the significance of its findings do not imply that simply changing residents' work schedules along the lines of the authors' intervention schedule would guarantee a similar 36 percent reduction in resident-caused serious medical errors across the spectrum of U.S. medical residency programs. It is not known to what extent the results of Landrigan and colleagues (2004) can be used to represent other medical or surgical subspecialties, and how well the results from this single center represent effects in other teaching hospitals. The committee has a number of concerns about how replicable and generalizable the results of this important and seminal study were:

- *Would the 36 percent reduction in serious medical errors hold up under an actual long-term implementation in this same setting?* It is often the case that a long-term implementation of a managerial or technical intervention, such as described in the study, loses its efficacy over time. The long-term effectiveness would likely depend on the dedication and intensiveness of the ongoing supervision and management of the intervention. The committee notes that notwithstanding its beneficial impact, the 16-hour intervention schedule was not continued at original study sites when the experiment ended. Thus, its long-term efficacy cannot be determined. This cautionary observation is not uniquely applicable to this intervention. The effect of scaling up from a laboratory trial to a full implementation is almost always fraught with difficulties, and it is not uncommon that upon full implementation the results are less dramatic than estimated from the initial trial.
- *Would or could this model be replicated with similar results in other clinics of the same type in other hospitals?* The committee notes that this study focused on interns, not residents in general, and was conducted in intensive care environments—where many hospitals do not assign interns to work at all. Moreover, cultures and systems differ from hospital to hospital, indeed from service to service—and culture and systems can be either a major enabler or a major barrier to the effectiveness of an intervention. The reported experiences of Dr. Peter Pronovost in exporting his anti-line infection checklist methodology from the original ICU setting at Johns Hopkins to other hospitals and services gives some basis for both optimism and caution about the exportation of safety-oriented managerial innovations in medicine. Dr. Pronovost's work has demonstrated that while dissemination is possible, it has been painfully slow and difficult (Gawande, 2007; Pronovost et al., 2006).
- *Does the 36 percent reduction in serious medical errors apply to other residency services and to residents in other years of training?* The studies focus on the residents in the first year of graduate medical training (interns), those with the least experience and the greatest propensity for error, and therefore, the results are unlikely to be indicative of the error rates of more senior residents. ICUs were an appropriate site for this research precisely because ICUs may be more vulnerable to fatigue-driven errors and are environs in which patients experience more frequent AEs. On the other hand, the typical ICU has a redundancy that could facilitate the interception of errors before they affect the patient—as seen in this study. Residents who are less supervised on the service floors of hospitals

may make the same or fewer errors, but there may be fewer protections to keep errors from reaching the patient.

- *Do reduced serious errors translate into improved patient safety?* Making a medical error is the first step in a chain of events that can lead to harm to a patient. Errors are precursors of AEs, so reductions in errors would appear to hold the promise of improvements in patient safety. The reduction in serious medical errors found in this study is scientifically valid and is consistent with evidence from Chapter 7. However, no differences were found in PAEs and ICU mortality. For this reason, and for the reasons offered above, it is problematic to project the benefit to patients from this intervention. Larger-scale trials will be needed to evaluate these outcomes.

The detail of the error data reported in the Landrigan paper did not enable the committee to judge the causal, but potentially off-setting, roles of the two key factors of fatigue and handovers in the generation of errors. A motivation for this study was clearly the hypothesis that fatigued interns will make more errors. While the Lockley paper isolated sleep data on an intern-by-intern basis, the authors of the Landrigan paper did not report error data by intern, and unfortunately, there is no analysis of the timing of the serious medical errors that occurred. Moreover, the committee was interested in the question of whether certain individuals in this population of interns made the preponderance of errors either due to lack of knowledge and supervision or whether error incidence was related to their sleep patterns. The committee advocates a systems approach to error reduction, so the intent and spirit of this query is not on seeking out and blaming interns as individuals if the circumstances are beyond their individual control, but on the potential to detect whether error frequency was related to sleep patterns and how to address that through scheduling modifications.

A frequently offered counterargument to the benefits of shorter work schedules is that the concomitant increase in patient handovers could actually increase risk. The committee noted that there is also no discussion in the papers of whether any of the serious medical errors were attributable to handovers and communication failures. The authors' attempt to institute new and improved sign-out practices was met with resistance and eventually abandoned by the ICU (Landrigan et al., 2004).

Conclusions Relative to Further Restrictions of Duty Hours

The study by Landrigan and colleagues is a carefully conducted experiment that demonstrates remarkable improvements in the two services studied (Landrigan et al., 2004; Lockley et al., 2004; Rothschild et al., 2005). Notwithstanding the caveats raised above, it demonstrates that a substantial reduction in error rates appears possible through such duty hour interventions and increased opportunity for sleep. Together with earlier reports on errors made by fatigued physicians (e.g., Friedman et al., 1971; Grantcharov et al., 2001) and the literature on the impact of fatigue and sleep deprivation on human performance (see Chapter 7), this study lends critical support to the hypothesis that long work hours, including long consecutive duty periods that are accompanied by acute sleep loss, can put patients at risk for errors that could lead to harm. The fatigue associated with long work hours and subsequent propensity for errors is what Bernstein and Etchells (2005) call a "latent hazard." Furthermore, as reported in the Landrigan et al. (2004) study, interns worked beyond their scheduled hours; their colleagues recommend that any maximum hours prescribed in rules account for this inevitability (Lockley et al., 2004). Chapter 7 details additional evidence and suggests ways to reduce acute and chronic sleep deprivation among residents, and Chapter 8 addresses handovers of care because they constitute a period when errors may occur and shortening the length of duty periods increases the number of transitions.

Finally, the committee notes that error rates by residents were high even during the intervention schedule (Landrigan et al., 2004; Rothschild et al., 2005). For example, during the intervention period, interns committed 16.5 errors (PAEs) per 1,000 patient-days. Unit-wide PAE rates, committed by all staff in the unit, were even higher (and almost equal at 38.5 per 1000 patient-days) during both phases of the study. It is noteworthy that these high error rates occurred despite the fact that the subjects were participating in a research program whose ultimate aim was error reduction. The fact that error rates remained high under the intervention schedule suggests to the committee that factors besides work hours, workload, and schedules contribute substantially to the error rates of both interns and others working in the units. Similarly, another study has found that hospital-wide adverse drug events remained the same after duty hour reform (Mycyk et al., 2005). The committee's conclusion is that a vigorous, systematic effort must be made to identify the root causes of medical errors by residents and others in addition to any adjustments in duty hours.

OTHER CONTRIBUTORS TO ERROR

Resident reports teach us about their experiences with error, how they learn from them, and where systems change might most effectively address the potential for intercepting resident errors. Residents do not want to make mistakes and often feel great anguish upon making an error, and the poorer the patient outcome is, the more intense is their reaction (Engel et al., 2006). Residents see both positive and negative results from the 2003 duty hour reforms with respect to patient safety (Fletcher et al., 2008; Lin et al., 2006). On the positive side, well-rested residents find their clinical decision making is improved especially on post-call days, working conditions are better, and they have a generally improved sense of personal well-being. They report downsides including that hour limits are inflexible, patient care can be rushed under the compressed duty hours, treatment decisions are sometimes delayed, and information can be lost in handoffs, thus creating fragmented and less patient-centered care. From the resident's perspective, duty hours alone are not the only issue when it comes to making errors (Jagsi et al., 2005, 2008; Lin et al., 2006).

A systems view of AEs in hospitals and other nonmedical environments recognizes the organizational contribution to a chain of events that can lead to error rather than blaming the individual (Barach and Small, 2000; Leape, 1994; Shojania et al., 2002; Volpp and Grande, 2003). Residents often blame their inexperience and faulty judgment for making errors (e.g., did not ask for advice, missed patient warning signs, had never seen a patient with an atypical presentation of a certain condition, hesitated to act for too long) (Wu et al., 2003). Yet just as frequently they note job overload—too much work to do within the time allotted (Jagsi et al., 2008; Wu et al., 2003). Adverse events are "more likely when suboptimal working conditions occur" (Tibby et al., 2004, p. 1160). Vidyarthi and colleagues (2007), in their analysis of a cross-sectional survey of internal medicine residents (n = 125), found that a multifactorial work stress factor (fatigue, excessive workload, inadequate time, distractions, and stress) (mean = 2.92, SD = 0.67 on a 5-point Likert scale) contributes more often than an intellectual stress factor (inadequate knowledge, inadequate supervision) (mean = 2.39, SD = 0.54, $p < .0001$) to errors. Resident use of suboptimal care practices (e.g., working while fatigued, forgetting to transmit information during sign-out) was the only significant feature predictive of error ($p < .0001$). These internal medicine residents also report that they make cognitive errors more often than administrative errors or procedural ones. Other specialties make procedural errors more often (Jagsi et al., 2005).

Jagsi and colleagues (2008) later surveyed residents in 76 different residency programs at two major teaching hospitals before and after implementation (n = 684/801 residents) of the 2003 duty hour limits to look for contributors to error. In the post-duty hour reform period, similar proportions of residents respond as to what the contributing factors for errors are whether they are in programs that reduced their total weekly work hours (e.g., reduced by 5 or more hours) or made no change in work hours. The values, respectively, for the reduced hours group and the other programs follow: poor handoffs (63.5-61.6 percent), working too many hours (44.0-45.4 percent), carrying or admitting too many patients (47-51.8 percent), cross-covering too many patients (46.9-45.9 percent), or inadequate supervision (24.7-34.1 percent).

Studies of resident errors should identify how the work system itself contributes to resident errors. Rothschild et al. (2005) point out that most of the errors in which residents were involved occurred during treatments involving medications and in procedures (78 percent of incidents) and communication (13.7 percent), and these can be system-level problems not just individual performance issues. It is unreasonable to expect residents not to make mistakes in unreliable work settings. For example, medication vials that look almost identical increase the risk of a mistake. Improving systems (e.g., changing paging practices to decrease interruptions, improved handover procedures, computerized orders to avoid illegible handwriting, better supervision) can improve the performance of residents and improve patient safety (Volpp and Grande, 2003).

Wu says that residents need help: "although patients are the first and obvious victims of medical mistakes, doctors are wounded by the same errors; they are the second victims" (Wu, 2000, p. 358). West and colleagues confirm this observation, finding that errors appear to beget increased burnout and depression and that these, in turn, may set up a continuing cycle as burnt-out residents make errors more frequently (West et al., 2006). Fahrenkopf and colleagues also report that depressed pediatric residents make 6.2 times more medication errors than those who are not depressed (Fahrenkopf et al., 2008). Burnout in residents is discussed more fully in Chapter 5.

Learning from Errors

Wu and colleagues (2003, republished from 1991, p. 221) argue that mistakes can be "powerful formative experiences" and ideally should be used as teaching tools. They queried internal medicine residents (n = 114) at three large tertiary care facilities about the most significant medical mistake they *ever made* and how they responded to it. Mistakes were defined as "an act or omission for which the resident felt responsible that had serious or potentially serious consequences for the patient and would have been judged wrong by knowledgeable peers at the time it occurred." The most significant mistakes reported by residents fell into several categories (33 percent diagnosis, 29 percent prescribing, 21 percent evaluation, 11 percent procedural, 5 percent communication) and the majority occurred in the first year of residency. Residents perceived that 90 percent of the patients involved had adverse outcomes as a result of their mistake (e.g., physical discomfort, additional procedure, prolonged hospital stay, death).

In June 2003, Jagsi and colleagues surveyed medical and surgical residents doing clinical training in 15 specialties at two major teaching hospitals about their exposure to errors made during the delivery of patient care by themselves or others (Jagsi et al., 2005). More than half of the surveyed medical and surgical residents (55 percent) reported that they had cared for a patient who had experienced an AE sometime during their training, with the residents' most recent AE "exposure" (median time since last event = 21 days) being related to procedures (31 percent), adverse drug events (21 percent), and infections (11 percent). The categories of error are consistent with medical records review studies (Gawande et al., 1999; Leape et al., 1991; Neale et al., 2001; Thomas and Brennan, 2000; Thomas et al., 2000a). Eighteen percent of these residents reported exposure to an AE in the past week in a patient that they cared for,[5] and about one-third of these residents felt that they had, at least in part, been responsible (Jagsi et al., 2005). The percentage of those who report AEs caused by mistakes that they felt at least partially responsible for varied by specialty (surgical 10.9 percent, medical 4.7 percent, hospital based such as radiology or anesthesiology 3.4 percent), procedural specialty (yes 8.0 percent, no 3.7 percent), and year of training (PGY-1 8.2 percent, PGY-2 or more 5.4 percent).

This high level of self-reported exposure in this study illustrates the key role residents could play in the reduction of errors if error reporting and system quality improvement were integrated into residency programs. In Chapter 8, the committee recommends changes in error-reporting systems to enhance the opportunity for teaching and learning when errors occur.

Conclusion About Other Factors

The committee concludes that a number of factors can contribute to resident errors (whether errors of commission or omission) and that it is not just a matter of hours worked or length of shift. Because first-year residents tend to work longer hours than residents in other years, more frequently violate duty hours, and appear to be more vulnerable to making mistakes—and yet can be reluctant to reach out for help—the committee has recommended in Chapter 4 the particular need to increase supervision for these trainees. Additionally, the committee has concluded in Chapters 3 and 4 that excessive workload creates pressure to violate work hours and can limit learning.

The resident self-report studies discussed in this section examine the experiences of residents at a small number of major teaching institutions. As noted earlier in this chapter, clearly, residents make mistakes during patient

[5]Note that these are not considered rates of "resident-committed errors" because the study questioned exposure to events and thus could be double counting errors due to cross-coverage of patients by different residents.

care and these can result in harm to patients, but research studies to date do not allow us to determine with precision the frequency and the severity of those mistakes across all specialties or how often they lead to adverse patient effects that would be preventable. However, first-year residents appear particularly vulnerable to these mistakes or near misses although they occur with residents of all training years, and the types of mistakes (diagnosis, delays in treatment, and performing procedures) are ones that better supervision would help address (Jagsi et al., 2005; Wu et al., 2003). Many of the perceived causes of the mistakes that residents make appear avoidable not only by better supervision but also by workload reduction, more rest, better handovers, and other changes in the work environment.

SUMMARY

This chapter has examined five questions that are central to the debate on the scope of resident errors while in training, the extent to which duty hour reforms have already made a difference, and the potential contribution of further duty hour reductions.

1. *Do residents make errors that contribute to patient harm?* Residents do make errors that contribute to patient harm (Jagsi et al., 2005, 2008; Landrigan et al., 2004; Rothschild et al., 2005; Wu et al., 2003). However, data are too limited to determine what portion of errors in training facilities are due to residents and what portion of errors result in preventable adverse events that contribute to patient harm.
2. *Is resident fatigue from long duty hours among the most significant risks to patient safety?* There is evidence that residents can experience fatigue under the current ACGME duty hours (2003) and that fatigue may derive from a number of factors, one of which is lengthy duty hours. There is also evidence that schedules that induce fatigue can result in increased medical errors by residents, which are a potential risk to patients' safety. The one randomized controlled trial of duty hour reduction reported to date found that serious medical errors (including medication and diagnostic errors) and non-intercepted serious errors were significantly higher with longer duty hours and less sleep (Landrigan et al., 2004). However, they did not find a statistically significant difference in patient safety as directly measured by PAEs (Landrigan et al., 2004). Consequently, while resident fatigue might pose a risk to patient safety, it is not possible to determine the extent of this risk.
3. *Did the 2003 reduction in resident duty hours affect patient safety?* The national studies of mortality, at the very least, show that

there is no evidence of widespread harm occurring after implementation of the limits (i.e., 2003 duty hour restrictions did not lead to an increase in mortality rates for the common conditions studied) and there may be modest improvements for medical if not surgical patients (Landrigan et al., 2004; Prasad, 2008; Shetty and Bhattacharya, 2007; Volpp et al., 2007a,b). The results from national studies as well as smaller institution-specific studies indicate how difficult it is to scientifically substantiate the conventional wisdom that reduced hours would clearly result in improved patient care. Based on the available data, the committee concludes that movement toward the 80-hour week has not had an adverse effect on patient outcomes. It also recognizes that all training programs in the country have not actually achieved compliance with the 80-hour week consistently.

4. *Would further reductions in resident duty hours improve patient safety?* At this point, no study indicates that 80 hours or some other lower duty hour total is optimal for patient safety. A number of studies of individual programs have found that they have been able to accommodate to the 80-hour week, even in surgical programs, without sacrificing educational or patient outcomes or increasing error (e.g., de Virgilio et al., 2006; Vaughn et al., 2008). The study by Landrigan and colleagues tested in an ICU setting an intervention with a shorter workweek, shorter shift lengths, and more sleep for interns. This study suggests that further reductions in resident work hours could potentially improve conditions for patient safety by reducing errors although the reduction in PAEs was not statistically significant. As noted by Landrigan et al. (2004, p. 1844), "Therefore, it remains to be determined whether the decrease in the rate of serious medical errors by interns will translate into a reduction in the rate of adverse events." Although Landrigan and colleagues conducted a well-designed study, there are a number of questions about its generalizability to other settings, specialties, and years of training. Chapter 7 examines evidence from the human performance literature on the contribution of shift length, night work, and amount of sleep in order to help identify the factors that contribute to diminished performance and to identify opportunities for preventing and mitigating fatigue.
5. *What factors in the resident work and learning environment contribute to error?* Numerous factors can contribute to resident errors. The causes of resident errors as well as those of other clinical staff are not one-dimensional but include multiple factors in addition to fatigue: a work and learning environment with insufficient staffing and heavy workload, inadequate supervision, mental

health (e.g., burnout, depression), level of skills and knowledge, complexity of patient's clinical condition, communication problems between team members, language barriers with patients, and inherent system failures (Carayon and Gurses, 2008; Dean et al., 2002; Fahrenkopf et al., 2008; West et al., 2006; Wu et al., 2003).

The committee encourages additional research on the questions in this chapter. Identifying ways to prevent resident fatigue and the risks it poses to patient safety requires a more systematic understanding of the extent to which fatigued residents are causing patient harm and, if so, under what conditions. For example, the following information would help identify how to best protect patients from errors by residents: When during shifts are errors made? Are many errors made by a few residents or are all residents equally likely to commit errors? What types of errors are made, and how serious and preventable are they? To what extent are errors corrected by other clinicians and systems, and to what extent could more be prevented by the committee's recommendations for changes in supervision, handovers, and protected sleep? Larger samples of residents from a greater variety of programs and institutions would provide a better population estimate for identifying best practices to prevent risks to patient and resident safety. Notwithstanding some of the excellent research that has been done in recent years, multi-institutional studies would also have the power to detect changes in preventable adverse errors and mortality as a function of changes in duty hours and any resultant increases in handovers, and would provide data on what kinds of situations need to be targeted to reduce risks to patients and residents.

While the research studies discussed in this chapter concerning residents, duty hours, and patient safety generally have limitations and are less conclusive about the effects of duty hours on patient safety, the research discussed in Chapter 7 presents strong evidence that sleep deprivation, which can result from some aspects of current duty hours, can cause fatigue, which contributes to reduced well-being, increased errors, and accidents. The evidence presented in the next chapter provides the basis for the committee's recommendations concerning changes in duty hours to prevent fatigue.

REFERENCES

AHRQ (Agency for Healthcare Research and Quality). 2002. Medical schools and residency programs should provide more training on preventing adverse drug reactions. *Research Activities* 263:8.

———. 2007. *National healthcare quality report—chapter 3: Patient safety.* Rockville, MD: U.S. Department of Health and Human Services.

Barach, P., and S. D. Small. 2000. Reporting and preventing medical mishaps: Lessons from non-medical near miss reporting systems. *BMJ* 320(7237):759-763.

Bates, D. W., D. J. Cullen, N. Laird, L. A. Petersen, S. D. Small, D. Servi, G. Laffel, B. J. Sweitzer, B. F. Shea, and R. Hallisey. 1995. Incidence of adverse drug events and potential adverse drug events. Implications for prevention. ADE Prevention Study Group. *JAMA* 274(1):29-34.

Beckmann, U., C. Bohringer, R. Carless, D. M. Gillies, W. B. Runciman, A. W. Wu, and P. Pronovost. 2003. Evaluation of two methods for quality improvement in intensive care: Facilitated incident monitoring and retrospective medical chart review. *Critical Care Medicine* 31(4):1006-1011.

Berstein, M., and E. E. Etchells. 2005. Does reducing interns' work hours reduce the rate of medical errors? *CMAJ: Canadian Medical Association Journal* 172(4).

Bhavsar, J., D. Montgomery, J. Li, E. Kline-Rogers, F. Saab, A. Motivala, J. B. Froehlich, V. Parekh, J. Del Valle, and K. A. Eagle. 2007. Impact of duty hours restrictions on quality of care and clinical outcomes. *American Journal of Medicine* 120(11):968-974.

Brady, J., K. Ho, and C. Clancy. 2008. Progress slows in improving patient safety for all populations. *Patient Safety & Quality Healthcare* (July/August):6-7.

Brennan, T. A., L. E. Hebert, N. M. Laird, A. Lawthers, K. E. Thorpe, L. L. Leape, A. R. Localio, S. R. Lipsitz, J. P. Newhouse, and P. C. Weiler. 1991. Hospital characteristics associated with adverse events and substandard care. *JAMA* 265(24):3265-3269.

Carayon, P., and A. P. Gurses. 2008. Nursing workload and patient safety—A human factors engineering perspective. In *Patient safety and quality: An evidence-based handbook for nurses*. Rockville, MD: Agency for Healthcare Research and Quality. Original edition, AHRQ Publication No. 08-0043.

Chang, B. K. 2007. Presentation to the Committee on Optimizing Graduate Medical Trainee (Resident) Hours and Work Schedules to Improve Patient Safety, December 3, 2007, Washington, DC.

Classen, D. C., R. C. Lloyd, L. Provost, F. A. Griffin, and R. Resar. 2008. Development and evaluation of the Institute for Healthcare Improvement Global Trigger Tool. *Journal of Patient Safety* 4(3):169-177.

Commonwealth Fund. 2008. *Why not the best?* New York: The Commonwealth Fund.

de Virgilio, C., A. Yaghoubian, R. J. Lewis, B. E. Stabile, and B. A. Putnam. 2006. The 80-hour resident workweek does not adversely affect patient outcomes or resident education. *Journal of Surgical Education* 63(6):435-439.

Dean, B., M. Schachter, C. Vincent, and N. Barber. 2002. Causes of prescribing errors in hospital inpatients: A prospective study. *Lancet* 359(9315):1373-1378.

DeBuono, B. A., and W. M. Osten. 1998. The medical resident workload: The case of New York State. *JAMA* 280(21):1882-1883.

DeFrances, C. J., C. A. Lucas, V. C. Buie, and A. Golosinskiy. 2008. 2006 National Hospital Discharge Survey. Hyattsville, MD: National Center for Health Statistics.

Dingell, J. D., J. Barton, B. Stupak, and E. Whitfield. 2007. Letter to William Munier, Acting Director, Agency for Healthcare Research and Quality. Washington, DC: U.S. House of Representatives Committee on Energy and Commerce.

Eastridge, B. J., E. C. Hamilton, G. E. O'Keefe, R. V. Rege, R. J. Valentine, D. J. Jones, S. Tesfay, and E. R. Thal. 2003. Effect of sleep deprivation on the performance of simulated laparoscopic surgical skill. *American Journal of Surgery* 186(2):169-174.

Ellman, P. I., M. G. Law, C. Tache-Leon, T. B. Reece, T. S. Maxey, B. B. Peeler, J. A. Kern, C. G. Tribble, and I. L. Kron. 2004. Sleep deprivation does not affect operative results in cardiac surgery. *Annals of Thoracic Surgery* 78(3):906-911.

Ellman, P. I., I. L. Kron, J. S. Alvis, C. Tache-Leon, T. S. Maxey, T. B. Reece, B. B. Peeler, J. A. Kern, and C. G. Tribble. 2005. Acute sleep deprivation in the thoracic surgical resident does not affect operative outcomes. *Annals of Thoracic Surgery* 80(1):60-65.

Engel, K. G., M. Rosenthal, and K. M. Sutcliffe. 2006. Residents' responses to medical error: Coping, learning, and change. *Academic Medicine* 81(1):86-93.

Fahrenkopf, A. M., T. C. Sectish, L. K. Barger, P. J. Sharek, D. Lewin, V. W. Chiang, S. Edwards, B. L. Wiedermann, and C. P. Landrigan. 2008. Rates of medication errors among depressed and burnt out residents: Prospective cohort study. *BMJ* 336(7642):488-491.

Fischer, J. E. 2004. Continuity of care: A casualty of the 80-hour work week. *Academic Medicine* 79(5):381-383.

Fletcher, K. E., S. Q. Davis, W. Underwood, R. S. Mangrulkar, L. F. McMahon, Jr., and S. Saint. 2004. Systematic review: Effects of residents work hours on patient safety. *Annals of Internal Medicine* 141(11):851-857.

Fletcher, K. E., V. Parekh, L. Halasyamani, S. R. Kaufman, M. Schapira, K. Ertl, and S. Saint. 2008. Work hour rules and contributors to patient care mistakes: A focus group study with internal medicine residents. *Journal of Hospital Medicine* 3(3):228-237.

Forster, A. J., H. J. Murff, J. F. Peterson, T. K. Gandhi, and D. W. Bates. 2003. The incidence and severity of adverse events affecting patients after discharge from the hospital. *Annals of Internal Medicine* 138(3):161-167.

Friedman, R. C., J. T. Bigger, and D. S. Kornfield. 1971. The intern and sleep loss. *New England Journal of Medicine* 285:201-203.

Gandhi, T. K., A. Kachalia, E. J. Thomas, A. L. Puopolo, C. Yoon, T. A. Brennan, and D. M. Studdert. 2006. Missed and delayed diagnoses in the ambulatory setting: A study of closed malpractice claims. *Annals of Internal Medicine* 145(7):488-496.

Garbutt, J., A. D. Waterman, J. M. Kapp, W. C. Dunagan, W. Levinson, V. Fraser, and T. H. Gallagher. 2008. Lost opportunities: How physicians communicate about medical errors. *Health Affairs* 27(1):246-255.

Gawande, A. 2007. The checklist: If something so simple can transform intensive care, what else can it do? *New Yorker*, December 17, pp. 86-101.

Gawande, A. A., E. J. Thomas, M. J. Zinner, and T. A. Brennan. 1999. The incidence and nature of surgical adverse events in Colorado and Utah in 1992. *Surgery* 126(1):66-75.

Grantcharov, T. P., L. Bardram, P. Funch-Jensen, and J. Rosenberg. 2001. Laparoscopic performance after one night on call in a surgical department: Prospective study. *BMJ* 323(7323):1222-1223.

Griffin, F. A., and D. C. Classen. 2008 (August). Detection of adverse events in surgical patients using the trigger tool approach. *Quality and Safety in Health Care* 17:253-258.

Hayward, R. A., and T. P. Hofer. 2001. Estimating hospital deaths due to medical errors: Preventability is in the eye of the reviewer. *JAMA* 286(4):415-420.

HHS (U.S. Department of Health and Human Services). 2008. *Hospital compare.* http://www.hospitalcompare.hhs.gov/Hospital/Static/About-Overview.asp?dest=NAV|Home|About|Overview#TabTop (accessed July 28, 2008).

Horwitz, L. I., M. Kosiborod, Z. Lin, and H. M. Krumholz. 2007. Changes in outcomes for internal medicine inpatients after work-hour regulations. *Annals of Internal Medicine* 147(2):97-103.

Howard, D. L., J. H. Silber, and D. R. Jobes. 2004. Do regulations limiting residents' work hours affect patient mortality? *Journal of General Internal Medicine* 19(1):1-7.

Howard, S. K., D. M. Gaba, M. R. Rosekind, and V. P. Zarcone. 2002a. The risks and implications of excessive daytime sleepiness in resident physicians. *Academic Medicine* 77(10):1019-1025.

Howard, S. K., M. R. Rosekind, J. D. Katz, and A. J. Berry. 2002b. Fatigue in anesthesia: Implications and strategies for patient and provider safety. *Anesthesiology* 97(5):1281-1294.

IHI (Institute for Healthcare Improvement). 2008. *100,000 lives campaign general strategy/overview calls.* http://www.ihi.org/IHI/Programs/Campaign/100kCampaignStrategyOverviewCalls.htm (accessed August 12, 2008).

IOM (Institute of Medicine). 2000. *To err is human: Building a safer health system.* Washington, DC: National Academy Press.

———. 2006. *Preventing medication errors: Quality chasm series.* Washington, DC: The National Academies Press.

Jacques, C. H., J. C. Lynch, and J. S. Samkoff. 1990. The effects of sleep loss on cognitive performance of resident physicians. *Journal of Family Practice* 30(2):223-229.

Jagsi, R., B. T. Kitch, D. F. Weinstein, E. G. Campbell, M. Hutter, and J. S. Weissman. 2005. Residents report on adverse events and their causes. *Archives of Internal Medicine* 165(22):2607-2613.

Jagsi, R., D. F. Weinstein, J. Shapiro, B. T. Kitch, D. Dorer, and J. S. Weissman. 2008. The Accreditation Council for Graduate Medical Education's limits on residents' work hours and patient safety: A study of resident experiences and perceptions before and after hours reductions. *Archives of Internal Medicine* 168(5):493-500.

Jakubowicz, D. M., E. M. Price, H. J. Glassman, A. J. G. Gallagher, N. Mandava, W. P. Ralph, and M. P. Fried. 2005. Effects of a twenty-four hour call period on resident performance during simulated endoscopic sinus surgery in an Accreditation Council for Graduate Medical Education-compliant training program. *Laryngoscope* 115:143-146.

Joint Commission. 2007. *Sentinel event policy and procedures.* http://www.jointcommission.org/SentinelEvents/PolicyandProcedures/ (accessed July 25, 2008).

Kaafarani, H. M. A., K. M. F. Itani, L. A. Petersen, J. Thornby, and D. H. Berger. 2005. Does resident hours reduction have an impact on surgical outcomes? *Journal of Surgical Research* 126(2):167-171.

Kahol, K., M. J. Leyba, M. Deka, V. Deka, S. Mayes, M. Smith, J. J. Ferrara, and S. Panchanathan. 2008. Effect of fatigue on psychomotor and cognitive skills. *American Journal of Surgery* 195(2):195-204.

Kaldjian, L. C., E. W. Jones, B. J. Wu, V. L. Forman-Hoffman, B. H. Levi, and G. E. Rosenthal. 2008. Reporting medical errors to improve patient safety: A survey of physicians in teaching hospitals. *Archives of Internal Medicine* 168(1):40-46.

Kennedy, R. 1998. Residents' work hours termed excessive in hospital study. *New York Times*, May 1, p. 1.

Kiernan, M., J. Civetta, C. Bartus, and S. Walsh. 2006. 24 hours on-call and acute fatigue no longer worsen resident mood under the 80-hour work week regulations. *Current Surgery* 63(3):237-241.

Laine, C., L. Goldman, J. R. Soukup, and J. G. Hayes. 1993. The impact of a regulation restricting medical house staff working hours on the quality of patient care. *JAMA* 269(3):374-378.

Landrigan, C. P., J. M. Rothschild, J. W. Cronin, R. Kaushal, E. Burdick, J. T. Katz, C. M. Lilly, P. H. Stone, S. W. Lockley, D. W. Bates, and C. A. Czeisler. 2004. Effect of reducing interns' work hours on serious medical errors in intensive care units. *New England Journal of Medicine* 351(18):1838-1848.

Leape, L. L. 1994. Error in medicine. *JAMA* 272(23):1851-1857.

Leape, L. L., and D. M. Berwick. 2005. Five years after *To Err Is Human*: What have we learned? *JAMA* 293(19):2384-2390.

Leape, L. L., T. A. Brennan, N. Laird, A. G. Lawthers, A. R. Lacolio, B. A. Barnes, L. Hebert, J. P. Newhouse, P. C. Weiler, and H. Hiatt. 1991. The nature of adverse events in hospitalized patients. Results of the Harvard Medical Practice Study II. *New England Journal of Medicine* 324(6):377-384.

Leonard, C., N. Fanning, J. Attwood, and M. Buckley. 1998. The effect of fatigue, sleep deprivation and onerous working hours on the physical and mental wellbeing of pre-registration house officers. *Irish Journal of Medical Science* 167(1):22-25.

Lin, G. A., D. C. Beck, and J. M. Garbutt. 2006. Residents' perceptions of the effects of work hour limitations at a large teaching hospital. *Academic Medicine* 81(1):63-67.

Lockley, S. W., J. W. Cronin, E. E. Evans, B. E. Cade, C. J. Lee, C. P. Landrigan, J. M. Rothschild, J. T. Katz, C. M. Lilly, P. H. Stone, D. Aeschbach, C. A. Czeisler, and Harvard Work Hours Health and Safety Group. 2004. Effect of reducing interns' weekly work hours on sleep and attentional failures. *New England Journal of Medicine* 351(18):1829-1837.

McCannon, C. J., A. D. Hackbarth, and F. A. Griffin. 2007. Miles to go: An introduction to the 5 million lives campaign. *The Joint Commission Journal on Quality and Patient Safety* 33(8):477-484.

Mycyk, M. B., M. R. McDaniel, M. A. Fotis, and J. Regalado. 2005. Hospitalwide adverse drug events before and after limiting weekly work hours of medical residents to 80. *American Journal of Health-System Pharmacists.*

Neale, G., M. Woloshynowych, and C. Vincent. 2001. Exploring the causes of adverse events in NHS hospital practice. *Journal of the Royal Society of Medicine* 94(7):322-330.

NTSB (National Transportation Safety Board). 2008. *NTSB training center.* http://www.ntsb.gov/TC/CourseInfo/IM303_2009.htm (accessed October 24, 2008).

Petersen, L. A., T. A. Brennan, A. C. O'Neil, E. F. Cook, and T. H. Lee. 1994. Does housestaff discontinuity of care increase the risk for preventable adverse events? *Annals of Internal Medicine* 121(11):866.

Petersen, L. A., E. J. Orav, J. M. Teich, A. C. O'Neil, and T. A. Brennan. 1998. Using a computerized sign-out program to improve continuity of inpatient care and prevent adverse events. *Joint Commission Journal on Quality Improvement* 24(2):77-87.

Poulose, B. K., W. A. Ray, P. G. Arbogast, J. Needleman, P. I. Buerhaus, M. R. Griffin, N. N. Abumrad, R. D. Beauchamp, and M. D. Holzman. 2005. Resident work hour limits and patient safety. *Annals of Surgery* 241(6):847-856.

Prasad, M. 2008. The effect of resident work-hour regulations on ICU mortality. Paper read at 2008 American Thoracic Society International Conference, May 21, 2008, Toronto, ON, Canada.

Pronovost, P., D. Needham, S. Berenholtz, D. Sinopoli, H. Chu, S. Cosgrove, B. Sexton, R. Hyzy, R. Welsh, G. Roth, J. Bander, J. Kepros, and C. Goeschel. 2006. An intervention to decrease catheter-related bloodstream infections in the ICU. *New England Journal of Medicine* 355(26):2725-2732.

Regenbogen, S. E., C. C. Greenberg, D. M. Studdert, S. R. Lipsitz, M. J. Zinner, and A. A. Gawande. 2007. Patterns of technical error among surgical malpractice claims: An analysis of strategies to prevent injury to surgical patients. *Annals of Surgery* 246(5):705-711.

Robbins, J., and F. Gottlieb. 1990. Sleep deprivation and cognitive testing in internal medicine house staff. *Western Journal of Medicine* 152(1):82-86.

Rosekind, M. R., K. B. Gregory, D. L. Miller, E. L. Co, and J. V. Lebacqz. 1994. *Aircraft accident report: Uncontrolled collision with terrain, American International Airways Flight 808, Douglas DC-8, N814CK, U.S. Naval Air Station, Guantanamo Bay, Cuba, August 18, 1993.* Washington, DC: National Transportation Safety Board. http://human-factors.arc.nasa.gov/zteam/PDF_pubs/G_Bay/GauntanamoBay.pdf (accessed March 16, 2009).

Rothschild, J. M., C. P. Landrigan, J. W. Cronin, R. Kaushal, S. W. Lockley, E. Burdick, P. H. Stone, C. M. Lilly, J. T. Katz, C. A. Czeisler, and D. W. Bates. 2005. The critical care safety study: The incidence and nature of adverse events and serious medical errors in intensive care. *Critical Care Medicine* 33(8):1694-1700.

Saxena, A. D., and C. F. P. George. 2005. Sleep and motor performance in on-call internal medicine residents. *Sleep* 28(11):1386-1391.

Schneider, J. R., J. J. Coyle, E. R. Ryan, R. H. Bell, Jr., and D. A. DaRosa. 2007. Implementation and evaluation of a new surgical residency model. *Journal of the American College of Surgeons* 205(3):393-404.
Shetty, K. D., and J. Bhattacharya. 2007. Changes in hospital mortality associated with residency work-hour regulations. *Annals of Internal Medicine* 147(2):73-80.
Shojania, K. G., H. Wald, and R. Gross. 2002. Understanding medical error and improving patient safety in the inpatient setting. *Medical Clinics of North America* 86(4):847-867.
Singh, H., E. J. Thomas, L. A. Petersen, and D. M. Studdert. 2007. Medical errors involving trainees: A study of closed malpractice claims from 5 insurers. *Archives of Internal Medicine* 167(19):2030-2036.
Thomas, E. J., and T. A. Brennan. 2000. Incidence and types of preventable adverse events in elderly patients: Population based review of medical records. *BMJ* 320(7237):741-744.
Thomas, E. J., D. M. Studdert, J. P. Newhouse, B. I. W. Zbar, K. M. Howard, E. J. Williams, and T. A. Brennan. 1999. Costs of medical injuries in Utah and Colorado. *Inquiry* 36(3):255-264.
Thomas, E. J., D. M. Studdert, H. R. Burstin, E. J. Orav, T. Zeena, E. J. Williams, K. M. Howard, P. C. Weiler, and T. A. Brennan. 2000a. Incidence and types of adverse events and negligent care in Utah and Colorado. *Medical Care* 38(3):261-271.
Thomas, E. J., D. M. Studdert, W. B. Runciman, R. K. Webb, E. J. Sexton, R. M. Wilson, R. W. Gibberd, B. T. Harrison, and T. A. Brennan. 2000b. A comparison of iatrogenic injury studies in Australia and the USA I: Context, methods, casemix, population, patient and hospital characteristics. *International Journal for Quality in Health Care* 12(5):371-378.
Tibby, S., J. Correa-West, A. Durward, L. Ferguson, and I. Murdoch. 2004. Adverse events in a paediatric intensive care unit: Relationship to workload, skill mix and staff supervision. *Intensive Care Medicine* 30(6):1160-1166.
Vaughn, D. M., C. L. Stout, B. L. McCampbell, J. R. Groves, A. I. Richardson, W. K. Thompson, M. L. Dalton, and D. K. Nakayama. 2008. Three-year results of mandated work hour restrictions: Attending and resident perspectives and effects in a community hospital. *American Surgeon* 74(6):542-546.
Veasey, S., R. Rosen, B. Barzansky, I. Rosen, and J. Owens. 2002. Sleep loss and fatigue in residency training: A reappraisal. *JAMA* 288(9):1116-1124.
Vidyarthi, A. R., A. D. Auerbach, R. M. Wachter, and P. P. Katz. 2007. The impact of duty hours on resident self reports of errors. *Journal of General Internal Medicine* 22(2):205-209.
Volpp, K. G. M., and D. Grande. 2003. Residents' suggestions for reducing errors in teaching hospitals. *New England Journal of Medicine* 348(9):851-855.
Volpp, K. G., A. K. Rosen, P. R. Rosenbaum, P. S. Romano, O. Even-Shoshan, A. Canamucio, L. Bellini, T. Behringer, and J. H. Silber. 2007a. Mortality among patients in VA hospitals in the first 2 years following ACGME resident duty hour reform. *JAMA* 298(9):984-992.
Volpp, K. G., A. K. Rosen, P. R. Rosenbaum, P. S. Romano, O. Even-Shoshan, Y. Wang, L. Bellini, T. Behringer, and J. H. Silber. 2007b. Mortality among hospitalized Medicare beneficiaries in the first 2 years following ACGME resident duty hour reform. *JAMA* 298(9):975-983.
Weinger, M. B., and S. Ancoli-Israel. 2002. Sleep deprivation and clinical performance. *JAMA* 287(8):955-957.
West, C. P., M. M. Huschka, P. J. Novotny, J. A. Sloan, J. C. Kolars, T. M. Habermann, and T. D. Shanafelt. 2006. Association of perceived medical errors with resident distress and empathy: A prospective longitudinal study. *JAMA* 296(9):1071-1078.
Wu, A. W. 2000. Medical error: The second victim. *BMJ* 320(7237):726-727.
Wu, A. W., S. Folkman, S. J. McPhee, and B. Lo. 2003. Do house officers learn from their mistakes? *Quality and Safety in Health Care* 12(3):221-226.

7

Strategies to Reduce Fatigue Risk in Resident Work Schedules

The scientific literature makes clear that risks of fatigue-related errors and accidents derive from multiple interacting variables of work and sleep. This chapter discusses the literature on sleep and human performance and recommends specific adjustments to the current Accreditation Council for Graduate Medical Education (ACGME) resident duty hours to enhance the prevention and mitigation of resident fatigue as an unsafe condition, thereby improving performance and the safety of both patients and residents. The major rationales for the recommendations are the following: (1) work duration should be limited because human performance degrades after 16 hours of wakefulness whether one is working or not; (2) sufficient time for sleep needs to be incorporated into daily and weekly work schedules to prevent acute and chronic sleep deprivation, respectively, and to allow recovery from accumulated sleep debt; and (3) when extended duty periods are considered an essential aspect of resident training and continuity of care, a protected sleep period should be provided during that period to reduce the effects of acute sleep loss and to enhance performance. Because of the diversity of specialty and hospital needs, the committee leaves some flexibility for programs, but urges that adequate protected sleep periods be maintained, and that fatigue prevention and mitigation be a matter of professionalism that requires attention by residents, attending physicians and all those charged with maintaining patient safety.

The consensus committee was charged by the Agency for Healthcare Research and Quality (AHRQ) to "1) synthesize current evidence on medical resident schedules and healthcare safety and 2) develop strategies to enable optimization of work schedules to improve safety in the healthcare work environment." This chapter deals with scientific evidence that fatigue is an unsafe condition that can occur relative to the timing and duration of work and sleep opportunities, which are fundamental components of residents' schedules. The chapter reviews scientific literature on fatigue, its consequences, and its prevention, and provides recommendations and evidenced-based justifications for ways to reduce fatigue as a safety risk while residents are training intensively by working long hours.

Many of the recommendations focus on ensuring residents obtain adequate sleep, which research has shown is among the most fundamental biological needs, to counter fatigue and promote learning and memory. The focus on resident fatigue prevention in the recommendations for duty hour adjustments in the latter sections of this chapter is a response to AHRQ's charge that the committee develop strategies to enable optimization of work schedules to improve safety in the healthcare work environment. Other chapters in this report recommend additional ways in which safety can be enhanced through supervision, appropriate workload, teamwork, and system changes. This chapter takes an evidenced-based approach to developing recommended changes in only those aspects of resident duty hours that are most likely to result in fatigue as an unsafe condition that can pose risks to both patients and residents. Thus, to retain the training value and flexibility in scheduling required by different specialties and rotations within specialties, while preventing and mitigating sleep loss that contributes to fatigue-related errors and accidents, the recommendations derived from this chapter relative to duty hours are focused more on providing predictable and protected time for sleep and recovery sleep than on limiting total work hours.

FATIGUE, WORK HOURS, AND SLEEP LOSS

In healthy individuals, fatigue is a general term used to describe feelings of tiredness, reduced energy, and the increased effort needed to perform tasks effectively and avoid errors. It occurs as performance demands increase because of work intensity and work duration, but it is also a product of the quantity and quality of sleep and the time of day work occurs (Dinges, 2001). All of these factors are relevant aspects of residency training duty hours (Buysse et al., 2003). Some current aspects of resident duty hours can interfere with normal sleep patterns and lead to sleep deprivation, with the extent of this deprivation differing according to the workload demands and schedule of each residency specialty.

Risks of fatigue-related errors and accidents in relation to work schedules derive not from the single factor of the total hours of work in a week, but from multiple interrelated and interacting aspects of work, rest, and sleep. These include but are not limited to (1) the duration of work periods within a single day and over time, (2) the time of day at which work occurs, (3) variation in the timing of work within and between weeks, (4) the duration of sleep obtained on work days and on non-work days, (5) the frequency and duration of days off from work, (6) the different vulnerabilities of workers to fatigue from these factors, and (7) the volume and intensity of work (Dinges, 1995; Drake et al., 2004; Folkard et al., 2005; Rosa, 2001; Van Dongen, 2006). It is not surprising that concern about the negative effects of sleep deprivation on residents is one of the primary reasons duty hour restrictions have been implemented by the Accreditation Council for Graduate Medical Education (ACGME, 2003).

This chapter contains a synthesis of the current evidence about fatigue, performance, and safety risks posed by different work-rest-sleep factors applicable to current ACGME duty hour rules and possible adjustments to those rules. First, the chapter focuses on acute sleep deprivation and ways to prevent the development of acute sleep loss (e.g., shorter duty periods) or ways to mitigate the effects of acute sleep loss by use of sleep during extended duty periods that may be required for patient care and education. The chapter then examines ways to prevent and address the accumulation of chronic partial sleep loss in residents. After reviewing this evidence, the committee proposes strategies to enable optimization of resident work schedules to improve safety in the healthcare work environment while taking into account the learning and experience that residents must achieve during their training.

NEED FOR SLEEP

A sizeable scientific literature exists on the extent to which preventing fatigue, and its associated cognitive performance deficits, depends heavily upon the extent to which acute (daily) and chronic (weekly) sleep needs are met. Moreover, reviews of the risks posed by residency duty hours have emphasized that prevention of sleep deprivation in residents is the most important way to reduce fatigue risks to patient and resident safety (Baldwin and Daugherty, 2004; Buysse et al., 2003; Cavallo and Mallory, 2004; Gaba and Howard, 2002; Howard et al., 2002; Landrigan et al., 2007; Lockley et al., 2006; Parshuram, 2006; Veasey et al., 2002; Weinger and Ancoli-Israil, 2002). Below the committee reviews evidence concerning the benefits to human performance, and potentially to patient safety, from increased sleep time.

Before the 2003 duty hour reforms, first-year residents reported sleeping an average of 5.7 (standard deviation [SD] .90) hours per night and second-year residents reported an average of 5.98 (SD .98) hours (Baldwin and Daugherty, 2004). However, self-reported sleep times tend to overestimate actual physiological sleep obtained (Jean-Louis et al., 2000). No national study of resident sleep hours is available post duty hour reform in 2003. Furthermore, the average number of hours reported by residents may mask the actual degree of sleep deprivation because the schedule of work (e.g., night work, extended duty shifts) may significantly influence the amount of uninterrupted sleep possible.

Reducing work hours does not necessarily result in a corresponding increase in sleep hours. An examination of work hours and the amount of sleep obtained by residents shows that there is a statistically significant but only moderate correlation between residents' work hours and their sleep hours, with approximately 15 percent (Baldwin and Daugherty, 2004; Baldwin et al., 2003) to 33 percent (Lockley et al., 2004) of common variance between work time and sleep time (see Figure 7-1).

The varied elements influencing fatigue interact in complex ways that make it difficult to attribute risk of reduced resident performance to reduced hours of work. For example, the study by Landrigan and colleagues (2004) restricted the work of interns to approximately 16 hours at a time and eliminated extended duty periods (24 hours or more), which resulted in an average of 19.5 hours less work per week than the traditional schedule, but it also resulted in 5.8 hours more sleep per week. A survey conducted by Baldwin and Daugherty (2004) of residents prior to the 2003 duty hour limits also revealed an inverse relationship between average weekly work hours reported by residents and average weekly sleep time (Figure 7-1A). Thus, it is not possible to isolate the distinct effects of shift length, total work hour limitation, increased sleep time, and/or other consequences of adjusting the duty hour limits (e.g., increased staffing) on the reduction of medical errors found by Landrigan and colleagues (2004). Although the separate contribution of increased sleep time to error reduction cannot be measured, there is ample reason (reviewed below) that sleep could be the primary way in which fatigue and its risks were mitigated in the residents studied by Landrigan and colleagues.

Circadian Influence

Acute sleep loss begins when an individual remains awake beyond 16 to 18 hours or into the habitual nocturnal period for sleep (Van Dongen and Dinges, 2005). Extensive research has shown the brain's circadian system ceases to oppose the physiologic pressure for sleep after 16-18 hours of being awake. This results in steady increases between midnight and 6-10 a.m. (when the body is biologically programmed to sleep) in sleepiness and sleep propensity, lapses of attention and memory, and a wide range of other cognitive performance deficits (Van Dongen and Dinges, 2005). This natural pressure for sleep occurs when someone is awake at night, whether or not the person is working. Morning hours (4-10 a.m.) are a peak time

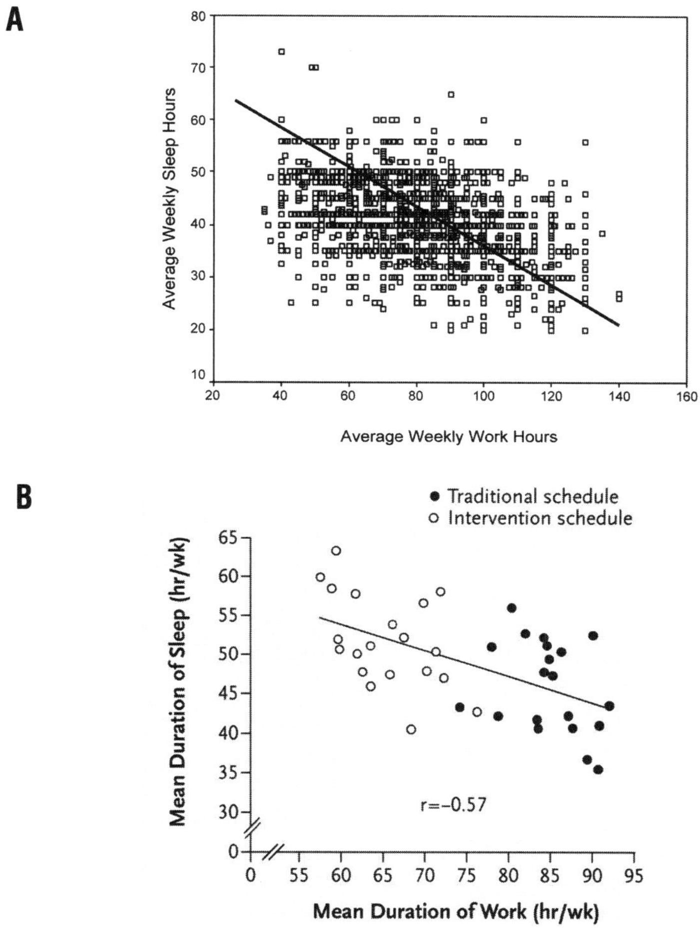

FIGURE 7-1 Relationship of residents' average weekly sleep to average weekly hours of work.

NOTE: Figure A is survey data from n = 3,406 residents. Scatterplot of reported average weekly work hours of sleep with reported average weekly work hours, PGY (postgraduate year) 1 and PGY2 combined (regression line plotted, $r = -.39$). Figure B is sleep log data in n = 20 interns. Relationship between the duration of work and the duration of sleep for 20 interns during the traditional schedule and the intervention schedule. The duration of work and the duration of sleep were inversely correlated ($r = -0.57$, $p < 0.001$) during the traditional intervention schedules, with the best-fit regression predicting a 19.2-minute loss of sleep per week for every additional hour of work per week.

(A) Reprinted, with permission, from Baldwin and Daugherty, 2004. Copyright 2004 by the American Academy of Sleep Medicine.

(B) Reprinted, with permission, from Lockley et al., 2004. Copyright © 2004 Massachusetts Medical Society. All rights reserved.

for drowsy driving accidents (Pack et al., 1995) and other industrial catastrophes (Dinges, 1995; Mitler et al., 1988). However, our innate circadian biology begins promoting wakefulness during the day. Performance impairments from a night without sleep actually decline somewhat by 6-10 p.m. (i.e., at 36-40 hours of being awake) relative to the peak for poor performance earlier in the day between 6 and 10 a.m. (i.e., 24-28 hours awake) (Van Dongen and Dinges, 2005). This circadian modulation of sleepiness and alertness was part of the justification for ACGME's settling on a 30-hour rather than a 24-hour extended duty period (ACGME, 2003, 2004), although some programs scheduled these extended duty periods in a non-circadian fashion (e.g., starting the 30 hours at noon).

EFFECTS OF ACUTE SLEEP DEPRIVATION ON HUMAN PERFORMANCE

Findings on the effects of 30-hour extended duty periods on the performance of physicians (Philibert, 2005) and the reduction in adult intensive care unit (ICU) medical errors when intern duty periods were limited to approximately 16 hours (Landrigan et al., 2004) received much attention in the medical community. These results, however, were not surprising given data that have accumulated over the past 100 years about the effects of sleep deprivation on attention, memory, and a range of cognitive functions (Dinges and Kribbs, 1991; Durmer and Dinges, 2005; Harrison and Horne, 2000; Kleitman, 1963; Patrick and Gilbert, 1896). There is a substantial scientific literature on the cognitive and functional deficits induced when healthy (non-physician) adult volunteers remain awake for 24 to 40 hours (Durmer and Dinges, 2005; Harrison and Horne, 2000; IOM, 2006; Philibert, 2005). In addition, neuroimaging studies have confirmed that a night without sleep results in changes in brain functions that are associated with unstable and inaccurate performance on a wide range of cognitive tasks including attention, working memory, and executive functions such as problem solving and decision making (Bell-McGinty et al., 2004; Chee and Choo, 2004; Chee et al., 2006, 2008; Chuah et al., 2006; Drummond et al., 1999, 2000, 2005; Habeck et al., 2004; Lim et al., 2007; Portas et al., 1998; Thomas et al., 2000; Wu et al., 2006).

Although the majority of healthy adults exposed to wakefulness extended from 16-18 hours to 24-30 hours experience performance-impairing fatigue, there are substantial differences among individuals in the onset and magnitude of cognitive changes induced by a night without sleep (Doran et al., 2001; Leproult et al., 2003; Van Dongen et al., 2004). There is as yet no reliable objective biomarker for differential vulnerability to the effects of sleep loss, although a recent study suggests one possible genetic candidate (Groeger et al., 2008; Viola et al., 2007). Approaches designed to help individuals become aware of their own personal vulnerability to the cognitive effects of sleep loss, combined with information on how to prevent these effects, could form the basis of a more personalized fatigue management system (Dinges, 2004; Van Dongen et al., 2007). ACGME is encouraged to look into developing these systems based on the approaches developed in regulated transportation industries for their applicability to residency.

ACUTE SLEEP DEPRIVATION AND RESIDENT PERFORMANCE

Current ACGME duty hours set an upper limit on duty hours of 24 hours with an additional 6 hours to allow adequate time for patient follow-up, didactic learning, and patient handovers (ACGME, 2003, 2004). ACGME proposed this 30-hour work limit (also referred to as a "long call" schedule or an extended duty period [Knauth, 2007]) in order "to address the effects of acute sleep loss" (ACGME, 2003). However, this limit does not adequately protect against acute sleep loss (Dinges, 2005; Philibert, 2005). An ACGME meta-analysis of 60 studies on the effects of sleep deprivation in 959 physicians found that "Sleep loss of less than 30 hours reduced physicians' overall performance by nearly 1 standard deviation and clinical performance by more than 1.5 standard deviations" (Philibert, 2005, p. 1392). These very large effect sizes on a large sample of resident physicians leave little doubt that acute total sleep loss of 30 hours diminishes resident performance. Thus, it was concluded that allowing residents to stay awake for 30 hours on duty "may not completely guard against the negative effect of sleep loss on cognitive and clinical performance" (Philibert, 2005, p. 1392).

Both realistic patient simulator studies (Howard et al., 2003) and field studies of residents working extended duty periods (24 or more hours) have often found performance deficits post-call relative to pre-call (Eastridge et al., 2003; Friedman et al., 1971; Kahol et al., 2008; Leonard et al., 1998). However, some studies have not found such deficits (e.g., Jakubowicz et al., 2005; Jensen et al., 2004; Uchal et al., 2005). It is not possible to determine whether the latter studies had inadequate power to detect statistically significant differences as a function of work time. Since all of these field studies vary greatly in the rigor of their study designs and methods, what factors specifically contributed to the different outcomes cannot be determined. As in the more well-designed study by Landrigan and colleagues (2004), factors other than work duration per se (e.g., differences in degree of sleep deprivation) may have contributed to those findings in which extended duty hours reduced resident performance. The prospective controlled study of Landrigan and colleagues suggests that performance deficits associated with extended duty periods (24 + 6 hours) could adversely affect patient safety (Landrigan et al., 2004), and that the increased sleep time resulting from elimination of these long on-call duty periods might have helped to decrease attentional failures stemming from sleep loss during overnight work hours (Lockley et al., 2004). In contrast, retrospective studies, such as one by Ellman (2005), have not shown differences in patient outcomes.

The findings of the meta-analysis conducted by ACGME on the effects of sleep deprivation on physicians (Philibert, 2005) are consistent with previous reviews of the adverse effects of work periods beyond 24 hours on resident performance (Asken and Raham, 1983; Veasey et al., 2002), and with the report on the beneficial effects of limiting ICU interns' continuous work periods to "approximately 16 hours" (Landrigan et al., 2004, p. 1839). Furthermore, a nationwide web-based survey of 2,737 interns found that extended work duration was associated with an increased risk of percutaneous injuries to interns (Ayas et al., 2006), and the more frequently interns experienced an extended work period of 24-30 hours, the more fatigue-related errors they reported (Barger et al., 2006b; see also Gander et al., 2000). Thus, considerable evidence as reviewed in this chapter now exists to suggest that the 2003 ACGME extended duty hour limit for residents of 24 + 6 continuous hours (ACGME, 2003) is likely to result in increased risks (via performance errors) to both patients and residents than shorter-duration work periods.

The same 2,737 interns queried for information on fatigue-related errors (Barger et al., 2006b) and percutaneous injuries (Ayas et al., 2006) during extended duty periods were also asked about motor vehicle incidents and crashes. Sleep deprivation from extended shifts contributed to significantly elevated risks of motor vehicle crashes, near-miss incidents, and incidents involving involuntary sleep while driving home from the hospital after an extended duty period (post-call) compared to drives home after non-extended work periods. The odds ratio for sleep-deprived drivers was 2.3 for crashes and 5.9 for near-miss incidents (Barger et al., 2005). An earlier smaller survey of residents found comparable results (Marcus and Loughlin, 1996). The willingness of residents to drive when they are drowsy may be, in part, associated with the effects of sleep deprivation on judgment. Recent studies have found that sleep loss can result in greater risk taking (Killgore et al., 2006; McKenna et al., 2007; Roehrs et al., 2004; Venkatraman et al., 2007).

Collectively, the research on resident physicians indicates that sleep loss associated with having to stay awake for an extended period (up to 30 hours or more)—rather than the performance of work per se—is likely the primary reason that neurobehavioral and cognitive performance degrades during residents' extended duty periods. This conclusion is consistent with findings from laboratory studies of healthy adults showing that the adverse effects on cognitive performance of remaining awake 24 or more hours are found even when people are awake without working (Dinges and Kribbs, 1991; Harrison and Horne, 1999, 2000).

ACUTE SLEEP LOSS PLUS INEXPERIENCE IN FIRST-YEAR RESIDENTS

The seminal studies by the Harvard Work Hours, Health and Safety Group on resident duty hours in relation to both patient safety (Barger et al., 2006b; Landrigan et al., 2004) and resident safety (Ayas et al., 2006; Barger et al., 2005) focused on interns (first-year resident physicians) who are the least experienced and work more hours, and therefore are more prone to errors if not properly supervised. Interns are also more sleep deprived than other resident physicians. A national random sample survey that obtained data on sleep from 3,604 first-year (interns) and second-year residents during 1998-1999 found that interns reported obtaining significantly less sleep than second-year residents (Baldwin and Daugherty, 2004). Interns also had significantly more prolonged sleep deprivation and longer single periods without sleep than second-year residents. Additionally, this study found that residents averaging 5 or fewer hours of sleep per night were more likely to report serious accidents or injuries, conflict with other professional staff, use of alcohol, use of medications to stay awake, noticeable weight change, working in an impaired condition, and having made significant medical errors. Residents reporting more instances of inadequate supervision and occasions of working while impaired, as well as those who believed that they should have taken time off for illness but did not, also reported less sleep time and more sleep deprivation (Baldwin and Daugherty, 2004). While these results pre-date the current ACGME resident duty hour limits and consequently may not generalize to residents today, they are consistent with the more recent studies by the Harvard Work Hours, Health and Safety Group indicating that sleep deprivation in interns (the least experienced residents) poses an unsafe condition.

WORK DURATION AND RISK

Continuous time spent performing work (referred to as time on task) may also increase the risk of accidents, but this is less clearly understood and documented than the contribution of acute sleep deprivation to the risk of accidents. Reviews across industries of the relative risk of accidents as a function of work hour duration (with attempts to adjust for exposure) generally conclude that the risk of accidents can begin to increase as time working exceeds 8 hours, and especially when it exceeds 12 hours, although the increases in risk after 12 hours of work are not always consistent or large (Caruso et al., 2004; Knauth, 2007; Nachreiner, 2001).

There are very few data, however, to inform work-hour guidelines in health care. A retrospective analysis of 411 recorded medical staff exposures to biological fluid at a university hospital with an emergency medicine residency program found a statistically significant increase (after adjusting for the number of workers per exposure) in exposures during 9-12 hours on duty (Macias et al., 1996). Two studies involving a total of 895 U.S. hospital staff nurses found an increase in self-reported errors and near-errors when work shifts were extended to 12.5 hours or longer (Rogers et al., 2004; Scott et al., 2006), although a smaller (and likely underpowered) study of nurses in Japan who were allowed to nap while working reported that 16-hour night shifts did not result in greater fatigue or difficulties concentrating than 8-hour night shifts (Takahashi et al., 1999).

A report by the National Institute of Occupational Safety and Health on overtime and extended work shifts has concluded that factors other than simply work duration per se contribute to the relationship between work duration and risk (Caruso et al., 2004). These factors are similar to the list at the beginning of this chapter and include shift start time, total hours worked in a week, rotation of work shifts between day and night work, and workload (e.g., Macdonald and Bendak, 2000). Moreover, there is much less information on the effects of work durations beyond 12 hours (Caruso et al., 2004), prompting the National Occupational Research Agenda Long Work Hours Team to propose a framework for future studies of long work hours, "including determinants, outcomes, and moderating factors of long work hours, suggesting that studies need to include more clear and complete descriptions of work schedules, worker characteristics, and the work environment, and need to consider a wider range of possible health, safety, social and economic outcomes for workers, families, employers, and the community. Additional studies are needed on vulnerable employee groups and those critical to public safety. More studies are also needed to develop interventions and test their effectiveness" (Caruso et al., 2006, p. 930).

Maximal Hour Limits per Shift

The evidence reviewed above supports the conclusion that performance is compromised by remaining awake beyond 16 hours (i.e., acute sleep deprivation). Therefore the extended duty shifts (24 + 6 hours) permitted in the current ACGME resident duty hour limits (ACGME, 2003) promote conditions for fatigue-related errors that pose risks to both patients and residents (Ayas et al., 2006; Barger et al., 2006a; Landrigan et al., 2004; Lockley et al., 2007). Limiting continuous work time to 16 hours would reduce these risks. A 16-hour continuous work limit is also reasonable in light of studies that equate the effects on performance from being awake more than 16 hours to the effects of 0.05 to 0.10 percent blood alcohol concentration (Arnedt et al., 2005; Dawson and Reid, 1997; Lamond and Dawson, 1999; Williamson and Feyer, 2000). While 16 hours of continuous work reflects a clear limit relative to safety, there is no compelling evidence that risks to patient safety increase from 8 to 12 hours of work, but some data from nurses suggest that risk may increase after 12 hours of work, although the work-related factors that contribute to this risk are unknown (Bollschweiler et al., 2001; Rogers et al., 2004).

In conclusion, the reviews of safe work hour limits and sleep deprivation indicate that either extended duty periods must be eliminated to improve patient safety relative to resident performance (Landrigan et al., 2004, 2007), or if extended duty periods of 24 + 6 hours are to remain an essential feature of resident training, provision for sleep following 16 hours of work will be needed, before the extended work period continues to 30 hours. The mitigating effects of sleep are discussed below.

PREVENTION OF ACUTE SLEEP DEPRIVATION

This section reviews what is known about sleep obtained by residents during extended duty periods. As a fundamental biological function, sleep both stabilizes waking performance and enhances the ability to learn and remember (Huber et al., 2004; Lim and Dinges, 2008; Stickgold, 2005; Stickgold et al., 2000; Walker and Stickgold, 2006). Continued advances in neurobiology have identified circadian timing and homeostatic mechanisms in the brain (Fuller et al., 2006) that require sleep to be obtained *daily* in adequate quantity and quality to prevent the physiological and behavioral effects of sleep deprivation. Prevention of sleep deprivation in residents is regarded as among the most essential ways to manage fatigue and its risks (Buysse et al., 2003; Dawson and McCulloch, 2005; Gaba and Howard, 2002; Gabow et al., 2006; Horrocks et al., 2006; Landrigan et al., 2007; Parshuram, 2006; Veasey et al., 2002).

The committee observed that ensuring residents obtain adequate sleep during their scheduled workweeks is a feature currently missing in the ACGME duty hours and resident training culture. Reduced sleep periods are common in many residency programs (Baldwin and Daugherty, 2004). Since current efforts to educate residents about sleep and fatigue management are by themselves not sufficient to increase sleep durations (Arora et al., 2007), requirements for protected sleep periods should be a priority in any new ACGME duty hour limits. Residents should practice good sleep hygiene and learn the importance of avoiding fatigue-related errors by obtaining essential sleep both daily and weekly as a matter of professional responsibility.

Although reduction of resident duty hours alone is one way to achieve more sleep (Lockley et al., 2004, 2006, 2007), it is an indirect and inefficient way to increase sleep given the moderate correlation between resident work hours and sleep time (see Figure 7-1) (Baldwin and Daugherty, 2004; Lockley et al., 2004). Reducing work hours could limit the time available for educational training experiences of residents (Ludmerer and Johns,

2005), without necessarily increasing their sleep time. The most direct way to increase residents' sleep time—to prevent fatigue risks due to sleep deprivation, and if necessary, allow residents to stay for extended duty periods up to 30 hours for educational purposes—would be to specifically mandate and protect periods of sleep for residents during extended duty. This approach has a long history in fatigue management in operational scenarios from other industries.

Protected Sleep Periods to Counter Fatigue During Extended Duty

The use of limited sleep periods typically between 10 minutes and 3 hours duration (i.e., naps) has been studied in many non-medical work contexts to manage fatigue risks and prevent performance errors. Studies have evaluated the use of planned naps (also called power naps) prior to and during night work, as well as during extended work periods up to 30+ hours (Akerstedt et al., 1989; Bonnefond et al., 2004; Bonnet, 1991; Caldwell and Caldwell, 1998; Dinges, 1989, 1992; Dinges et al., 1987, 1988; Driskell and Mullen, 2005; Kubo et al., 2007; Naitoh, 1992; Naitoh and Angus, 1989; Rosekind et al., 1994, 1995, 1997; Schweitzer et al., 2006; Takeyama et al., 2005). Strategic use of naps and longer sleep periods has been advocated as a fatigue countermeasure for residents during extended duty shifts (Veasey et al., 2002), and planned naps have been studied as fatigue countermeasures for physicians and nurses working 12-hour night shifts (Smith-Coggins et al., 2006). The vast majority of studies have found that naps and longer sleep periods (e.g., 4-5 hours) can help mitigate some of the effects of fatigue during night shifts and extended duty periods, suggesting that naps and longer sleep periods may be a valuable countermeasure to fatigue experienced by residents.

Although some residents take ad hoc naps during 24 + 6 hours extended duty periods when work demands permit, napping during extended duty periods is not addressed by the current ACGME duty hours. Studies indicate that some residents can obtain 1-3 hours of sleep when they nap on call under the current ACGME duty hours. These naps are usually during the nighttime (i.e., between 9 p.m. and 8 a.m.) (Arora et al., 2006, 2007; Lockley et al., 2004; Marcus and Loughlin, 1996). The sleep obtained can vary considerably by the year and subspecialty of the resident (Gabow et al., 2006). Although naps of 1 to 3 hours mitigate some of the effects of sleep deprivation, longer periods of sleep (4-8 hours) afford greater benefits for cognitive performance, even if some subjective fatigue and sleepiness may persist from sleep inertia (Belenky et al., 2003; Driskell and Mullen, 2005; Jewett et al., 1999; Van Dongen et al., 2003).

Benefits of Protected Sleep Periods

Reviews of studies that examined the effects of planned and protected nap sleep ranging in duration from 10 minutes to 4 hours in healthy adults concluded that naps and protected sleep periods can mitigate the effects of sleep loss on cognitive performance and subjective fatigue (Dinges and Broughton, 1989; Stampi, 1992). The longer the sleep, the greater are the benefits for performance, and the longer the benefits last. Thus, a recent meta-analysis revealed that a 15-minute nap could benefit fatigue and performance for a few hours, while a 4-hour sleep period could benefit performance for up to 10 hours (Driskell and Mullen, 2005). Relative to no sleep or a brief nap, 4 hours of protected sleep obtained at night during an extended duty period after about 16 hours of work could reduce fatigue and improve cognitive performance during a terminal portion of up to 10 hours of the 30-hour extended duty period.

Use of Protected Sleep Time by Residents During Extended Duty Periods

Two studies have evaluated the feasibility of deploying a protected pager-free sleep period at night during extended duty periods. One study designed to assess the effects of an on-duty protected nighttime period for sleep on first-year residents' sleep and fatigue during 24 + 6 hour work periods provided supplementary night float coverage to interns from 12:00 a.m. to 7:00 a.m. The residents on extended duty could finish their work, forward their pagers to a night-float resident physician, and sleep while the other resident cared for their patients. This was compared to a 2-week period when no night-float physician was provided while residents were on call. The nights with protected sleep time and night-float physician coverage resulted in significantly longer sleep (mean sleep = 3.50 hours) and less subjective fatigue during call and post-call than the unprotected nights (mean sleep = 2.37 hours) ($p < .001$). Protected sleep time, or the minutes that an intern's pager was "covered" by a night-float resident, was significantly associated with increased on-call sleep duration ($r = 0.69$). Thus, for each hour that an intern was covered by a night-float physician, the intern received 42 additional minutes of sleep (Arora et al., 2006). Sleep efficiency, which indicates how consolidated (unfragmented) sleep was, also improved for interns with the nap intervention.

Since use of protected night sleep schedule was at the discretion of interns (i.e., not mandated), adherence was relatively low. The authors reported that "interns on the nap schedule forwarded their pagers to the night-float physician on only 22% of available opportunities. In lieu of forwarding their pagers, interns with the nap schedule preferred to forward the care of only their cross-cover patients and to retain care of their own patients" (Arora et al., 2006, p. 795). Interns indicated they preferred the nap schedule when they were on call because it improved sleep, decreased fatigue, and/or gave them a greater ability to focus on care for their own patients without the additional workload associated with cross-covering patients. In other words, interns recognized the advantages of the night float—protected sleep schedule for increased sleep time and reduced fatigue—but they tended not to use it because of concerns about their patients and discontinuity of care (i.e., potential for risks posed by two transitions of care—one at the start and one at the end of their protected sleep period) (Arora et al., 2006). This suggests that teaching residents to work with the night-float physician and nurses as a team, oriented toward patient safety, might help them achieve the protected sleep they need without concerns that they have to remain awake for the sake of their patients.

A study published a decade earlier on the effects of a 4-hour (2 a.m.-6 a.m.) protected sleep time during a 36-hour on-call period found that sleep obtained during the protected period averaged 3.54 hours, but this was not significantly different from the mean sleep of 3.74 hours during the unprotected nights (Richardson et al., 1996). There were also no differences in measures of alertness and performance, despite impressions of attending physicians and supervising residents that the night-float system improved the alertness of the interns (Richardson et al., 1996). However, sleep efficiency (a measure of the consolidation of sleep) and slow wave sleep were markedly increased on the protected sleep nights due to fewer interruptions, indicating that the sleep was likely more restorative. Thus protecting the night sleep period by turning over pager calls to a night-float senior resident resulted in less sleep fragmentation from interruptions—and increased its recuperative value (Bonnet and Arand, 2003; Levine et al., 1987; Wesensten et al., 1999). In fact, there is evidence that merely the *perceived risk* of sleep interruption (e.g., wearing a pager while attempting to sleep) can fragment sleep, even if no interruption actually occurs (Richardson et al., 1996; Torsvall and Akerstedt, 1988). The study by Richardson and colleagues (1996) also illustrates that the amount of time off for a resident to spend in bed for sleep does not equate to actual sleep acquired. Their interns spent just under an average of 5 hours (295.4 minutes) in bed, but the sleep they obtained averaged 3.67 hours (220.1 minutes)—this is 75 percent sleep efficiency (i.e., total sleep time divided by total time in bed), which is the only estimate of the proportion of time residents could be expected to sleep during a protected nocturnal 5-hour sleep period in the hospital.

Thus, studies of a protected 4- to 7-hour sleep period during an extended duty schedule (Arora et al., 2006; Richardson et al., 1996) that allowed interns to decide whether they wanted to sleep during the protected night nap schedule found that adherence to the protected sleep period schedule was lower than expected. "On only 56 percent of the nights during which coverage was provided did interns make full use of the available time for sleep, i.e. by spending at least 4 hours in bed trying to sleep. The reasons cited in the diaries and in subsequent interviews suggested that with a guarantee of time for sleep at a later point, the covered interns preferred to catch up on incomplete work" (Richardson et al., 1996, p. 724).

The protected sleep schedule investigation of Richardson and colleagues (1996) used a less robust (between-subjects) study design and longer duty hours (36 hours) than the more recent report by Arora and colleagues (2006), which used a within-subjects design and the current ACGME duty hours (30 hours). Nevertheless, the two investigations support the conclusion that providing a protected nighttime period of 4-7 hours for sleep during an extended duty period (on-call schedule) can result in sleep that is less fragmented and somewhat longer than the 1-3 hours typically reported during residents' extended duty periods (Arora et al., 2006, 2007; Lockley et al., 2004; Marcus and Loughlin, 1996). Since less fragmented, longer duration of sleep is more recuperative (Bonnet and Arand, 2003; Wesensten et al., 1999), protected nocturnal sleep periods during extended work schedules offer a means to prevent the severe fatigue-related performance deficits that can occur when remaining awake for 24-30 hours (Landrigan et al., 2004; Philibert, 2005).

IMPROVING ADHERENCE TO USE OF PROTECTED SLEEP PERIODS

While a protected nighttime sleep of up to 4-5 hours duration appears feasible as a way to prevent acute sleep deprivation in resident physicians during an extended duty period, the limited data available indicate that adherence to such a schedule was relatively poor (22-56 percent). Thus, resident adherence to a protected nocturnal sleep during a 30-hour extended duty period remains a major challenge, especially if they are given the option of deciding when to use the night-float residents or other mechanisms for coverage and obtain at least 4 hours of sleep. Resident unwillingness to obtain sleep when a protected period with pager handoff to others is available appears to be due to their concerns about patients they admitted, continuity of care (i.e., increased hand-offs), and their own workload (Arora et al., 2006; Richardson et al., 1996). Other factors, such as availability of sleep-conducive environments, staff willingness to not interrupt the resident's sleep, and attending physicians' encouragement of interns and other residents' use of the protected nighttime sleep opportunity may also have played a role in adherence in these studies.

The barriers to resident use of protected nocturnal sleep periods during extended work periods could be overcome (1) by requiring residents to use the protected sleep periods for sleep and nothing else (especially if they are being paid and are required to remain in the hospital and work after the rest period), (2) by using more experienced physicians to cover patients for them so residents will have less cause for concern about their patients, (3) by sequestering residents in sleep-conducive environments, (4) by better management of resident workload, and (5) by ensuring that all hospital personnel avoid disrupting the resident's sleep. A negative aspect of a protected sleep period at night is the possibility of sleep inertia, which refers to a period of grogginess and cognitive performance deficits immediately after awakening from naps and abbreviated sleep periods (Tassi and Muzet, 2000). Ironically, sleep inertia is more severe when sleep is more consolidated (i.e., less fragmented) and deeper, the type of sleep which is associated with better recuperation (Dinges, 1990; Dinges et al., 1985). Sleep inertia usually is overcome within 15-30 minutes by physical activity, environmental and social stimulation, and caffeine (Van Dongen et al., 2001). Once sleep inertia dissipates, alertness and cognitive performance can return to levels that reveal the benefits of the sleep for performance (Dinges, 1990; Dinges et al., 1988; Driskell and Mullen, 2005; Jewett et al., 1999). If 4-5 hours of sleep are obtained in the protected nocturnal period, improved alertness and performance generally will remain for the final 9-10 hours of a 30-hour extended duty period, thus reducing fatigue-related risks, improving learning while providing follow-up care, allowing attendance at didactic sessions, and improving the quality of handovers.

PREVENTION OF CHRONIC SLEEP DEPRIVATION

Chronic sleep deprivation occurs when the quantity and quality of sleep being obtained across days is insufficient to prevent daytime sleepiness, elevated sleep propensity, cognitive deficits, and other neurobehavioral problems (e.g., drowsy driving) produced by repeated days of inadequate recovery sleep (Dinges et al., 2005). Residents are most likely to experience chronic sleep deprivation when their daily recovery sleep opportunities are restricted to 7 hours or less in duration (Howard et al., 2002), regardless of the reason for the sleep restriction (e.g., unable to sleep during the daytime following a night shift, inadequate time off from scheduled work, moonlighting, non-work activities that reduce sleep time). The following section contains studies of healthy adults, but not specifically resident populations.

Total Daily Sleep Restriction and Cognitive Performance

Chronic sleep restriction can result in cumulative fatigue manifesting as reduced alertness and cognitive deficits in a sleep dose-response manner (Banks and Dinges, 2007; Dinges et al., 2005). Experimental studies of healthy adults chronically restricted to less than 8 hours time in bed at night

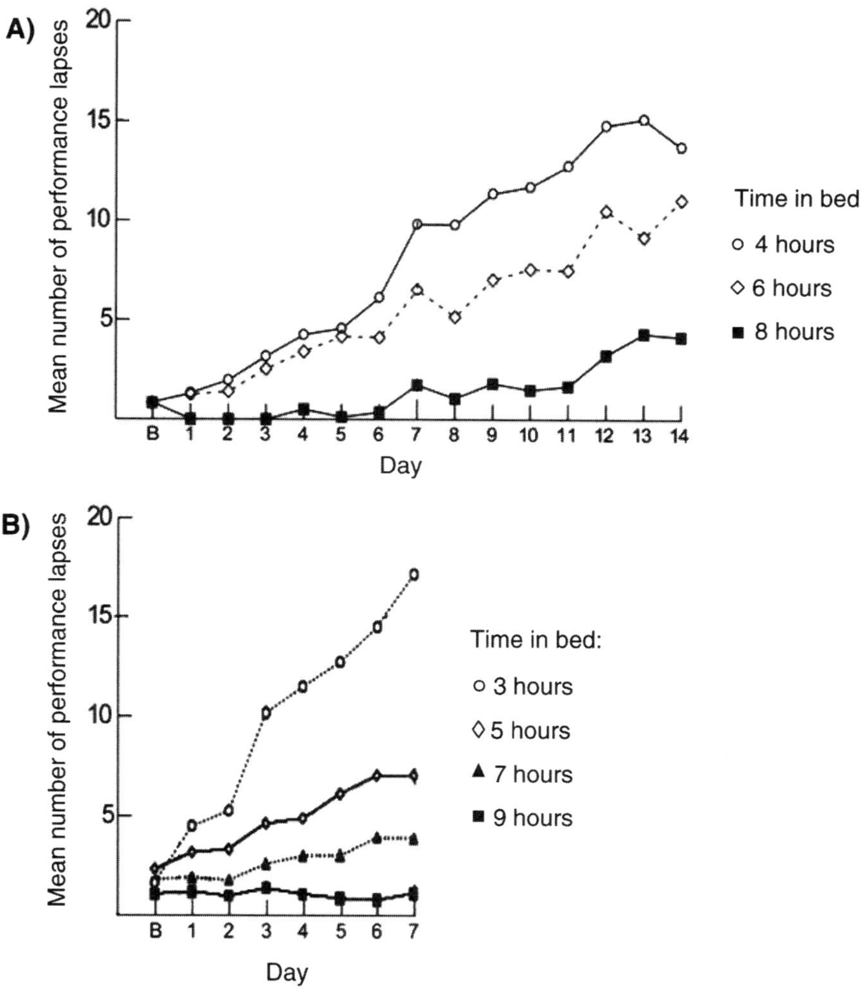

FIGURE 7-2 Repeated nights of sleep loss result in cumulative cognitive impairment. Higher number of performance lapses indicate poorer performance and more unstable alertness.
NOTE: B on the x-axis = baseline day.
SOURCE: IOM, 2006, based on (A) Van Dongen et al., 2003, and (B) Belenky et al., 2003.

reveal statistically reliable near-linear cumulative increases in cognitive deficits across days of sleep restriction (see Figure 7-2) (Belenky, 2003; Dinges et al., 1997; Van Dongen et al., 2003). The magnitudes of cognitive deficits found during sleep restriction were sleep dose-dependent between 3 hours and 7 hours per night, both within and between subjects. The less sleep provided each day, the more rapidly performance deficits accumulated over days (Belenky et al., 2003; Van Dongen et al., 2003). By the seventh day of nightly sleep restriction to between 4 and 6 hours time in bed, deficits in attention and unstable alertness (Doran et al., 2001; Dorrian et al., 2005) due to inadequate sleep were substantial. They reached levels equivalent to those found following one night of total sleep deprivation. Between days 9 and 12 of such restriction, performance deficits were equivalent to those found after two consecutive nights without any sleep (Van Dongen et al., 2003). These analyses took into account inter-individual differences in response to sleep loss to ensure the effects were not due solely to the sample populations' containing more vulnerable individuals.

A recent large-scale experimental study of chronic sleep restriction using split-sleep schedules (i.e., nighttime sleep plus a daytime nap) also found that reductions in total daily sleep resulted in a near-linear accumulation of impairment regardless of whether sleep was scheduled as a consolidated nocturnal sleep period or split into a nocturnal anchor sleep period and a daytime nap (Mollicone et al., 2008). Thus, the three largest sleep-dose response experiments published to date indicate that cumulative cognitive performance deficits occur in response to chronic restriction of sleep to less than 7-8 hours' time in bed for sleep. This occurs even if subjects' ratings and self-reports indicate they believe that are not very sleepy or fatigued (Banks and Dinges, 2007; Belenky et al., 2003; Dinges et al., 2005; Mollicone et al., 2008; Van Dongen et al., 2003). It is important to remember that time in bed does not equate to hours of sleep; hours of actual sleep will be less (Belenky et al., 2003; Richardson et al., 1996; Van Dongen et al., 2003).

APPROACHES TO PREVENT CHRONIC SLEEP LOSS

Resident duty hours should protect against chronic sleep restriction over consecutive days and weeks of work. A number of work-hour factors are important for the prevention of cumulative sleep deprivation including provision of (1) adequate sleep time to recover after each work day, (2) additional sleep time to recover from extended duty periods (24 + 6 hours), (3) sleep time to recover from repeated days of restricted sleep due to night shift work, (4) sleep time to recover from work hours that result in sleep restriction due to averaging hours over weeks, and (5) off-duty periods with sufficient time for sleep to recover and still have time for personal (quality of life) activities. In the following section the adequacy of the current ACGME duty hour limits is assessed relative to the prevention of chronic sleep deprivation through these means.

Current ACGME Limits Relative to Chronic Sleep Deprivation

Current ACGME duty hours set an upper limit of 80 hours of work per week averaged over 4 weeks (this limit includes all in-house moonlighting) "to safeguard against the negative effects of chronic sleep loss" (ACGME, 2003, 2004). Because the 80 hours is an average, it permits weeks in which work hours could be in excess of 90 hours (Lockley et al., 2004). ACGME has tried to modulate wide swings in weekly hours by saying that the average should be within a rotation and not across rotations (e.g., not allowing a 40-hour ambulatory care rotation to be averaged with a 100-hour surgical one).

The consequences of having people work 80 hours a week on average have not been extensively studied. However, a report by the National Institute for Occupational Safety and Health concluded that "a pattern of deteriorating performance on psychophysiological tests as well as injuries while working long hours was observed across study findings, particularly with very long shifts and when 12-hour shifts combined with more than 40 hours of work a week" (Caruso et al., 2004, p. iv). Not every study, however, found that 84-hour workweeks adversely affect fatigue, performance, and health more than 40-hour workweeks do (e.g., studies of construction workers working 12 hours a day for 7 consecutive days) (Persson et al., 2003, 2006), although more carefully designed studies are needed on long work hours (Caruso et al., 2006). Nevertheless, the reduction of duty hours for ICU interns from a mean of 84.9 hours per week to a mean of 65.4 hours per week was associated with significant reductions in resident medical errors and potentially patient harm (Landrigan et al., 2004). However, as noted previously, it is not known if these benefits were the result of reductions in work duration, increased sleep time, or both, and what role concomitant factors—such as time of day of work—had in the results.

Obtaining Daily Sleep of Sufficient Duration

There are several reasons why the current ACGME limit of 80 hours per week may not protect against chronic sleep loss. Permitting the 80 hours to be an average over 4 weeks provides training flexibility within and between subspecialties, but it also results in the possibility of weeks in which residents can work more than 80 hours. In addition to the potential safety risks posed by overly long work hours (Caruso et al., 2004), the ACGME duty hours stipulate that "adequate time for rest and personal activities must be provided. This should consist of a 10-hour period provided between all daily duty periods after in-house call" (ACGME, 2003). The 10 hours off duty each day for recovery from acute fatigue is consistent with what is federally mandated for most transportation industries.

As noted above, scientific evidence indicates that the recovery potential of sleep in healthy individuals depends heavily on the duration of sleep—whether it is acquired in one continuous period or in a split-sleep period (Banks and Dinges, 2007; Belenky et al., 2003; Dinges et al., 2005; Mollicone et al., 2008; Van Dongen et al., 2003). Residents must have an adequate period for obtaining at least 7-8 hours of sleep per day to avoid chronic sleep restriction. Although studies have not addressed this, it is likely that for residents off duty outside the hospital for 10 hours, sleep will occupy no more than 80 to 95 percent (i.e., 8.0-9.5 hours) of the 10-hour nocturnal period (Ohayon et al., 2004), which leaves 0.5-2.0 hours for residents to transit to and from the hospital, eat, and have time for personal hygiene, domestic activities, and psychosocial requirements.

If the sleep period for residents is during the daytime (following night work), sleep time will be reduced by 2-4 hours due to circadian interface with daytime sleep as well as sleep disturbance from environmental factors such as noise and light (Akerstedt, 2003). If the off-duty period is less than 10 hours (which is possible under the current ACGME rules, which state "should provide" rather than "must provide" 10 hours off), recovery sleep time is reduced and the resident's cumulative sleep debt grows. As a result, a failure to mandate and enforce the 10-hour "adequate rest" rule poses a challenge to the prevention of chronic sleep restriction and its consequences for cumulative performance impairments in residents, even more so after night shifts and overnight call than day shifts.

There is evidence that the current 80-hour work limit and 10-hour rest advisory are associated with reduced average daily sleep times. The intervention study by Landrigan and colleagues (2004) that restricted the work of interns to approximately 16 hours and eliminated extended duty periods (24 hours or more), reduced weekly hours from an average of 84.9 hours to an average of 65.4 hours (mean decrease in weekly work time of 19.5 hours), and increased sleep time from an average of 45.9 hours to an average of 51.7 hours (mean increase in weekly sleep time of 5.8 hours) (Lockley et al., 2004). Thus, restriction of weekly work hours from 80+ hours to less than 80 hours resulted in a daily average sleep time increase from a mean of 6.6 ± 0.8 hours per day to a mean of 7.4 ± 0.9 hours per day (Lockley et al., 2004). In terms of chronic sleep deprivation, this mean increase of 0.8 hours (48 minutes) more sleep per day is important, because it moves the daily sleep average above the minimum threshold of 7 hours per night that experimental studies have found are needed to prevent cognitive deficits from chronic sleep restriction (see Figure 7-2) (Belenky et al., 2003; Van Dongen et al., 2003). It is also consistent with survey findings of 3,604 residents taken before the ACGME duty hours went into effect, showing that the less sleep residents reported obtaining, the more they reported feeling sleep deprived. When sleep was less than 5 hours per night, they were significantly more likely to report serious accidents or injuries, working in an impaired condition, and having made significant medical errors (Baldwin and Daugherty, 2004).

Although reducing the weekly work hours of ICU interns in the Landrigan et al. study (2004) resulted in more sleep on average being obtained by the interns, there was, as noted previously, only a moderate correlation between work hours and sleep time (e.g., 33 percent of the variance in common) (Lockley et al., 2004). A similar outcome was obtained from the national random sample survey of 3,604 first- and second-year residents in the 1998-1999 training year (Baldwin and Daugherty, 2004). Figure 7-1 shows the relationship of resident sleep time to work hours found in both of these studies.

Although it is clear from these figures that work-hour reductions can result in increased sleep time, there is considerable variability in resident sleep time that remains unexplained, whether comparing a large cohort of residents from multiple subspecialties (Baldwin and Daugherty, 2004) or the same individuals within the same subspecialty and in the same hospital (Lockley et al., 2004). For example, in the study by Lockley and colleagues (2004), about one-third (7/20) of the interns averaged less than 7 hours of reported sleep a night when work hours were reduced during the study (see Figure 7-1B). Thus factors other than just work hours influence resident sleep time. The committee believes that identification of these factors, and ensuring that resident sleep time is protected and optimized, offer important avenues for prevention of fatigue-related resident errors and their risks to patient and resident safety.

RECOVERY SLEEP

Recovery Sleep Following Extended Duty Hours

The current ACGME rules have no minimum off-duty requirement that ensures residents obtain adequate recovery sleep following an extended duty period (24 + 6 hours) in the hospital. Instead, ACGME says only that time for rest and personal activities must be provided and should be 10 hours, which is identical to what is suggested following any work period with a shorter length (ACGME, 2003). However, the limited experimental data that exist regarding the amount of sleep needed to recover from an extended period of wakefulness involving a night of sleep deprivation suggests that 9 hours of time in bed *at night* is needed for recovery sleep. Six hours of time in bed does not result in full recovery even after five consecutive nights of 6 hours' time in bed (Jay et al., 2007; Lamond et al., 2007).

In light of these data and the common experience of people needing to sleep longer after a period of total or partial sleep deprivation, resident fatigue could be reduced following extended duty periods (24 + 6 hours) if the mandatory time off duty is greater than the 10 hours suggested by ACGME. It may not be realistic to expect that residents will be able to obtain 9 hours recovery sleep in a 10-hour off-duty period, due to the time required by other non-work-related waking activities as well as the influence of circadian biology on sleep duration (Czeisler et al., 1980; Strogatz et al., 1986). Residents may end a 30-hour extended-duty work period in midday, then due to circadian influences over sleep propensity remain awake until the evening near habitual bedtime (Lavie, 1986; Strogatz et al., 1987). The addition of a 5-hour break for a 4-hour protected sleep period at night in the hospital (with a night float covering for the resident) during the extended (24 + 6 hours) duty period will not change these recovery sleep dynamics. The committee believes that a recovery period of at least 14 hours is necessary following the end of an extended duty period (that includes a 5-hour protected nighttime sleep). This recovery period should include a prohibition on starting work again before 6 a.m. the next morning. The combination of a minimum of 14 hours off duty and a start time of not before 6 a.m. the next day will ensure that residents can acquire at least 8-9 hours recovery sleep *during the nocturnal period* after an extended duty period in the hospital.

Recovery Sleep Following Night Shifts

As noted at the beginning of this chapter, a number of factors interact to influence the relationship of work and fatigue. The underlying assumption that cumulative work time is the predominant factor is not always substantiated. The time of day at which work takes place—especially work during the night shift when sleep propensity is highest—has proven to be one of the factors that can be more important to the risk of fatigue than the duration of work (Caruso et al., 2006; Feyer and Williamson, 1995).

Circadian biology has a profound effect on the night shift worker by increasing fatigue and decreasing performance at night during the work period and decreasing sleep duration during the daytime (Akerstedt, 2003; Rosa, 2001), and there are substantial inter-individual differences in the severity of these responses (Sack et al., 2007; Van Dongen, 2006). As a result, repeated, consecutive night shifts are associated with a growing sleep debt (due to daytime circadian restriction of sleep by 2-4 hours below sleep obtained at night), and with reduced productivity and increasing night-shift errors (Akerstedt, 2003; Folkard et al., 2005; Rosa, 2001). Studies of residents experiencing night shifts find results consistent with what has been observed in other areas. Daytime sleep duration after night shifts was significantly shorter than nighttime sleep duration following day shifts, and residents had decreased mood and alertness during night shifts compared to during day shifts (Cavallo et al., 2002, 2003).

A review of the risks posed by repeated day shifts versus repeated night shifts concluded that, on average, the risk of an incident on the night shift was more than double that on the day shift, and the risk increased progressively with each consecutive night shift up to four nights (Folkard et al., 2005). Thus, there is evidence that accident risk increases markedly over successive night shifts, but much less so over successive morning or day shifts. The committee notes therefore that using night shifts in place of extended duty periods (24 + 6 hours) will necessitate strategies to prevent the risks posed by fatigue at night and reduced daytime sleep.

Night shifts pose a performance risk to residents, and possibly a risk to patient safety, that could be prevented and mitigated by use of countermeasures (e.g., prophylactic naps prior to night-shift work) (Bonnefond et al., 2004; Bonnet et al., 2005; Horrocks et al., 2006; Knauth and Hornberger, 2003), as well as adequate time for recovery sleep following each night shift. ACGME duty hours currently have no special provisions for the duration of night-shift work—which is typically 12 hours—or for the duration of off-duty recovery time for sleep after night shifts (except for the 10 hours of "adequate rest" requirement). Permitting residents a 12-hour recovery period (rather than 10 hours) between consecutive night shifts would help ensure they either avoid or minimize any chronic sleep debt from cumulative sleep restriction across consecutive night shifts. It would do so by allowing more time for them to obtain both a sleep period in the morning immediately post night shift, and a prophylactic nap before the next nights.

There is evidence that night shifts that include backwards rotation in time (i.e., start earlier each consecutive day) are more fatiguing and disruptive of sleep, alertness, and performance than are nights shifts that begin at the same time each day or later each day (Akerstedt, 2003; Driscoll et al., 2007; Rosa, 2001). Since 12-hour night shifts are common in resident work schedules, a 12-hour off-duty period between night shifts would be appropriate to maintain a circadian work-rest schedule and avoid backward rotation between consecutive night shifts. Therefore both the need for sleep and synchronization with circadian rhythms would benefit from allowing residents a 12-hour off-duty recovery period following each night shift. The committee notes that such a schedule could present a challenge to the scheduling of an overlap of shifts to allow for handovers, which are discussed in the next chapter, unless the complementary day shift is longer than 12 hours or other members of the team have schedules that overlapped between shifts.

The scientific literature indicates that night-shift work also requires more days off for recovery due to the repeated challenge it poses to the circadian system and the sleep restriction it engenders, which occurs during the typical four to five nights worked by residents. A review of recovery following different work schedules suggested that as many as 3-4 days may be necessary for recovery after night-shift work due to the disturbances of normal circadian rhythms (Akerstedt et al., 2000). Therefore, recovery from the cumulative sleep loss and circadian perturbation of consecutive night shifts requires at least two nocturnal sleep periods when sleep can naturally occur uninterrupted and extend up to at least 9 hours of time in bed in order for recovery to occur (Jay et al., 2007; Lamond et al., 2007).

Recovery Sleep Following Six Consecutive Days of Work

ACGME duty hours currently require 1 day off in 7, averaged over a 4-week period. This averaging can result in residents working for 14 or even 21 consecutive days to get a consolidated 2-day weekend off. On weeks when residents do not get a day off from work, there is no possibility for ad libitum recovery sleep and little time for non-professional activities other than sleep. Without a day off in 7, as permitted by averaging days off, residents have a greater likelihood of chronic sleep restriction that can progress to levels of severe performance impairment. An experiment in healthy adults revealed that restriction of nightly sleep to between 4 and 6 hours resulted in cumulative performance deficits that reached levels equivalent to 48-64 hours of total sleep deprivation after 9 consecutive days without a day off for extra recovery sleep (Van Dongen et al., 2003). These analyses took into account inter-individual differences in response to sleep loss to ensure that the effects were not due solely to more vulnerable individuals. The findings are shown in Figure 7-2A. They suggest that prolonged work hours without a day off weekly can lead to substantial performance risks.

A review of recovery from work on various schedules concluded that those who work long shifts require more frequent days off for recovery (Akerstedt et al., 2000). Recent studies of recovery from chronic sleep restriction indicate that at least 1 full recovery day (24 hours) for ad libitum sleep time is needed to ensure that performance deficits do not become excessive and pose a risk to performance when work schedules that induce sleep restriction exceed a 6-day period (Belenky et al., 2003; Van Dongen et al., 2003). This can only be accomplished in residents if the 1 day a week is guaranteed in duty hour limits.

To ensure that residents do not continue to accumulate fatigue and performance deficits from chronic sleep restriction due to a single ad libitum recovery day each week, duty hour limits should also ensure that residents have at least one 48-hour period free of duty each month. This will reduce the practice of residents trading work days to consolidate 2 days off in a single weekend.

ADJUSTMENTS TO THE 2003 RESIDENT DUTY HOUR LIMITS

After reviewing the literature on work duration, sleep, and human performance in this chapter, the committee concludes that there are specific adjustments to the current ACGME resident duty hours that would help prevent and reduce resident fatigue, and thereby enhance resident performance and reduce risks to both patients and residents. The major rationales behind the recommendations are the following: (1) work duration should be limited because human performance degrades after 16 hours whether one is working or not; (2) sufficient time for sleep should be incorporated into daily and weekly work schedules to prevent acute and chronic sleep deprivation, respectively, and allow recovery from accumulated sleep debt; and (3) when extended duty periods are considered an essential aspect of resident training and continuity of care, a protected sleep period should be provided during the extended duty period. These changes will reduce the chances of residents providing patient care when their performance may be less than optimal due to fatigue.

The committee has concluded that the best way to prevent the problem of resident fatigue was to ensure adequate sleep is obtained by residents, and this can be done without changing the overall total of allowable hours of work in a week, retaining its putative training value. Hence the recommendations below, derived from the evidence in this chapter, are focused more on providing predictable and protected time for recovery sleep than on work hours per se. Appendix B provides sample monthly schedules for an individual resident comparing the application of current ACGME duty hour limits and the committee's recommendations that illustrate enhanced regularity of days off and protected sleep during extended duty periods under the latter.

Focusing on resident off-duty time (for sleep), more so than on-duty time for work, to prevent fatigue-related risks is a novel way to view the necessary balance between the need for long resident duty hours inherent in their intensive training, and the goal of preventing fatigue as a condition of risk. New scientific evidence on the biological causes of fatigue (e.g., sleep need, circadian rhythms) has guided the committee's recommendations (below) to focus on the need to enhance recovery sleep opportunities for residents, something emphasized in many of the reviews on how to reduce resident fatigue (Baldwin and Daugherty, 2004; Buysse et al., 2003; Cavallo and Mallory, 2004; Gaba and Howard, 2002; Landrigan et al., 2007; Lockley et al., 2006; Parshuram, 2006; Veasey et al., 2002). The committee emphasizes that it will be necessary to educate both residents and faculty on the need for taking advantage of these sleep opportunities as a patient safety and quality-of-care issue.

Recommendation 7-1: ACGME should adopt and enforce requirements for residency training that adhere to the following principles: duty hour limits and schedules should promote the prevention of sleep loss and fatigue; additional measures should mitigate fatigue when it is unavoidable (e.g., during night work and extended duty periods); and schedules should provide for predictable, protected, and sufficient uninterrupted recovery sleep to relieve acute and chronic sleep loss, promote resident well-being, and balance learning requirements. Programs should design resident schedules using the following parameters:

- **Duty hours must not exceed 80 per week, averaged over 4 weeks.**
- **Scheduled continuous duty periods must not exceed 16 hours unless a 5-hour uninterrupted continuous sleep period is provided between 10 p.m. and 8 a.m. This period must be free from all work and call, and used by the resident for sleep in a safe and sleep-conducive environment. The 5-hour period for sleep must count toward total weekly duty hour limits. Following the protected sleep period, a resident may continue the extended duty period up to a total of 30 hours, including any previous work time and the sleep period.**
- **Residents should not admit new patients after 16 hours during an extended duty period.**
- **Extended duty periods (e.g., 30 hours that include a protected 5-hour sleep period) must not be more frequent than every third night with *no* averaging.**
- **After completing duty periods, residents must be allowed a continuous off-duty interval of**
  - **A minimum of 10 hours following a daytime duty period that is not part of an extended duty period,**
  - **A minimum of 12 hours following a night float or night shift work that is not part of an extended duty period, and**
  - **A minimum of 14 hours following an extended duty period, and residents should not return to service earlier than 6 a.m. the next day.**
- **Night-float or night-shift duty must not exceed four consecutive nights and must be followed by a minimum of 48 continuous hours off duty after three or four consecutive nights.**
- **At least one 24-hour off-duty period must be provided per 7-day period without averaging; one additional (consecutive) 24-hour period off duty must be provided to ensure at least one continuous 48-hour period off duty per month.**

- In exceptional circumstances requiring the resident's physical presence to ensure patient safety or to engage in a critical learning opportunity, program faculty may permit, but not require, residents to remain on duty beyond the scheduled time; programs must record for ACGME review the nature of each exception allowed. These exceptions are not to become routine practice. Residency Review Committees should determine at the time of program re-accreditation whether the documented exceptions to scheduled duty hours warrant citation.
- The ACGME should develop criteria for granting individual programs waivers from one or more of the above scheduling parameters; such criteria should be formulated *only* to accommodate rare, well-documented circumstances in which patient safety and/or educational requirements of specific programs outweigh the advantages of full compliance with the committee's recommendations and cannot be addressed by means other than the requested waiver(s); programs that are granted waivers (if any) and the nature of those waivers should be posted on the public access portion of the ACGME website. Included in the application for waiver should be a long-term plan that articulates how the program will work to avoid a permanent need for the requested waiver. All waivers should be monitored and reviewed on an annual basis to determine suitability for renewal.
- Programs should provide annual formal education for residents and staff on the adverse effects of sleep loss and fatigue and on the importance of and means for their prevention and mitigation.
- Sponsoring institutions and programs should ensure that their practices promote and ensure that residents take the required sleep during extended duty periods.

ADDITIONAL CONSIDERATIONS UNDERPINNING RECOMMENDATION 7-1

As noted above, the intent of the committee's recommendations for changes to resident duty hours was to prevent fatigue when possible and to provide measures to relieve both acute and chronic sleep deprivation, recognizing that some fatigue may be inevitable when attempting to provide service in the hospital 24 hours a day, 7 days a week. Because of the diversity of specialty education needs, program sizes, and patient populations, the committee believes some flexibility in duty hour rules is needed for programs to design their own resident training schedules within certain limits supported by the evidence in this chapter.

Maximum Hours of Work per Week

The committee has retained the current 80-hour per week duty hour maximum averaged over 4 weeks (Table 7-1) rather than reduce it or eliminate averaging. This preserves flexibility for each specialty and program site to have what they determine are sufficient hours to achieve their learning goals just as they have now under the 2003 rules. As is currently the case, the committee does not believe all specialties and rotations will require this lengthy workweek. Programs with certain constraints related to number of residents or patient characteristics may require longer hours than other programs. While the 80-hour limit does not have an empirical evidence base demonstrating that it is better or worse for educational outcomes or patient safety than other limits (e.g., 72 or 56 hours), it has been in place for nearly 20 years in some specialties and New York State, and the rule has been in place nationally for 5 years. Board certification results are just beginning to be available nationally across specialties to monitor the achievement of residents (see Chapter 4). Without additional documentation of the actual hours worked by different residency programs and their outcomes, the committee was reluctant to reduce the 80-hour framework. Furthermore reducing hours of work could limit the time for education and training experiences of residents without resulting in increased hours of sleep since, as noted earlier in this chapter, reducing total work hours alone is an inefficient and indirect way to increase sleep time (Baldwin and Dougherty, 2004; Lockley et al., 2004; Ludmerer and Johns, 2005). Studies on the effects of long workweeks are limited in health care and other industries (Landrigan et al., 2004; Persson et al., 2003, 2006), and when performance improves with a shorter workweek, it is not clear if it is due to hours of service or hours of additional sleep obtained.

Any of the Residency Review Committees, which set educational standards for each specialty in conjunction with ACGME, may choose to create more restrictive duty hour limits if it considers changes to be necessary for its particular circumstances, such as the severity of patient cases and the constancy of high-intensity work. For example, this has been done in emergency medicine, which limits shift length to 12 hours, totaling 60 hours per week, plus 12 hours for education (ACGME, 2007b); the committee does not recommend any change in the hours for emergency medicine.

Maximum Shift Length

Although the scientific evidence base establishes that human performance begins to deteriorate after 16 hours of wakefulness, the committee did not believe that limiting all shifts to a maximum of 16 hours would leave sufficient flexibility to address the educational needs of all specialties.

TABLE 7-1 Comparison of IOM Committee Adjustments to Current ACGME Duty Hour Limits

	2003 ACGME Duty Hour Limits	IOM Recommendation
Maximum hours of work per week	80 hours, averaged over 4 weeks	No change
Maximum shift length	30 hours (admitting patients up to 24 hours then 6 additional hours for transitional and educational activities)	• 30 hours (admitting patients for up to 16 hours, plus 5-hour protected sleep period between 10 p.m. and 8 a.m. with the remaining hours for transition and educational activities) • 16 hours with no protected sleep period
Maximum in-hospital on-call frequency	Every third night, on average	Every third night, no averaging
Minimum time off between scheduled shifts	10 hours after shift length	• 10 hours after day shift • 12 hours after night shift • 14 hours after any extended duty period of 30 hours and not return until 6 a.m. of next day
Maximum frequency of in-hospital night shifts	Not addressed	4 night maximum; 48 hours off after 3 or 4 nights of consecutive duty
Mandatory time off duty	• 4 days off per month • 1 day (24 hours) off per week, averaged over 4 weeks	• 5 days off per month • 1 day (24 hours) off per week, no averaging • One 48-hour period off per month
Moonlighting	Internal moonlighting is counted against 80-hour weekly limit	• Internal and external moonlighting is counted against 80-hour weekly limit • All other duty hour limits apply to moonlighting in combination with scheduled work
Limit on hours for exceptions	88 hours for select programs with a sound educational rationale	No change
Emergency room limits	12-hour shift limit, at least an equivalent period of time off between shifts; 60-hour workweek with additional 12 hours for education	No change

So extended duty periods of up to 30 hours, the current limit, are allowed with the inclusion of a 5-hour sleep period to address acute sleep deprivation (Table 7-1). The committee anticipates that its new recommendations regarding sleep within extended duty periods and the need for protected and predictable off-duty times for sleep will result in residents being less likely to work in excess of 80 hours per week because the frequency of 30-hour duty period may no longer be averaged (Table 7-1), and when those duty periods occur, a 5-hour period for sleep is incorporated (e.g., see schedules B-1a, B-1b, B-2a, and B-2b in Appendix B as examples).

The committee is not mandating that duty periods be 16 hours, but rather is setting 16 hours as the upper limit for continuous work without a protected sleep period. Only after a 5-hour sleep period can work continue. ACGME currently suggests a 10-hour separation between shifts of any length regardless of whether the work occurs in the day or night time. The committee recognizes that this required 10 hours off between daytime work periods will likely result in routine shift lengths shorter than 16 hours (e.g., 10 to 14 hours per day).

Although under the proposed changes 14-hour days could routinely be scheduled along with the 10-hour interval between shifts to achieve a schedule less than 80 hours a week, the committee is cognizant that this may be a draining schedule whether every day is 14 hours or 14-hour days are in combination with extended duty periods. Bertrand Bell, in developing the rationale behind the 80-hour week for New York State commented, "It is reasonable for residents to work a 10-hour day for 5 days a week. It is humane for people to work every fourth night" (Bell, 2003, p. 40). Samples of schedules that might be drawn up following the committee's recommendations are compared with schedules following current ACGME duty hour limits (see Appendix B); other variations are possible.

Maximum Time Off Between Scheduled Shifts

The committee believes 10 hours is a minimum rest break for residents to receive adequate sleep after daytime work periods and recommends that this become mandatory. Variable amounts of time 10, 12, and 14 hours are recommended as the minimum time off after day, night, and extended duty periods (Table 7-1), respectively, as guided by the evidence for the periods of recovery necessary depending on the time of work during the day and our circadian biology. The committee decided on a requirement of 14 hours off after an extended duty period coupled with a requirement that a resident not return to service before 6 a.m. on the following day. This combination was selected to allow flexibility in the start and ending times of extended duty periods (e.g., 7 a.m. to 1 p.m. on the next day, or 10 a.m. on day 1 until 4 p.m. on day 2) yet ensure that residents are able to have a nocturnal period of sleep before returning to work after an extended duty period. As noted earlier, it is estimated that a person needs 9 hours of time in bed at night to gain sufficient recovery sleep after being up the previous night (Jay et al., 2007; Lamond et al., 2007).

Maximum In-Hospital Frequency of Night Shift

The committee also recommends that 48 hours be given off after three or four consecutive night shifts to provide opportunities for sleep to offset the increasing performance deficits associated with multiple nights of work (Akerstedt, 2003; Folkard et al, 2005; Rosa, 2001). The ACGME does not currently limit the number of consecutive night shifts (Table 7-1).

Mandatory Time Off Duty

The committee recommends 5 rather than 4 days off per month and eliminating averaging of the days off per week to ensure sufficient time weekly for recovery sleep (Table 7-1). Elimination of averaging days off will help prevent schedules where residents may go 2 or 3 weeks without a day off, whether scheduled by the program director or due to residents trading days to have multiple consecutive days off. Offering a fifth day per month that creates a consolidated 48-hour period further enhances recovery from sleep debt and gives residents 2 consecutive days off. The committee recognizes that scheduling a Saturday-Sunday break for every resident, although it may be preferable for resident personal life, may not always be possible, but encourages programs to have at least one of the days always be a weekend day (e.g., Friday-Saturday or Sunday-Monday) so that residents are available when other family and friends are off work to provide some work-life balance.

Protected Sleep Period During Extended Duty Period

The committee debated the best course for continuity of patient care, educational purposes, and addressing fatigue when deliberating on whether to maintain the 30-hour extended duty period. It also debated the appropriate length of a rest period that would be observed by residents yet maximally address acute fatigue so the residents would perform well for the duration of the duty period. A 5-hour period with the expectation of a 4-hour period of sleep best fit the evidence on the amount of sleep required that would improve performance for the balance of the duty period (Driskell and Mullen, 2005). Shorter nap periods, such as 2 hours, provide some relief but are inadequate. Longer periods (e.g., 7-10 hours), while desirable appeared impractical for two reasons: (1) there would be too many sleep hours counted against duty hour limits, and (2) residents would prefer to and would be inclined to leave the hospital to sleep in their own beds for such a long break, driving home while overtired and not obtaining sufficient sleep before they returned regardless of how the official schedule is drawn. Thus, the committee recommends incorporating a 5-hour sleep period in any duty period over 16 hours and recommends that this sleep period be counted as part of total duty hours.

Given a 5-hour protected sleep period, the committee understands that a resident would only likely obtain up to about 4 hours of sleep (e.g., Arora et al., 2006; Richardson et al., 1996). However a benefit of a mandated rather than optional sleep period is that it is predictable and protected and will improve performance for the balance of the extended duty period (Driskell and Mullen, 2005). The committee recognizes that some residents have in the past preferred to use the protected time to catch up on paperwork (Arora et al., 2006). Consequently it both recommends training for residents in sleep hygiene, so that they will understand the importance of that sleep break, and reductions in resident workloads to make them more manageable within reduced duty hours. Workload limits are discussed in more detail in Chapter 3, but the committee also recommends here confining admissions to the first 16 hours that a resident is on duty during an extended duty period. The extended duty period is one whose limit is frequently violated (see Chapter 2) and limiting admissions will help residents complete their work before the end of their shift.

Educating Residents and Faculty on Benefits of Sleep

The committee does not endorse any specific educational program on the consequences of sleep loss and fatigue management but agrees that residents, their supervisors, and institutional leaders all need to be aware of the patient and resident safety implications of acute and chronic sleep deprivation and act on solutions that will work in their training environment. Both AHRQ and ACGME recommend that residents receive sleep education (ACGME, 2003; Jha et al., 2001), and the education should emphasize obtaining sound sleep and preventing sleep loss as primary tactics for reducing the likelihood of fatigue and its risks (e.g., Horrocks et al., 2006). ACGME might advance the spread of best practices in sleep education by developing or endorsing existing educational material (e.g., high-quality materials developed by individual residency programs), Duke University LIFE Curriculum (Learning to Address Impairment and Fatigue to Enhance Patient Safety), or SAFER—Sleep Alertness, and Fatigue Education in Residency (SAFER Task Force, 2007). A number of articles have emphasized that changing residency culture that perceives long hours as the mark of dedication and professionalism will require more than a single education session, just as it requires more than duty hour regulation (Arora et al., 2008; Owens et al., 2008; Parthasarathy et al., 2007).

Limit on Admitting Patients During Extended Duty Period

Current ACGME rules allow extended duty periods of 30 hours with no provision for sleep, but they do limit when residents may admit patients to the first 24 hours. The balance of the 30 hours under current practice is 6 hours for educational purposes, completion of patient care duties, and handovers of care responsibilities. The 30-hour duty period is frequently violated, with residents often citing insufficient time to complete their workload, patient care responsibilities, and educational requirements. The committee now recommends that residents on an extended duty period only admit patients during the first 16 hours before they take the 5-hour sleep period and then continue on until the end of the 30 hours; the committee believes that allowing residents to admit patients only in the first 16 hours will help them complete their work and permit them to leave the hospital within the time constraints of a 30-hour duty period. Caring for fewer patients would also provide more time for an in-depth evaluation of each patient which has educational advantages as well. Even after the sleep period, residents would now have up to 9 hours to obtain needed test results on these new admissions and hence would have a greater likelihood of receiving the necessary information to make management decisions about a patient they admitted before handing the patient over to another team member. Having a mandatory sleep period will necessitate that either a night float, hospitalist service, or some other mechanism be available to admit new patients and cover the sleeping residents' patients after residents reach 16 hours on duty.

Adherence to Recommendations and Opportunities for Innovation

In Chapter 2 the committee recommended that ACGME continue to monitor duty hours. The ACGME in employing the substantial compliance model should be firmly intolerant of any systematic duty hour violations. The committee expects that there will not be routine violations, either scheduled or ad hoc, of duty hours. This expectation applies whether it involves a duty period of 10, 16, or 30 hours, or any other variation. The committee expects that there will be rare occasions when the stability of the patient is in question or an exceptional learning opportunity will present itself, and that the application of duty hour rules should be flexible enough to accommodate these instances. However, the committee believes these occasions should be well documented at the local level, reviewed for reasonable cause by the ACGME, and the type and number of exceptions made public by ACGME on its website to determine if these instances can provide guidance for future duty hour adjustments in general and on a specialty specific basis. Having the frequency and the reasons for exceptions available will increase transparency and help address concerns about conflict of interest in the ACGME monitoring process (CIR/SEIU, 2007). Consequently, residency programs must document all such incidents and provide explanations of the "reasonable" violations. Additionally, it is also expected that unusual learning opportunities will mean that supervision is present during those times to mitigate resident fatigue.

ACGME currently also has a process for allowing broader exemptions to duty hour rules for specific programs and for purposes of innovation. ACGME should publish criteria for granting such exemptions, which should be rare, and methods for monitoring them closely for possible increased risks to patient or resident safety. ACGME should list on its website which programs have exemptions and the reasons for the exemptions. The committee encourages the development of innovative projects for improving patient and resident safety including those involving scheduling. These projects must have a rigorous data collection and analysis framework that will advance learning and safety under different duty hour scenarios beyond what is currently known, given that the committee has offered recommendations rooted in a strong evidence base from the human performance and sleep literature. Currently the ACGME is testing scheduling approaches and limits, including different napping strategies during extended duty periods, limiting the separation between shifts to 8 hours, and investigating whether duty hour limits should apply to residents in their last year of training.[1] The committee encourages ACGME and the respective Residency Review Committees to document why they need to continue 30-hour duty periods as it was the most contentious part of the duty hour debate before the committee.

Implementation of all of the committee's duty hour recommendations should include a national evaluation of the following: the changes individual programs make; the extent to which the recommended changes to ACGME duty hour limits actually result in increased sleep for residents and maintenance of alertness and performance during work; the costs of implemented changes; the effect on labor supply and patient coverage; and which specific schedules with protected sleep programs more or less effectively promote sleep and alertness. ACGME and other stakeholders should foster research studies across multiple institutions to examine the effects of duty hour changes and practices (e.g., improved handovers) on (1) serious medical errors and preventable adverse events, (2) resident safety, and

[1]Personal communication, I. Philibert, Accreditation Council for Graduate Medical Education, May 2, 2008.

(3) resident educational outcomes and well-being. In the interest of both resident and patient safety, it will also be valuable and cost effective to learn from research sponsored by the National Institutes of Health, the National Aeronautics Space Administration, the Department of Defense, and other agencies about ways to determine objectively when residents are impaired by fatigue, since perception of one's own fatigue is often inaccurate (Van Dongen et al., 2003).

RESIDENT MOONLIGHTING

There is no recent national assessment of the degree to which residents and fellows are moonlighting. Anecdotal reports say that duty hour cutbacks created their own demand for moonlighting within teaching hospitals to cover services around the clock;[2] however, this would apply primarily to residents beyond the first year. Two studies, one institution-specific and one specialty-specific, conflict on whether moonlighting by residents has increased or remained at the same level since the 2003 limits (Cull et al., 2006; Dola et al., 2006). Currently, ACGME only requires that "in-house" or internal moonlighting for patient care be considered part of the 80-hour weekly limit on duty hours; that moonlighting requires prospective, written permission from the program director; and that resident performance be monitored to ensure no adverse effects that may lead to withdrawal of permission (ACGME, 2003, 2007a). Despite requirements for advance permission, these processes are not always followed; leading to the conclusion that more residents may be moonlighting than their supervisors know (Cull et al., 2006). At this time ACGME Residency Review Committees or individual programs may add additional requirements such as forbidding moonlighting during on-call months or forbidding moonlighting altogether. The Federation of State Medical Boards and legislative proposals have advocated more restrictions on moonlighting (Federation of State Medical Boards, 2007; GovTrack.us, 2005).

The committee concludes that all moonlighting for patient care, whether at the training facility (internal moonlighting) or elsewhere (external moonlighting), should come within the 80-hour weekly limit and that all other duty hour parameters should apply. Currently the 80-hour limit applies only to internal moonlighting. Having this provision in the resident contract (if the program chooses to permit moonlighting) gives a clearer expectation that the resident's primary duty is to the limits set by his or her training program and to patient safety. Residents have a responsibility to be ready for work and not take on too many additional obligations if these activities could interfere with their capacity to learn and to provide safe patient care.

---

[2]Personal communication, David Meltzer, University of Chicago, August 12, 2008.

Additionally, sponsoring institutions must not require that residents and fellows engage in moonlighting to cover cutbacks in hours on other services. So that moonlighting does not undermine the intent of limiting duty hours, the committee recommends the following.

Recommendation 7-2: The ACGME should immediately amend its current requirements on moonlighting by

- **Requiring that any internal and external moonlighting for patient care adhere to the duty hour limits listed above (e.g., 80 hours and all other limits), even if the program has received an exception to schedule longer hours, and**
- **Requiring that sponsoring institutions, if they choose to permit moonlighting, include provisions in resident contracts that (1) a resident must request prospective, written permission from the program director for moonlighting; and (2) resident performance will be monitored to ensure that there is no adverse effect of moonlighting on resident performance.**

REFERENCES

ACGME (Accreditation Council for Graduate Medical Education). 2003. *Common program requirements for duty hours.* Chicago, IL: ACGME.

———. 2004. *Report of the workgroup on resident duty hours and the learning environment.* Chicago, IL: ACGME.

———. 2007a. *ACGME institutional requirements: Checklist, effective: July 1, 2007.* http://www.acgme.org/acWebsite/irc/IRCheckList-07.pdf (accessed February 28, 2008).

———. 2007b. *ACGME program requirements for graduate medical education in emergency medicine.* Chicago, IL: ACGME.

Akerstedt, T. 2003. Shift work and disturbed sleep/wakefulness. *Occupational Medicine* 53(2):89-94.

Akerstedt, T., L. Torsvall, and M. Gillberg. 1989. Shift work and napping. In *Sleep and alertness: Chronobiological, behavioral and medical aspects of napping*, edited by D. F. Dinges and R. J. Broughton. New York: Raven Press. Pp. 205-220.

Akerstedt, T., G. Kecklund, M. Gillberg, A. Lowden, and J. Axelsson. 2000. Sleepiness and days of recovery. *Transportation Research Part F* 3:251-261.

Arnedt, J. T., J. Owens, M. Crouch, J. Stahl, and M. A. Carskadon. 2005. Neurobehavioral performance of residents after heavy night call vs after alcohol ingestion. *JAMA* 294(9):1025-1033.

Arora, V., C. Dunphy, V. Y. Chang, F. Ahmad, H. J. Humphrey, and D. Meltzer. 2006. The effects of on-duty napping on intern sleep time and fatigue. *Annals of Internal Medicine* 144(11):792-798.

Arora, V. M., E. Georgitis, J. N. Woodruff, H. J. Humphrey, and D. Meltzer. 2007. Improving sleep hygiene of medical interns—Can the sleep, alertness, and fatigue education in residency program help? *Archives of Internal Medicine* 167(16):1738-1744.

———. 2008. Improving sleep hygiene—Reply. *Archives of Internal Medicine* 168(11):1230.

Asken, M. J., and D. C. Raham. 1983. Resident performance and sleep-deprivation—A review. *Journal of Medical Education* 58(5):382-388.

Ayas, N. T., L. K. Barger, B. E. Cade, D. M. Hashimoto, B. Rosner, J. W. Cronin, F. E. Speizer, and C. A. Czeisler. 2006. Extended work duration and the risk of self-reported percutaneous injuries in interns. *JAMA* 296(9):1055-1062.

Baldwin, D. C., Jr., and S. R. Daugherty. 2004. Sleep deprivation and fatigue in residency training: Results of a national survey of first- and second-year residents. *Sleep* 27(2):217-223.

Baldwin, D. C., Jr., S. R. Daugherty, R. Tsai, and M. J. Scotti. 2003. A national survey of residents' self-reported work hours: Thinking beyond specialty. *Academic Medicine* 78(11):1154-1163.

Banks, S., and D. F. Dinges. 2007. Behavioral and physiological consequences of sleep restriction. *Journal of Clinical Sleep Medicine* 3(5):519-528.

Barger, L. K., B. E. Cade, N. T. Ayas, J. W. Cronin, B. Rosner, F. E. Speizer, and C. A. Czeisler. 2005. Extended work shifts and the risk of motor vehicle crashes among interns. *New England Journal of Medicine* 352(2):125-134.

Barger, L. K., N. T. Ayas, B. E. Cade, J. W. Cronin, B. Rosner, F. E. Speizer, and C. A. Czeisler. 2006a. Impact of extended-duration shifts on medical errors, adverse events, and attentional failures. *PLoS Medicine* 3(12):e487.

———. 2006b. Impact of extended-duration shifts on medical errors, adverse events, and attentional failures. *PLoS Medicine* 3(12):2440-2448.

Belenky, G., N. J. Wesensten, D. R. Thorne, M. L. Thomas, H. C. Sing, D. P. Redmond, M. B. Russo, and T. J. Balkin. 2003. Patterns of performance degradation and restoration during sleep restriction and subsequent recovery: A sleep dose-response study. *Journal of Sleep Research* 12:1-12.

Bell, B. M. 2003. Reconsideration of the New York state laws rationalizing the supervision and the working conditions of residents. *Einstein Journal of Biological Medicine* 20(1):36-40.

Bell-McGinty, S., C. Habeck, H. J. Hilton, B. Rakitin, N. Scarmeas, E. Zarahn, J. Flynn, R. DeLaPaz, R. Basner, and Y. Stern. 2004. Identification and differential vulnerability of a neural network in sleep deprivation. *Cerebral Cortex* 14(5):496-502.

Bollschweiler, E., A. Krings, K. H. Fuchs, G. Pistorius, T. Bein, U. Otto, E. Muhl, U. Backes-Gellner, and A. H. Hölscher. 2001. Alternative shift models and the quality of patient care. An empirical study in surgical intensive care units. *Langenbeck's Archives of Surgery* 386:104-109.

Bonnefond, A., P. Tassi, J. Roge, and A. Muzet. 2004. A critical review of techniques aiming at enhancing and sustaining worker's alertness during the night shift. *Industrial Health* 42(1):1-14.

Bonnet, M. H. 1991. The effect of varying prophylactic naps on performance, alertness and mood throughout a 52-hour continuous operation. *Sleep* 14(4):307-315.

Bonnet, M. H., and D. L. Arand. 2003. Clinical effects of sleep fragmentation versus sleep deprivation. *Sleep Medicine Reviews* 7(4):297-310.

Bonnet, M. H., T. J. Balkin, D. F. Dinges, T. Roehrs, N. L. Rogers, and N. J. Wesensten. 2005. The use of stimulants to modify performance during sleep loss: A review by the sleep deprivation and stimulant task force of the American Academy of Sleep Medicine. *Sleep* 28(9):1163-1187.

Buysse, D. J., B. Barzansky, D. Dinges, E. Hogan, C. E. Hunt, J. Owens, M. Rosekind, R. Rosen, F. Simon, S. Veasey, and F. Wiest. 2003. Sleep, fatigue, and medical training: Setting an agenda for optimal learning and patient care. *Sleep* 26(2):218-225.

Caldwell, J. A., Jr., and J. L. Caldwell. 1998. Comparison of the effects of zolpidem-induced prophylactic naps to placebo naps and forced rest periods in prolonged work schedules. *Sleep* 21(1):79-90.

Caruso, C. C., E. M. Hitchcock, R. B. Dick, J. M. Russo, and J. M. Schmit. 2004. *Overtime and extended work shifts: Recent findings on illnesses, injuries, and health behaviors.* Cincinnati, OH: National Institute for Occupational Safety and Health, Centers for Disease Control and Prevention, U.S. Department of Health and Human Services.

Caruso, C. C., T. Bushnell, D. Eggerth, A. Heitmann, B. Kojola, K. Newman, R. R. Rosa, S. L. Sauter, and B. Vila. 2006. Long working hours, safety, and health: Toward a national research agenda. *American Journal of Industrial Medicine* 49(11):930-942.

Cavallo, A., and M. L. Mallory. 2004. Sleep deprivation, residency training, and ACGME rules: Practical guidelines. *Journal of Pediatrics* 145(6):717-718.

Cavallo, A., J. Jaskiewicz, and M. D. Ris. 2002. Impact of night-float rotation on sleep, mood, and alertness: The resident's perception. *Chronobiology International* 19(5):893-902.

Cavallo, A., M. D. Ris, and P. Succop. 2003. The night float paradigm to decrease sleep deprivation: Good solution or a new problem? *Ergonomics* 46(7):653-663.

Chee, M. W. L., and W. C. Choo. 2004. Functional imaging of working memory after 24 hr of total sleep deprivation. *Journal of Neuroscience* 24(19):4560-4567.

Chee, M. W. L., L. Y. M. Chuah, V. Venkatraman, W. Y. Chan, P. Philip, and D. F. Dinges. 2006. Functional imaging of working memory following normal sleep and after 24 and 35 h of sleep deprivation: Correlations of fronto-parietal activation with performance. *Neuroimage* 31(1):419-428.

Chee, M. W. L., C. H. Tan, Z. Hui, S. Parimal, S. D. Weissman, V. Zagorodnov, and D. F. Dinges. 2008. Lapsing during sleep deprivation is associated with distributed changes in brain activation. *Journal of Neuroscience* 28(21):5519-5528.

Chuah, L. Y. M., V. Venkatraman, D. F. Dinges, and M. W. L. Chee. 2006. The neural basis of inter-individual variability in inhibitory efficiency following sleep deprivation. *Journal of Neuroscience* 26(27):7156-7162.

CIR/SEIU (Committee of Interns and Residents-Service Employees International Union) Healthcare. 2007. Presentation by L. Toni Lewis to the Committee on Optimizing Graduate Medical Trainee (Resident) Hours and Work Schedules to Improve Patient Safety, December 3, 2007, Washington, DC.

Cull, W. L., H. J. Mulvey, E. A. Jewett, E. L. Zalneraitis, C. E. Allen, and R. J. Pan. 2006. Pediatric residency duty hours before and after limitations. *Pediatrics* 118(6):e1805-e1811.

Czeisler, C., E. D. Weitzman, M. C. Moore-Ede, J. C. Zimmerman, and R. S. Knauer. 1980. Human sleep—Its duration and organization depend on its circadian phase. *Science* 210(4475):1264-1267.

Dawson, D., and K. McCulloch. 2005. Managing fatigue: It's about sleep. *Sleep Medicine Reviews* 9(5):365-380.

Dawson, D., and K. Reid. 1997. Fatigue, alcohol and performance impairment. *Nature* 388:235.

Dinges, D. F. 1989. Napping. In *Neuroscience year*, edited by G. Adelman. Boston: Birkhäuser. Pp. 113-114.

———. 1990. Are you awake? Cognitive performance and reverie during the hypnopompic state. In *Sleep and cognition*, edited by R. R. Bootzin, J. F. Kihlstrom, and D. L. Schacter. Washington, DC: American Psychological Association. Pp. 159-175.

———. 1992. Adult napping and its effects on ability to function. In *Why we nap. Evolution, chronobiology, and functions of polyphasic and ultrashort sleep*, edited by C. Stampi. Boston: Birkhäuser. Pp. 119-134.

———. 1995. An overview of sleepiness and accidents. *Journal of Sleep Research* 4(Suppl 2):4-14.

———. 2001. Stress, fatigue, and behavioral energy. *Nutrition Reviews* 59(1):S30-S32.
———. 2004. Critical research issues in development of biomathematical models of fatigue and performance. *Aviation Space and Environmental Medicine* 75(3):A181-A191.
———. 2005. Sleep deprivation, impairment of resident performance, and medical error. *Journal of Bone and Joint Surgery* 87a(11):2581-2582.
Dinges, D. F., and R. J. Broughton, eds. 1989. *Sleep and alertness: Chronobiological, behavioral and medical aspects of napping*. New York: Raven Press.
Dinges, D. F., and N. B. Kribbs. 1991. Performing while sleepy: Effects of experimentally-induced sleepiness. In *Sleep, sleepiness and performance*, edited by T. Monk. Chichester: John Wiley & Sons. Pp. 97-128.
Dinges, D. F., M. T. Orne, and E. C. Orne. 1985. Assessing performance upon abrupt awakening from naps during quasi-continuous operations. *Behavior Research Methods, Instruments and Computers* 17(1):37-45.
Dinges, D. F., M. T. Orne, W. G. Whitehouse, and E. C. Orne. 1987. Temporal placement of a nap for alertness: Contributions of circadian phase and prior wakefulness. *Sleep* 10(4):313-329.
Dinges, D. F., W. G. Whitehouse, E. Carota Orne, and M. T. Orne. 1988. The benefits of a nap during prolonged work and wakefulness. *Work and Stress* 2(2):139-153.
Dinges, D. F., F. Pack, K. Williams, K. A. Gillen, J. W. Powell, G. E. Ott, C. Aptowicz, and A. I. Pack. 1997. Cumulative sleepiness, mood disturbance, and psychomotor vigilance performance decrements during a week of sleep restricted to 4-5 hours per night. *Sleep* 20(4):267-277.
Dinges, D. F., N. L. Rogers, and M. D. Baynard. 2005. Chronic sleep deprivation. In *Principles and practice of sleep medicine*, 4th ed., edited by M. H. Kryger, T. Roth, and W. C. Dement. Philadelphia, PA: W. B. Saunders. Pp. 67-76.
Dola, C., L. Nelson, J. Lauterbach, S. Degefu, and G. Pridjian. 2006. Eighty hour work reform: Faculty and resident perceptions. *American Journal of Obstetrics & Gynecology* 195(5):1450-1456.
Doran, S. M., H. P. Van Dongen, and D. F. Dinges. 2001. Sustained attention performance during sleep deprivation: Evidence of state instability. *Archives Italiennes de Biologie* 139(3):253-267.
Dorrian, J., N. L. Rogers, and D. F. Dinges. 2005. Psychomotor vigilance performance: A neurocognitive assay sensitive to sleep loss. In *Sleep deprivation: Clinical issues, pharmacology and sleep loss effects*, edited by C. Kushida. New York: Marcel Dekker. Pp. 39-70.
Drake, C. L., T. Roehrs, G. Richardson, J. K. Walsh, and T. Roth. 2004. Shift work sleep disorder: Prevalence and consequences beyond that of symptomatic day workers. *Sleep* 27(8):1453-1462.
Driscoll, T. R., R. R. Grunstein, and N. L. Rogers. 2007. A systematic review of the neurobehavioural and physiological effects of shiftwork systems. *Sleep Medicine Reviews* 11(3):179-194.
Driskell, J. E., and B. Mullen. 2005. The efficacy of naps as a fatigue countermeasure: A meta-analytic integration. *Human Factors* 47(2):360-377.
Drummond, S. P. A., G. G. Brown, J. L. Stricker, R. B. Buxton, E. C. Wong, and J. C. Gillin. 1999. Sleep deprivation-induced reduction in cortical functional response to serial subtraction. *NeuroReport* 10:3745-3748.
Drummond, S. P., G. G. Brown, J. C. Gillin, J. L. Stricker, E. C. Wong, and R. B. Buxton. 2000. Altered brain response to verbal learning following sleep deprivation. *Nature* 403(6770):655-657.
Drummond, S. P. A., A. Bischoff-Grethe, D. F. Dinges, L. Ayalon, S. C. Mednick, and M. J. Meloy. 2005. The neural basis of the psychomotor vigilance task. *Sleep* 28(9):1059-1068.

Durmer, J. S., and D. F. Dinges. 2005. Neurocognitive consequences of sleep deprivation. *Seminars in Neurology* 25(1):117-129.

Eastridge, B. J., E. C. Hamilton, G. E. O'Keefe, R. V. Rege, R. J. Valentine, D. J. Jones, S. Tesfay, and E. R. Thal. 2003. Effect of sleep deprivation on the performance of simulated laparoscopic surgical skill. *American Journal of Surgery* 186(2):169-174.

Ellman, P. I., I. L. Kron, J. S. Alvis, C. Tache-Leon, T. S. Maxey, T. B. Reece, B. B. Peeler, J. A. Kern, and C. G. Tribble. 2005. Acute sleep deprivation in the thoracic surgical resident does not affect operative outcomes. *Annals of Thoracic Surgery* 80(1):60-65.

Federation of State Medical Boards. 2007. *State-specific requirements for initial medical licensure.* http://fsmb.org/usmle_eliinitial.html (accessed March 17, 2008).

Feyer, A. M., and A. M. Williamson. 1995. Work and rest in the long-distance road transport industry in Australia. *Work and Stress* 9(2-3):198-205.

Folkard, S., D. A. Lombardi, and P. T. Tucker. 2005. Shiftwork: Safety, sleepiness and sleep. *Industrial Health* 43(1):20-23.

Friedman, R. C., J. T. Bigger, and D. S. Kornfield. 1971. The intern and sleep loss. *New England Journal of Medicine* 285:201-203.

Fuller, P. M., J. J. Gooley, and C. B. Saper. 2006. Neurobiology of the sleep-wake cycle: Sleep architecture, circadian regulation, and regulatory feedback. *Journal of Biological Rhythms* 21(6):482-493.

Gaba, D. M., and S. K. Howard. 2002. Patient safety: Fatigue among clinicians and the safety of patients. *New England Journal of Medicine* 347(16):1249-1255.

Gabow, P. A., A. Karkhanis, A. Knight, P. Dixon, S. Eisert, and R. K. Albert. 2006. Observations of residents' work activities for 24 consecutive hours: Implications for workflow redesign. *Academic Medicine* 81(8):766-775.

Gander, P. H., A. Merry, M. M. Millar, and J. Weller. 2000. Hours of work and fatigue-related error: A survey of New Zealand anaesthetists. *Anaesthesia Intensive Care* 28(2):178-183.

GovTrack.us. 2005. *S. 1297—109th congress: Patient and Physician Safety and Protection Act of 2005.* http://www.govtrack.us/congress/bill.xpd?bill=h109-1297&tab=summary (accessed March 29, 2008).

Groeger, J. A., A. U. Viola, J. C. Y. Lo, M. Von Schantz, S. N. Archer, and D. J. Dijk. 2008. Early morning executive functioning during sleep deprivation is compromised by a PERIOD3 polymorphism. *Sleep* 31(8):1159-1167.

Habeck, C., B. C. Rakitin, J. Moeller, N. Scarmeas, E. Zarahn, T. Brown, and Y. Stern. 2004. An event-related fMRI study of the neurobehavioral impact of sleep deprivation on performance of a delayed-match-to-sample task. *Cognitive Brain Research* 18(3):306-321.

Harrison, Y., and J. A. Horne. 1999. One night of sleep loss impairs innovative thinking and flexible decision making. *Organic Behavioral Human Decision Processes* 78(2):128-145.

———. 2000. The impact of sleep deprivation on decision making: A review. *Journal of Experimental Psychology: Applied* 6(3):236-249.

Horrocks, N., R. Pounder, and R. C. P. Working Group. 2006. Working the night shift: Preparation, survival and recovery—A guide for junior doctors. *Clinical Medicine* 6(1):61-67.

Howard, S. K., D. M. Gaba, M. R. Rosekind, and V. P. Zarcone. 2002. The risks and implications of excessive daytime sleepiness in resident physicians. *Academic Medicine* 77(10):1019-1025.

Howard, S. K., D. M. Gaba, B. E. Smith, M. B. Weinger, C. Herndon, S. Keshavacharya, and M. R. Rosekind. 2003. Simulation study of rested versus sleep-deprived anesthesiologists. *Anesthesiology* 98(6):1345-1355.

Huber, R., M. F. Ghilardi, M. Massimini, and G. Tononi. 2004. Local sleep and learning. *Nature* 430(6995):78-81.

IOM (Institute of Medicine). 2006. *Sleep disorders and sleep deprivation: An unmet public health problem.* Washington, DC: The National Academies Press.

Jakubowicz, D. M., E. M. Price, H. J. Glassman, A. J. G. Gallagher, N. Mandava, W. P. Ralph, and M. P. Fried. 2005. Effects of a twenty-four hour call period on resident performance during simulated endoscopic sinus surgery in an Accreditation Council for Graduate Medical Education-compliant training program. *Laryngoscope* 115(1):143-146.

Jay, S. M., N. Lamond, S. A. Ferguson, J. Dorrian, C. B. Jones, and D. Dawson. 2007. The characteristics of recovery sleep when recovery opportunity is restricted. *Sleep* 30(3):353-360.

Jean-Louis, G., D. F. Kripke, S. Ancoli-Israel, M. R. Klauber, and R. S. Sepulveda. 2000. Sleep duration, illumination, and activity patterns in a population sample: Effects of gender and ethnicity. *Biological Psychiatry* 47(10):921-927.

Jensen, A., R. Milner, C. Fisher, J. Gaughan, R. Rolandelli, and H. Grewal. 2004. Short-term sleep deficits do not adversely affect acquisition of laparoscopic skills in a laboratory setting. *Surgical Endoscopy and Other Interventional Techniques* 18(6):948-953.

Jewett, M. E., D. J. Dijk, R. E. Kronauer, and D. F. Dinges. 1999. Dose-response relationship between sleep duration and human psychomotor vigilance and subjective alertness. *Sleep* 22(2):171-179.

Jha, A. K., B. W. Duncan, and D. W. Bates. 2001. Fatigue—Sleepiness and medical errors. Chapter 46. In *Making health care safer: A critical analysis of patient safety practices*, edited by K. G. Shojania, B. W. Duncan, K. M. McDonald, and R. M. Wachter. Evidence Report/Technology Assessment No. 43. Pp. 519-533. Rockville, MD: Agency for Healthcare Research and Quality.

Kahol, K., M. J. Leyba, M. Deka, V. Deka, S. Mayes, M. Smith, J. J. Ferrara, and S. Panchanathan. 2008. Effect of fatigue on psychomotor and cognitive skills. *American Journal of Surgery* 195(2):195-204.

Killgore, W. D. S., T. J. Balkin, and N. J. Wesensten. 2006. Impaired decision making following 49 h of sleep deprivation. *Journal of Sleep Research* 15(1):7-13.

Kleitman, N. 1963. *Sleep and wakefulness*. Chicago, IL: University of Chicago Press.

Knauth, P. 2007. Extended work periods. *Industrial Health* 45(1):125-136.

Knauth, P., and S. Hornberger. 2003. Preventive and compensatory measures for shift workers. *Occupational Medicine* 53(2):109-116.

Kubo, T., H. Takeyama, S. Matsumoto, T. Ebara, K. Murata, N. Tachi, and T. Itani. 2007. Impact of nap length, nap timing and sleep quality on sustaining early morning performance. *Industrial Health* 45(4):552-563.

Lamond, N., and D. Dawson. 1999. Quantifying the performance impairment associated with fatigue. *Journal of Sleep Research* 8:255-262.

Lamond, N., S. M. Jay, J. Dorrian, S. A. Ferguson, C. Jones, and D. Dawson. 2007. The dynamics of neurobehavioural recovery following sleep loss. *Journal of Sleep Research* 16(1):33-41.

Landrigan, C. P., J. M. Rothschild, J. W. Cronin, R. Kaushal, E. Burdick, J. T. Katz, C. M. Lilly, P. H. Stone, S. W. Lockley, D. W. Bates, and C. A. Czeisler. 2004. Effect of reducing interns' work hours on serious medical errors in intensive care units. *New England Journal of Medicine* 351(18):1838-1848.

Landrigan, C. P., C. A. Czeisler, L. K. Barger, N. T. Ayas, J. M. Rothschild, and S. W. Lockley. 2007. Effective implementation of work-hour limits and systemic improvements. *Joint Commission Journal on Quality and Patient Safety* 33(11 Suppl):19-29.

Lavie, P. 1986. Ultrashort sleep-waking schedule. 3: Gates and forbidden zones for sleep. *Electroencephalography and Clinical Neurophysiology* 63(5):414-425.

Leonard, C., N. Fanning, J. Attwood, and M. Buckley. 1998. The effect of fatigue, sleep deprivation and onerous working hours on the physical and mental wellbeing of pre-registration house officers. *Irish Journal of Medical Sciences* 167(1):22-25.

Leproult, R., E. F. Colecchia, A. M. Berardi, R. Stickgold, S. M. Kosslyn, and E. Van Cauter. 2003. Individual differences in subjective and objective alertness during sleep deprivation are stable and unrelated. *American Journal of Physiology-Regulatory Integrative and Comparative Physiology* 284(2):R280-R290.

Levine, B., T. Roehrs, E. Stepanski, F. Zorick, and T. Roth. 1987. Fragmenting sleep diminishes its recuperative value. *Sleep* 10(6):590-599.

Lim, J., and D. F. Dinges. 2008. Sleep deprivation and vigilant attention. *Annals of the New York Academy of Sciences* 1129:305-322.

Lim, J., W. C. Choo, and M. W. L. Chee. 2007. Reproducibility of changes in behaviour and fMRI activation associated with sleep deprivation in a working memory task. *Sleep* 30(1):61-70.

Lockley, S. W., J. W. Cronin, E. E. Evans, B. E. Cade, C. J. Lee, C. P. Landrigan, J. M. Rothschild, J. T. Katz, C. M. Lilly, P. H. Stone, D. Aeschbach, and C. A. Czeisler. 2004. Effect of reducing interns' weekly work hours on sleep and attentional failures. *New England Journal of Medicine* 351(18):1829-1837.

Lockley, S. W., C. P. Landrigan, L. K. Barger, and C. A. Czeisler. 2006. When policy meets physiology—The challenge of reducing resident work hours. *Clinical Orthopaedics and Related Research* (449):116-127.

Lockley, S. W., L. K. Barger, N. T. Ayas, J. M. Rothschild, C. A. Czeisler, and C. P. Landrigan. 2007. Effects of health care provider work hours and sleep deprivation on safety and performance. *Joint Commission Journal on Quality and Patient Safety* 33(11 Suppl):7-18.

Ludmerer, K. M., and M. M. E. Johns. 2005. Reforming graduate medical education. *JAMA* 294(9):1083-1087.

Macdonald, W., and S. Bendak. 2000. Effects of workload level and 8- versus 12-h workday duration on test battery performance. *International Journal of Industrial Ergonomics* 26(3):399-416.

Macias, D. J., J. Hafner, J. C. Brillman, and D. Tandberg. 1996. Effect of time of day and duration into shift on hazardous exposures to biological fluids. *Academic Emergency Medicine* 3(6):605-610.

Marcus, C. L., and G. M. Loughlin. 1996. Effect of sleep deprivation on driving safety in housestaff. *Sleep* 19(10):763-766.

McKenna, B. S., D. L. Dickinson, H. J. Orff, and S. P. A. Drummond. 2007. The effects of one night of sleep deprivation on known-risk and ambiguous-risk decisions. *Journal of Sleep Research* 16(3):245-252.

Mitler, M. M., M. A. Carskadon, C. A. Czeisler, W. C. Dement, D. F. Dinges, and R. C. Graeber. 1988. Catastrophes, sleep, and public policy: Consensus report. *Sleep* 11(1):100-109.

Mollicone, D. J., H. P. A. Van Dongen, N. L. Rogers, and D. F. Dinges. 2008. Response surface mapping of neurobehavioral performance: Testing the feasibility of split sleep schedules for space operations. *Acta Astronautica* 63(7-10):833-840.

Nachreiner, F. 2001. Time on task effects on safety. *Journal of Human Ergology* 30(1-2): 97-102.

Naitoh, P. 1992. Minimal sleep to maintain performance: The search for sleep quantum in sustained operations. In *Why we nap: Evolution, chronobiology, and functions of polyphasic and ultrashort sleep*, edited by C. Stampi. Boston: Birkhäuser. Pp. 199-216.

Naitoh, P., and R. G. Angus. 1989. Napping and human functioning during prolonged work. In *Sleep and alertness: Chronobiological, behavioral, and medical aspects of napping*, edited by D. F. Dinges and R. J. Brougthon. New York: Raven Press. Pp. 221-246.

Ohayon, M. M., M. A. Carskadon, C. Guilleminault, and M. V. Vitiello. 2004. Meta-analysis of quantitative sleep parameters from childhood to old age in healthy individuals: Developing normative sleep values across the human lifespan. *Sleep* 27(7):1255-1273.

Owens, J. A., A. Avidan, D. Baldwin, and C. Landrigan. 2008. Improving sleep hygiene. *Archives of Internal Medicine* 168(11):1229-a-1230.

Pack, A., A. M. Pack, E. Rodgman, A. Cucchiara, D. F. Dinges, and C. W. Schwab. 1995. Characteristics of crashes attributed to the driver having fallen asleep. *Accident Analysis and Prevention* 27(6):769-775.

Parshuram, C. S. 2006. The impact of fatigue on patient safety. *Pediatric Clinics of North America* 53(6):1135-1153.

Parthasarathy, S., K. Hettiger, R. Budhiraja, and B. Sullivan. 2007. Sleep and well-being of ICU housestaff. *CHEST* 131(6):1685-1693.

Patrick, G. T. W., and J. A. Gilbert. 1896. On the effects of loss of sleep. *Psychological Review* 3(5):469-483.

Persson, R., P. Orbaek, H. Ursin, G. Kecklund, and K. Osterberg. 2003. Effects of the implementation of an 84-hour workweek on neurobehavioral test performance and cortisol responsiveness during testing. *Scandinavian Journal of Work Environment & Health* 29(4):261-269.

Persson, R., P. Orbaek, G. Kecklund, and T. Akerstedt. 2006. Impact of an 84-hour workweek on biomarkers for stress, metabolic processes and diurnal rhythm. *Scandinavian Journal of Work Environment & Health* 32(5):349-358.

Philibert, I. 2005. Sleep loss and performance in residents and nonphysicians: A meta-analytic examination. *Sleep* 28(11):1392-1402.

Portas, C. M., G. Rees, A. M. Howseman, O. Josephs, R. Turner, and C. D. Frith. 1998. A specific role for the thalamus in mediating the interaction of attention and arousal in humans. *Journal of Neuroscience* 18(21):8979-8989.

Richardson, G. S., J. K. Wyatt, J. P. Sullivan, E. J. Orav, A. E. Ward, M. A. Wolf, and C. A. Czeisler. 1996. Objective assessment of sleep and alertness in medical house staff and the impact of protected time for sleep. *Sleep* 19(9):718-726.

Roehrs, T., M. Greenwald, and T. Roth. 2004. Risk-taking behavior: Effects of ethanol, caffeine, and basal sleepiness. *Sleep* 27(5):887-893.

Rogers, A. E., W. T. Hwang, L. D. Scott, L. H. Aiken, and D. F. Dinges. 2004. The working hours of hospital staff nurses and patient safety. *Health Affairs* 23(4):202-212.

Rosa, R. R. 2001. Examining work schedules for fatigue: It's not just hours of work. In *Stress, workload, and fatigue*, edited by P.A. Hancock and P. A. Desmond. Mahwah, NJ: Lawrence Erlbaum Associates. Pp. 513-528.

Rosekind, M. R., R. C. Graeber, D. F. Dinges, L. J. Connell, M. S. Rountree, C. L. Spinweber, and K. A. Gillen. 1994. *Crew factors in flight operations IX: Effects of planned cockpit rest on crew performance and alertness in long-haul operations.* Moffett Field, CA: NASA Ames Research Center.

Rosekind, M. R., R. M. Smith, D. L. Miller, E. L. Co, K. B. Gregory, L. L. Webbon, P. H. Gander, and J. V. Lebacqz. 1995. Alertness management: Strategic naps in operational settings. *Journal of Sleep Research* 4(Suppl 2):62-66.

Rosekind, M. R., D. F. Neri, and D. F. Dinges. 1997. From laboratory to flightdeck: Promoting operational alertness. Paper read at Proceedings: The Royal Aeronautical Society's Symposium on Fatigue and Duty Time Limitations—An International Review, London, UK.

Sack, R. L., D. Auckley, R. R. Auger, M. A. Carskadon, K. P. Wright, M. V. Vitiello, and I. Zhdanova. 2007. Circadian rhythm sleep disorders: Part 1, Basic principles, shift work and jet lag disorders. *Sleep* 30(11):1460-1483.

SAFER Task Force. 2007. *Sleep, alertness, and fatigue in residency.* http://www.aasmnet.org/store/product/details.aspx?pid=116 (accessed June 1, 2007).

Schweitzer, P. K., A. C. Randazzo, K. Stone, M. Erman, and J. K. Walsh. 2006. Laboratory and field studies of naps and caffeine as practical countermeasures for sleep-wake problems associated with night work. *Sleep* 29(1):39-50.

Scott, L. D., A. E. Rogers, W. T. Hwang, and Y. Zhang. 2006. Effects of critical care nurses' work hours on vigilance and patients' safety. *American Journal of Critical Care* 15(1):30-37.

Smith-Coggins, R., S. K. Howard, D. T. Mac, C. Wang, S. Kwan, M. R. Rosekind, Y. Sowb, R. Ballise, J. Levis, and D. M. Gaba. 2006. Improving alertness and performance in emergency department physicians and nurses: The use of planned naps. *Annals of Emergency Medicine* 48(5):596-604.

Stampi, C., ed. 1992. *Why we nap. Evolution, chronobiology, and functions of polyphasic and ultrashort sleep*. Boston: Birkhäuser.

Stickgold, R. 2005. Sleep-dependent memory consolidation. *Nature* 437(7063):1272-1278.

Stickgold, R., L. James, and J. A. Hobson. 2000. Visual discrimination learning requires sleep after training. *Nature Neuroscience* 3(12):1237-1238.

Strogatz, S. H., R. E. Kronauer, and C. A. Czeisler. 1986. Circadian regulation dominates homeostatic control of sleep length and prior wake length in humans. *Sleep* 9(2): 353-364.

———. 1987. Circadian pacemaker interferes with sleep onset at specific times each day—Role in insomnia. *American Journal of Physiology* 253(1):R172-R178.

Takahashi, M., H. Fukuda, K. Miki, T. Haratani, L. Kurabayashi, N. Hisanaga, H. Arito, H. Takahashi, M. Egoshi, and M. Sakurai. 1999. Shift work-related problems in 16-h night shift nurses. (2): Effects on subjective symptoms, physical activity, heart rate, and sleep. *Industrial Health* 37(2):228-236.

Takeyama, H., T. Kubo, and T. Itani. 2005. The nighttime nap strategies for improving night shift work in workplace. *Industrial Health* 43(1):24-29.

Tassi, P., and A. Muzet. 2000. Sleep inertia. *Sleep Medicine Reviews* 4(4):341-353.

Thomas, M., H. Sing, G. Belenky, H. Holcomb, H. Mayberg, R. Dannals, H. Wagner, D. Thorne, K. Popp, L. Rowland, A. Welsh, S. Balwinski, and D. Redmond. 2000. Neural basis of alertness and cognitive performance impairments during sleepiness. I. Effects of 24 h of sleep deprivation on waking human regional brain activity. *Journal of Sleep Research* 9(4):335-352.

Torsvall, L., and T. Akerstedt. 1988. Disturbed sleep while being on-call—An EEG study of ships engineers. *Sleep* 11(1):35-38.

Uchal, M., J. Tjugum, E. Martinsen, X. G. Qiu, and R. Bergamaschi. 2005. The impact of sleep deprivation on product quality and procedure effectiveness in a laparoscopic physical simulator: A randomized controlled trial. *American Journal of Surgery* 189(6):753-757.

Van Dongen, H. P. A. 2006. Shift work and inter-individual differences in sleep and sleepiness. *Chronobiology International* 23(6):1139-1147.

Van Dongen, H. P. A., and D. F. Dinges. 2005. Circadian rhythm in sleepiness, alertness and performance. In *Principles and practice of sleep medicine*. 4th ed., edited by M. H. Kryger, T. Roth, and W. C. Dement. Philadelphia, PA: W. B. Saunders. Pp. 435-443.

Van Dongen, H. P. A., N. J. Price, J. M. Mullington, M. P. Szuba, S. C. Kapoor, and D. F. Dinges. 2001. Caffeine eliminates psychomotor vigilance deficits from sleep inertia. *Sleep* 24(7):813-819.

Van Dongen, H. P., G. Maislin, J. M. Mullington, and D. F. Dinges. 2003. The cumulative cost of additional wakefulness: Dose-response effects on neurobehavioral functions and sleep physiology from chronic sleep restriction and total sleep deprivation. *Sleep* 26(2):117-126.

Van Dongen, H. P., M. D. Baynard, G. Maislin, and D. F. Dinges. 2004. Systematic interindividual differences in neurobehavioral impairment from sleep loss: Evidence of trait-like differential vulnerability. *Sleep* 27(3):423-433.

Van Dongen, H. P. A., C. G. Mott, J. K. Huang, D. J. Mollicone, F. D. McKenzie, and D. F. Dinges. 2007. Optimization of biomathematical model predictions for cognitive performance impairment in individuals: Accounting for unknown traits and uncertain states in homeostatic and circadian processes. *Sleep* 30(9):1129-1143.

Veasey, S., R. Rosen, B. Barzansky, I. Rosen, and J. Owens. 2002. Sleep loss and fatigue in residency training: A reappraisal. *JAMA* 288(9):1116-1124.

Venkatraman, V., Y. M. L. Chuah, S. A. Huettel, and M. W. L. Chee. 2007. Sleep deprivation elevates expectation of gains and attenuates response to losses following risky decisions. *Sleep* 30(5):603-609.

Viola, A. U., S. N. Archer, L. M. James, J. A. Groeger, J. C. Y. Lo, D. Skene, M. von Schantz, and D. J. Dijk. 2007. PER3 polymorphism predicts sleep structure and waking performance. *Current Biology* 17:1-6.

Walker, M. P., and R. Stickgold. 2006. Annual review of psychology. *Sleep, Memory, and Plasticity* 57:139-166.

Weinger, M. B., and S. Ancoli-Israel. 2002. Sleep deprivation and clinical performance. *JAMA* 287(8):955-957.

Wesensten, N., T. J. Balkin, and G. Belenky. 1999. Does sleep fragmentation impact recuperation? A review and reanalysis. *Journal of Sleep Research* 8(4):237-245.

Williamson, A. M., and A.-M. Feyer. 2000. Moderate sleep deprivation produces impairments in cognitive and motor performance equivalent to legally prescribed levels of alcohol intoxication. *Occupational and Environmental Medicine* 57:649-655.

Wu, J. C., J. C. Gillin, M. S. Buchsbaum, P. Chen, D. B. Keator, N. K. Wu, L. A. Darnall, J. H. Fallon, and W. E. Bunney. 2006. Frontal lobe metabolic decreases with sleep deprivation not totally reversed by recovery sleep. *Neuropsychopharmacology* 31(12):2783-2792.

8

System Strategies to Improve Patient Safety and Error Prevention

System changes are needed in addition to enhanced supervision, workload adjustment, and fatigue prevention methods to enhance conditions for resident performance and patient safety. The committee recommends ways to make more effective handovers and error reporting an integral part of resident learning experiences to help achieve these goals. Teamwork coordination and clinical information technology can also foster increased learning, productivity, and patient safety. A transformation in the medical environment is needed so that a system-wide culture of safety develops and a system of blame is replaced with one of shared responsibility.

The committee's examination of graduate medical education has revealed that duty hours represent only one among many factors in residents' experiences that may affect patient safety and resident learning. Although the committee's deliberations about recommendations to help mitigate and prevent resident fatigue were central to its charge, it became apparent that additional changes at the system level could also help improve patient safety, resident education, and the quality of care. The committee further recognized that redesigning hospital practices or system processes may be necessary to facilitate redesign of graduate medical education or implementation of the proposed recommendations. The strategies discussed in this chapter and the recommendations of the committee are aimed at systems that not only improve resident work and learning, but also improve the delivery of care in teaching institutions by all staff. The need for these steps is apparent now, under the current duty hour limits, and will continue to be important after implementation of the committee's recommended changes to duty hours.

In order to implement such changes, an organization-wide approach is necessary to create an environment that involves all hospital workers in achieving the desired results of maximum safety and the provision of quality care. Adjustments that would assist in transforming the resident work environment, and the environment for all health workers, include improving communications skills among hospital staff, implementing team strategies to complete work more efficiently, and developing a safety culture that extends across hospital settings. Therefore, this chapter discusses organizational and systems strategies that can help to (1) establish a culture of safety, (2) improve handover processes, (3) use adverse event and error-reporting systems for resident learning, and (4) develop a team culture to improve communication and task performance among residents. These elements can both enhance a physician's education and contribute to patient safety.

LEARNING IN A CULTURE OF SAFETY

Creating a culture of safety and developing teamwork have been broadly addressed in previous Institute of Medicine (IOM) reports, namely the Quality Chasm series (IOM, 2000, 2003, 2004). The committee builds on those earlier reports, focusing attention on adopting strategies for teamwork development and error reporting to better serve the educational needs of residents while fostering safe patient care.

Culture of Safety and High Reliability

Definitions of the concept of a culture of safety vary, but organizations that establish a safety culture generally demonstrate the following characteristics (Singer et al., 2003):

- Safety is considered the highest priority of the organization.
- There are strongly shared values and behavioral norms throughout the organization that are centered around safety.
- Resources and incentives are available for the organization to pursue and implement a safety commitment.
- There is non-hierarchical and open communication among workers—particularly in safety-related scenarios.
- There are rare occurrences of errors, but open recognition and reporting of them is accomplished without blame for individuals.
- Organizational learning is highly valued.

High-reliability organizations (HROs) build on culture of safety elements to go beyond the norm and approach their goals of zero errors and avoidance of potential disasters, such as multiple deaths (Weick and Sutcliffe, 2001). Businesses in particularly risky industries that could have a catastrophic impact on the public, such as military operations, commercial airlines, and nuclear power generation, were among the first to adopt the continual processes needed to achieve high-reliability operations while producing minimal errors.

Although recognition of a safety culture and high-reliability components and practices (e.g., teamwork, blame-free error reporting) are becoming more common in health care, there has yet to be widespread adoption of these practices across the medical field (Patterson et al., 2004). Reasons for slow adoption by some institutions include resistance to organizational change (Carroll and Quijada, 2004) and insufficient resources to support safety culture practices (Patterson, 2007), although some experts note that a major investment of resources is not necessary (Hines et al., 2008). Tension can exist between the goals of a safety culture and individual residents, program directors, or departments, which is why leadership—at both the clinician and the executive levels—is a critical component in overcoming any resistance and establishing the importance of high reliability throughout an organization (Roberts et al., 2005). Leaders in healthcare settings accomplish this by aligning incentives and encouraging the ideas that drive a culture of safety, promoting the continued progression of system redesign and eventually sustaining the developments made (Roberts and Perryman, 2007). Suggesting that they be more active in establishing patient safety standards for clinical performance and that such practices become part of medical training is in line with recommendations from the IOM report *To Err Is Human* (IOM, 2000).

In an effort to take a lead in promoting a culture of safety for healthcare settings, the Agency for Healthcare Research and Quality (AHRQ) now encourages hospitals to adapt the concepts of high reliability to their organizations, along with the previously mentioned elements of safety culture (Hines et al., 2008). The introduction of high-reliability practices is still relatively new in the medical field, and the exact impact of the culture of safety on specific improvements in healthcare organizations has yet to be documented on a broad scale (Shojania, 2005).

However, it is known that error rates in hospital care tend to be far greater than those associated with HROs in other industries (e.g., airlines). The 44,000-98,000 estimated deaths in the United States related to medical errors are just one component of risks to patients. Many more nonfatal preventable events also harm patients, with impacts such as extended hospital stays, pain and suffering due to hospital acquired infections, or an adverse drug event. The frequency of such errors certainly indicates a need for improvement and is discussed in Chapter 6. In this chapter the focus is on organizational attitudes and culture: ensuring that safety is given the prominence it requires for the provision of high-quality care, that residents and all other workers in the hospital are comfortable discussing errors, and that efforts are made to correct or prevent situations in which errors occur.

To prevent such occurrences, hospital environments that promote communications by all levels and professions of workers should be supported. Encouraging questions about safety and blame-free reporting of errors would likely enhance the educational value of residents' training and their ability to learn from all of their colleagues and continuously improve quality of care through cooperative teamwork (IOM, 2001). This does not apply only to medical residents, but focusing attention on them may be a good place to introduce the culture change required for this shift to team mentality or shared responsibility and accountability in healthcare settings.

REDUCING ERRORS BY IMPROVING HANDOVERS

Handovers, or transitions in patient care are an area of medical practice that is considered a substantial source of errors and risks to patients, but one that can benefit from immediate attention through processes improvement. More commonly referred to as "handoffs," "transfers," or "sign-out" in the United States, the committee chose the term "handover" for this report because it better encompasses the goal of these pivotal moments, suggesting that they are intended as a handing over of responsibility for a patient from one healthcare provider to another and not simply a quick transcription of patient information at the end of one's time on duty. Continuity of care as described in Chapter 4 refers primarily to relationship building between physician and patient, and gaining thorough knowledge of a patient's condition in order to provide the best treatment. A resident's familiarity with a patient and his/her care is important, and discontinuity of care due to handing cases over to other residents has been shown to result in increased levels of preventable adverse events (Laine, 1993; Petersen, 1994).

However, to achieve stronger patient-physician relationships and serve its educational purpose, continuity of care relies heavily on the continuity of information itself. Information transferred during handovers may include a patient's name, bed location, blood pressure, diagnosis, and other critical data on patient status or treatment plan. A service that must be available 24 hours daily, such as health care, requires transferring this information and responsibility of tasks from one team member to another at some point or points during the day. In a hospital setting, for example, where teams of physicians, nurses, and residents are all responsible for a single patient, continuity of care involves a comprehensive handover of patient information from one provider or team to another so that clinical care can be maintained successfully among a healthcare team. Handovers take place among teams of nurses, teams of physicians, and teams of residents, as well as between those teams and between integrated care teams consisting of various types of clinicians. Handovers occur between emergency departments, different inpatient settings from surgical to postoperative care, and different hospitals, not to mention transitions out of hospitals to nursing homes or home care settings (Patterson et al., 2004). The act of transferring responsibility for patient care is not inherently a negative practice. However with each additional handover per patient, there is more opportunity for dilution or omission of information, which can lead to inaccuracies that affect patient care and outcomes (Arora et al., 2005, 2007; Horwitz et al., 2008; Petersen, 1994). Such communication breakdowns result in information gaps that intensify discontinuity of patient care and the potential for errors. These factors make them pivotal moments in the care continuum and an important aspect of preventing medical errors and ensuring patient safety (Saultz, 2003). For residents, these exchanges are also opportunities for professional interaction, learning how to assess patient care situations, and problem solving.

The next several sections discuss the role handovers play in the continuity of patient care, the impact they have on patient safety and resident education, how they are affected by the regulation of duty hours, and suggestions for redesigning handover processes to optimize patient safety and resident education.

Consequences of Transfers and Communication Failure for Patient Safety

Several studies, not specific to residents, highlight observed patient cases that point out the errors—at times fatal—caused by poor communication during handovers (Beach et al., 2003; Gandhi, 2005; Vidyarthi, 2004; Wachter, 2008; Wachter et al., 2006). An evaluation by the Joint Commission in 2005 of more than 3,000 root-cause analyses of reported error data revealed that nearly 70 percent of sentinel events in accredited healthcare entities result from communication failures (Joint Commission International Center for Patient Safety, 2006). The Joint Commission further stated that there is evidence that at least half of such communication failures occur during handovers. In a study by Gandhi and colleagues, poorly executed handovers contributed to 20 percent (36 of 181) of malpractice claims that resulted in serious harm or death to patients (Gandhi et al., 2006). Additionally, poor handover and follow-up practices at discharge are particularly likely to increase safety risks for patients (Forster et al., 2003; Moore et al., 2003). Poor discharge practices have been associated with higher readmittance rates or avoidable readmission of discharged patients (Halasyamani et al., 2006). Because this evidence is not specific to residents, it demonstrates the extent to which poor communication permeates the health system, posing safety risks to patients.

Among residents, however, communication failures are among the most common factors contributing to adverse patient events (Sutcliffe et al., 2004), and handovers are just one form of communication between residents and their medical team. In a study by Singh (2007), 19 percent (46) of cases with errors made by residents (including interns and fellows) that resulted in malpractice claims and led to medical injury of patients were attributed to poor handoffs. Another study by Arora et al. (2007) in which 27 percent (1,876) of medication entries in sign-out forms from handover procedures by interns contained either omissions of or commissions from notations in the original patient chart, 54 percent of them had the potential to cause moderate or severe harm to patients. Reducing possible mistakes during these moments can be crucial, and residents can achieve this largely by improving their handover processes, especially with guidance from their attending physician on the critical clinical information that best prepares the next shift of residents to anticipate and respond to changes in patients' conditions.

Variability of Handovers

A likely contribution to errors during handovers is the variability of the handover process across settings without specification of the information that needs to be provided. Handover procedures and type of information transferred can vary from hospital to hospital and program to program within hospitals, and often are not structured or uniform between or among provider teams. Some use fax systems, others written tools, and others electronic tools, allowing handovers to take place either face-to-face, in written form only, in verbal form only, or in multiple forms—amplifying the variability of the process and information that gets transmitted across teams and care units.

Solet and colleagues (2005) illustrate such variations in a single internal medicine residency program that provides medical training across four different hospitals that each used different methods to transfer patient information. Three different computerized systems were utilized among the four institutions, and only two of the four used a computerized system for handovers. The other two hospitals conducted written handovers, one of which had a free-style form, using no templates or standard format, with residents' writing up or communicating their notes as they wished. Other studies describe additional variations in handover processes and their differing degrees of effectiveness in communicating necessary patient information (Borowitz et al., 2008; Horwitz et al., 2006).

Impact of Duty Hour Regulations on Handovers and Continuity of Care

Although fewer duty hours or appropriately placed rest periods may help to reduce fatigue in residents, they raise serious concerns for continuity of care. Practice has shown that the number of hours worked and the number of handovers among patients are inversely related, meaning that the fewer hours residents spend in the hospital, the more often patient care has to be handed over to other residents (Horwitz et al., 2007b; Vidyarthi, 2004; Vidyarthi et al., 2006). Therefore, shorter shifts to comply with the 2003 duty hour regulations and periods of sleep within extended duty periods, as this committee recommends, can result in an increase of handovers.

In light of the error rates associated with handovers (Arora and Farnan, 2008; Fletcher et al., 2005), increasing their frequency requires that hospitals improve the process in order to maintain or improve the quality of care. In the United Kingdom, this same trend of increased handovers (because of adherence to the European Working Time Directive) has led its Department of Health to emphasize the effectiveness of handovers as an area of improvement for patient safety (Sabir et al., 2006). Here in the United States, the Joint Commission has recently established a National Patient Safety Goal specific to improving handover practices (which apply to all healthcare professionals, not only to residents) as part of its accreditation process (Joint Commission, 2007). Experts in the field suggest that many errors stemming from poor handovers are preventable or can be made less severe if hospitals take steps to improve communication and coordination of care (Kripalani, 2008), create better opportunities for interaction, and provide better guidance for the process. Examinations of hospital systems and resident programs have shown that structured and supervised handover procedures can dramatically decrease the rates of errors associated with them (Catchpole et al., 2007; Horwitz et al., 2006). It has also been observed that implementing such processes is possible within current spending levels and without having to pass new legislation (Coleman and Berenson, 2004), thus building a case for improving quality of care through improved handovers.

Instead of merely viewing more frequent handovers as an increased opportunity for error, they can be viewed as another opportunity for resident learning. They represent a chance to develop macro cognitive skills such as recognizing and analyzing early warning signs or anticipating any problems that might arise for patients on the next shift, understanding warning signs in patients' response to treatment, better planning for the care of patients, and improving communication and teamwork skills. Handovers are particularly crucial for all clinical staff to learn to navigate, and it is important that residents be familiar with effective strategies in order to apply them successfully in any setting. Suggestions for possible interventions and training follow in the next section.

HANDOVER INTERVENTIONS

The attention that handovers have received as a target area to improve patient safety is exemplified by the Joint Commission's decision to issue a "Patient Safety Goal" (effective January 1, 2006) requiring hospitals to standardize their handover approaches and communications as components of improving continuity of care (see Box 8-1) (Joint Commission, 2007). It is generally believed that providing some structure for handover procedures is the appropriate solution for improving outcomes. Other industries in high-risk or high-reliability environments have already identified aspects of

BOX 8-1
National Patient Safety Goal 2: Improve the Effectiveness of Communication Among Caregivers

Requirement 2E

Implement a standardized approach to "hand off" communications, including an opportunity to ask and respond to questions.

Rationale for Requirement 2E

The primary objective of a handoff is to provide accurate information about a patient's care, treatment, and services; current condition; and any recent or anticipated changes. The information communicated during a handoff must be accurate in order to meet patient safety goals.

In health care there are numerous types of patient handoffs, including but not limited to nursing shift changes; physicians' transferring complete responsibility for a patient; physicians' transferring on-call responsibility; temporary responsibility for staff leaving the unit for a short time; anesthesiologist's report to post-anesthesia recovery room nurse; nursing and physician handoffs from the emergency department to inpatient units, different hospitals, nursing homes, and home health care; and critical laboratory and radiology results sent to physicians' offices.

Implementation Expectations for Requirement 2E

The organization's process for effective handoff communication includes the following:

- Interactive communications allowing for the opportunity for questioning between the giver and receiver of patient information.
- Up-to-date information regarding the patient's care, treatment and services, condition, and any recent or anticipated changes.
- A process for verification of the received information, including repeat-back or read-back, as appropriate.
- An opportunity for the receiver of the handoff information to review relevant patient historical data, which may include previous care, treatment, and services.
- Interruptions during handoffs are limited to minimize the possibility that information would fail to be conveyed or would be forgotten.

SOURCE: Joint Commission, 2007.

handover processes, and several of their lessons or techniques are applicable to hospital settings (Patterson et al., 2004). Examples from such industries were indeed drawn upon to help formulate the Joint Commission requirements for these procedures.

Improving handovers provides an opportunity to restructure the way residents learn, possibly leading them to greater collaboration with peers and supervisors and helping them to build new skills that promote quality care. One-size-fits-all interventions or complete standardization of the process across all settings, however, is not feasible in a highly variable and complex system such as health care; therefore flexibility in adopting any suggested handover method would be crucial to its success (Patterson, 2008). Application of core components should be evaluated for each setting and care scenario to ensure that they are not used superfluously and do not hinder existing effective transfer methods (Patterson, 2008; Perry et al., 2008). It is expected that handover practices would be tailored somewhat to accommodate the differing needs of intensive care units compared to emergency rooms, surgical and internal medicine disciplines (Arora and Johnson, 2006), or outpatient and inpatient settings, but that core components would be instituted within a basic framework with consistent principles. Therefore, the basic elements that may help improve current medical handover processes presented in the following section are general suggestions. Limited data are available on the implementation of handover guidelines or their effectiveness, but the existing evidence suggests that following a somewhat structured protocol does improve resident communication (Chung and Ahmed, 2007) and patient outcomes (Catchpole et al., 2007).

General Guidance for Improving Handovers

One of the factors most consistently found in the research to help ensure successful handovers for residents is face-to-face interaction (Horwitz et al., 2007a; Parke and Mishkin, 2005; Solet et al., 2005). Solet et al. (2005) suggest that the combination of oral and written handoff is the most effective for transmitting patient information. Most residency programs do solely written sign-outs, and there are times when physicians can be available only via phone or e-mail to exchange crucial information. However, direct face-to-face communication enhances the comprehension of written orders and allows for greater expression of what points need emphasizing and those that are less urgent (Solet et al., 2005). This approach also allows residents to ask questions and clarify instructions, interactions that are helpful for learning and avoiding errors. Face-to-face communication also creates clearer transitions of responsibility and authority on a case, which some believe is equally important to recognize during the handover process (Behara et al., 2005). Because of the benefits of face-to-face interactions, finding locations in which they can occur with limited distractions or interruptions may be helpful (Perry et al., 2008; Singer and Dean, 2006). For example, some advocate going to a patient's bedside to perform transfers, which may have additional benefits associated with patient centeredness. Building in overlap time between shift schedules also helps set aside the time for this type of interaction, improving handover processes and increasing their educational value by providing the opportunity to ask questions and clarify treatment plans or other pertinent information (Afessa et al., 2005; Goldstein et al., 2004; Landrigan et al., 2004; Volpp and Landrigan, 2008). Each of these actions is very team oriented and often requires training because they are not easily instituted by written standards alone.

The literature further suggests that structuring the content of what is exchanged during handovers and using uniform language or terminology to communicate information assist in preventing omission of necessary information and help reduce confusion about what tasks are to be completed (Arora et al., 2005). To aid in this process several studies recommend framing content by using written checklists such as "I pass the baton" or "Signout," created by TeamSTEPPS™ (2007) and Horwitz et al. (2007a), respectively. These checklists outline specific information to exchange during handovers such as patient name, diagnosis, pending tests, allergies to medications, and so forth. Ideally, they would be as concise as possible without omitting relevant information.

More advanced tools that achieve this same goal are electronic sign-out systems. Electronic systems can improve handover content by providing structured, easy-to-access databases of patient information and creating formatted checklists of tasks that need to be considered for patient treatment. When residents record information electronically, they reduce paperwork and duplication. Electronic systems can also enhance the uniformity of terminology and procedures if multiple departments or an entire hospital uses the same electronic program, much like the Department of Veterans Affairs (VA) does with the system it recently adopted (Carpenter, 2008). The VA system combines sign-out strategies by importing patient data electronically but also includes a free-text entry segment that allows users to personally add treatment plans or anticipated tasks (Solet et al., 2005). Together, these factors can increase handover efficiency, reduce instances of content omission, and help resident and integrated teams have consistent and up-to-date information about their patients and care schedules. Although electronic systems have demonstrated improved resident performance and patient outcomes by reducing rates of adverse events and allowing residents more time to spend on direct patient care (Petersen et al., 1998; Van Eaton et al., 2005), very few residency programs or hospitals actually employ electronic sign-out systems. There are reports that roughly 18 percent of large residency programs have some form of electronic sign-out, as do 3 percent of smaller residency programs (Horwitz et al., 2006), and that less than 5 percent of U.S. hospitals have adopted such procedures (Okie, 2007). At least one study has shown that if electronic sign-out systems are cumbersome, residents may find ways to work around them or discard them altogether (Landrigan et al., 2004). Furthermore, if implemented or used improperly, electronic systems can have unintended consequences that undermine clinician communication or patient care (Ash et al., 2007; Campbell et al., 2006; IOM, 2006), making the need for training in these systems an important one. Further discussion of electronic system use among healthcare staff is addressed in more detail later in this chapter.

In addition to the identified key components of tested handover methods mentioned thus far, results from an observational study of residents during sign-out by Horwitz et al. (2007a) also illuminated the importance of residents' having supervision available during the process and having time to formulate clear plans to carry out their assigned tasks. Opportunities for learning could be increased by the presence of appropriate supervisors during the handover process. Learning how to hand over responsibility and information is important, as is learning what patient signs to look for and what types of information are critical to forward to another caregiver. A supervisor can help new residents anticipate a patient's future care needs.

All together, the above results fall in line with a 2005 study that interviewed 26 interns from a university teaching hospital. These interns suggested improvements in handover practices to help them make more informed and accurate decisions about patient care and reduce duplicative or unnecessary work. The recommendations included a request for face-to-face interactions; reviewing anticipated areas for care or troubleshooting; and having an accurate, updated, legible, written worksheet that includes standard patient content and medical information (Arora et al., 2005). Since these were the suggestions of first-year residents, it may be that having the structured format is more beneficial to residents as they first learn these processes (rather than after several years of experience), which underlines the educational benefits of using these methods for handovers.

Other components that can add structure to handover processes include agreeing on an end-of-shift time that allows for an overlap of shifts, establishing pre-handover routines, determining a set location for transfers to take place, requiring that outgoing residents inform incoming residents of all patients in the department, and conducting joint bedside visits (Singer and Dean, 2006).

Innovative Handover Strategies

A number of handover strategies currently being developed and practiced incorporate several of the components addressed above. A particular handover strategy used to improve patient care and help residents learn patient-centered techniques is bedside handover. Bedside strategies establish patient centeredness and visible continuity that reduces patient confusion or anxiety (Singer and Dean, 2006). For example, a pilot study in Ontario, Canada, involving nurses showed that implementing bedside handover helped catch incorrect patient armbands or intravenous solutions (via the bedside safety checks that were part of the handover intervention), preventing possible harm to the patient and helping to reach established patient safety goals. Patients seem to appreciate this approach and acknowledge that "they are reassured by knowing information about their care requirements has been communicated" (Alvarado et al., 2006, p. 78), promoting a culture of patient safety and team integration.

Another suggestion to promote team structure and shared information through bedside handover strategies is to share the care responsibilities of specific teammates more regularly with patients. For example, introductions of the care team at the bedside could indicate not only the names of the team members but their titles or roles on the team. If staff shifts change at times when no patient visits are planned, a record could be kept in the ward (or a note in each patient's record) to indicate which resident and attending are responsible for each patient at a particular time. Also, a schedule could be kept in the ward of when patient rounds with the attending physician could reasonably be expected, so patients and their families can be informed (Simmons and Gonzalez del Rey, 2008). These changes, suggested by patients, would be relatively easy to implement in facilities that do not already follow such protocols. Integrating patients more openly into the care team allows team culture to extend beyond the resident or integrated teams, adding a more personal view of the patient's perspective to the team. Patients familiar with this handover practice also suggest that hospital staff introduce themselves, use an understandable vocabulary when speaking to them, and include patients in discussions to maximize the value to patient and to promote team thinking (Simmons and Gonzalez del Rey, 2008).

As previously noted, handovers that occur at discharge create substantial risk to patients. Also known as a type of "care transition," "transfer of care," or "transitional care," these types of transfer have been defined as "a set of actions designed to ensure the coordination and continuity of health care as patients transfer between different locations or different levels of care in the same location" (Coleman and Berenson, 2004). Often they are transfers of patients to somewhere outside the hospital setting, usually to home settings or home care facilities where regular monitoring of a patient's condition is not necessarily possible. A number of strategies have been shown to be effective in increasing patient centeredness and reducing the occurrence of errors, such as Dr. Eric Coleman's Care Transitions Program and the Transforming Care at the Bedside program launched by the Robert Wood Johnson Foundation and the Institute for Healthcare Improvement

(Care Transitions Program, 2008; IHI, 2007). Since information in these cases passes from health professionals to patients or their families (instead of to other health professionals), residents need special training in how to present the information adequately and appropriately in a way that patients will understand, which is what the mentioned programs aim to do for all health professionals.

Handovers, from the perspective of patients, can appear to be a confusing interruption or discontinuity in their care, as responsibility for their care shifts from physician to physician. Lessons learned from the above studies could decrease the discontinuities that the patient experiences with handovers, regardless of their frequency. Likewise, applying some of the suggested methods can also help residents learn what information is most pertinent for quality care and patient safety during handovers and how to handle both the clinical and the relationship side of the process by interaction with their peers and supervisors as well as patients. Creating a formal protocol to transfer clinical information and patient care thoroughly and accurately, in any setting, can go a long way to help prevent or intercept errors, enhance workforce communication, provide educational opportunities for residents, and possibly assist to minimize the negative effects of increased shift work.

The committee concludes that whichever method or combination of methods is used to improve handovers, the key factor is that handovers be *structured*, while also conforming to the needs and capacity of particular departments or settings. Residents and all other participants in handover processes should be trained in how to perform effective handovers. Training other hospital staff in addition to residents will be particularly important for those who work on integrated care teams and those who hand over to other units. Establishing some basic, facility-wide principles and structures should assist all teams to work more seamlessly together and foster more open communication and accurate transfer of information and responsibility across hospital settings. Both clinical and executive leaders can help promote these system-wide practices. Successful handover processes should try to include the following components:

- Face-to-face interactions (whenever possible),
- A set location and time for handovers to occur,
- Minimal interruptions,
- Structured content (e.g., use of checklists) to ensure that all relevant information is transmitted,
- Uniform language or terminology,
- Sufficient time to interact and clarify questions or concerns (e.g., overlap in shift schedules), and
- Presence of a supervisor to oversee the process and answer additional questions.

Teaching the Handover Process

A lack of standard educational practices for teaching how to do handovers is another factor contributing to the degree of variability in conducting them. Evidence suggests that a formal curriculum including sessions on handovers does not exist in 92 percent of medical schools and that most medical students and residents alike learn handover procedures informally from other residents (Solet et al., 2005), highlighting the lack of attention this matter receives. A later study found that 60 percent of medical training programs (not including those of New York State) provided no lectures or workshops on sign-out skills (Horwitz et al., 2006). Therefore, the committee agrees with suggestions to improve handover education that include providing standard instructional materials, training faculty leaders to encourage shared responsibility and effective handover, encouraging or requiring faculty or resident leaders to properly supervise handover procedures, and teaching residents formal communication techniques (Solet et al., 2005). Additionally, materials on handovers could give case examples of how the type of information transferred can influence the outcomes of patient care. Because in addition to providing basic patient information, handovers "support macrocognitive functions, such as problem recognition, problem analysis, sensemaking, and planning" (Perry et al., 2008, p. 2), where residents have to determine future actions for a patient's care, anticipate any problems that might arise, and adequately communicate these things when necessary. Training materials that teach residents how to approach these actions would highlight the clinical lessons that can be gained from good handovers. Hospitals should consider each factor for incorporation into new education strategies for the improvement of handovers by residents.

Examples from the literature of effective curricula for training healthcare professionals in handover practices vary from providing a comprehensive series of classes over time to providing a one-time instructional conference (Alvarado et al., 2006; Horwitz et al., 2007a). When introducing new training or a new curriculum, however, the already extensive nature of residents' workload must be considered thoughtfully. Some studies found that programs had difficulty finding a time when sufficient numbers of residents were available to attend the proposed conferences (Horwitz et al., 2007a). As a result, only a small number of residents trained on the handover process. Given that resident schedules and workload are already so demanding, it is important either to find a time that works with their schedules (e.g., during orientation) or to make this lesson a priority and place it in the regular curriculum where appropriate. Education about these methods should also occur in real time, with patients under the resident's care to reinforce the lessons learned in general orientation on systems. For example, attendings could incorporate the discussion of what should be in each patient's sign-out during rounds and the nature of errors (omission or commission) that might occur without vigilance during these interactions. Alternately, computer-based or simulated lessons could be designed so that residents could learn whenever convenient.

Recommendation 8-1: Teaching hospitals should design, implement, and institutionalize structured handover processes to ensure continuity of care and patient safety.

- **Programs should train residents and teams in how to hand over their patients using effective communications.**
- **Programs should schedule an overlap in time when teams transition on and off duty to allow for handovers.**
- **The process should include a system that quickly provides staff and patients with the name of the resident currently responsible in addition to the name of the attending physician.**

Because of widespread concern across medical specialties that increasing handovers—a necessary consequence of restricting resident duty hours—will result in decreased continuity of care and increased risk to patient safety, systematic research is required on the effects of different handover techniques designed to prevent loss of continuity of care and risks to patients. Currently, we do not know if the relative risk of resident duty hours and fatigue mitigation as recommended by the committee, combined with good handover practices, results in better or worse patient safety outcomes. There should be detailed examination of specific elements of handovers—for example, the optimal time(s) required by residents for handovers of a specific number and severity of patients, when joint bedside visits would be recommended, minimum information transfer needed for all patients, availability of supervisors at handovers, the impact of face-to-face handovers and how handovers can be opportunities for intercepting errors.

TRAINING DOCTORS AND ERROR REPORTING

In addition to the latest, evidence-based best practices for patient care and structured handover procedures, new physicians must also learn and practice safety and quality improvement principles and methods. Throughout medical centers or hospitals there should be encouragement for residents to participate in ongoing quality improvement efforts and support for them to learn from constructive feedback. As part of its six core competencies the Accreditation Council for Graduate Medical Education (ACGME) requires residency programs to teach about quality improvement practices and produce residents who can "systematically analyze practice using quality improvement methods, and implement changes with the goal of practice improvement," and "participate in identifying system errors and implementing potential systems solutions," to improve care based on an understanding of resource allocation and integration of care delivery systems as well as individual patients' clinical needs (ACGME, 2007, pp. 1, 3). Some health researchers are finding that educating residents on quality improvement methods for patient care can have a beneficial effect on the outcomes of patients that they treat during training (Stevens et al., 2008; Warm et al., 2008). If the quality of education that residents receive during training affects the quality of care they give to future patients once they are working independently, then learning from their mistakes or those of others as part of that education can be valuable for future patient safety.

Teaching hospitals typically have error-reporting systems (as ACGME states they should in its competency requirements), but residents are often not fully integrated into the hospital's culture of safety and either do not know how to report errors or do not see the value of doing so. A serious barrier is that residents, regardless of whether they see the value of reporting errors, are often reluctant to report them because they fear retribution for asking questions, displaying ignorance, or facing legal consequences (Hines et al., 2008; Kaldjian et al., 2008).

The Joint Commission recently issued a Sentinel Event Alert concerning "intimidating and disruptive behaviors [that] can foster medical errors . . . and preventable adverse outcomes" that indicates that such disruptive behavior is not unusual (Joint Commission, 2008, p. 1). The Joint Commission mentioned examples of intimidating behavior, such as "reluctance or refusal to answer questions, return phone calls or pages; condescending language or voice intonation; and impatience with questions. Overt and passive behaviors undermine team effectiveness . . ." (Joint Commission, 2008, p. 1). The alert states that several surveys have found that a majority of healthcare workers have seen or experienced such behavior and one study found that "40 percent of clinicians have kept quiet or remained passive during patient care events rather than question a known intimidator" (Joint Commission, 2008, p. 1). Likewise, an AHRQ database comprised of voluntary survey responses by hospital staff on the efforts to create a patient safety culture in their institutions revealed that only 44 percent of respondents agreed that their hospital had a nonpunitive response to errors (AHRQ, 2008).

Among the core concepts of HROs is a perception of errors or near misses not as an occasion to point blame, but as an opportunity to improve system design and performance to achieve an even safer environment (Hines et al., 2008). This leads to creating a blame-free environment through a systemic response to errors, which could help reform the punitive culture often observed in healthcare settings that tend to inhibit open communication and, thus, learning. The careful design of an error-reporting system, analyses resulting from it, and feedback to those involved and to others who can learn from error-related events are critical to the success of the system (Kaplan and Rabin Fastman, 2003). An understanding of the errors in a system is the foundation for building a strong culture of safety. Information from error reporting and root-cause analyses of critical cases could also contribute significantly to residents' education.

Since the focus of most hospital error-reporting programs has been on system-wide problems rather than on the individual, and they frequently guarantee confidentiality, they often do not note characteristics of the individual who was involved in the event, such as profession, discipline, and training status. Without data on error patterns—including what type of caregiver was involved, at what training level, and whether there were errors of omission, commission, misdiagnosis, or work-around—it is more difficult to address educational deficiencies (Battles and Shea, 2001). However, if data were available on when during a work period the event occurred, it might show that events happened most frequently at the end of an extended work period and were possibly an indication of decreased attention due to fatigue. Error reports indicating the time and other specific circumstances of events might reveal patterns related to work shifts of individuals or teams, the transitions from team to team, and whether fatigue or communications failures were a significant factor. Currently, reporting does not capture such information in hospitals for use in residency programs.

While most individual institution reporting systems would have a limited volume of reports and insufficient power to draw statistically valid conclusions about certain events, they could be valuable to management and educators by identifying any problem. Just one report of a near miss could identify a critical situation in need of redesign and lead to significant quality improvement. Residents in particular can play an important role in improving health systems in this regard. Acting as the "spackle" in the busy training settings of the health profession, they know where gaps exist in the system, and helping to identify them can be an asset in any care setting as well. Not only would residents be a part of the solution to these problems, they would benefit from the educational benefits these systems can provide.

If more hospitals had robust error-reporting systems with sufficiently detailed data reported, and an atmosphere that encouraged all workers to participate, and if such data were consistently recorded from hospital to hospital and could be aggregated to a national level, or if there were an effective national reporting program in place, it might have been possible for this committee to assess whether errors by residents were a serious threat to patients and to what extent those errors could be attributed to fatigue and long work hours. However, data at that level do not currently exist.

The issues surrounding error-reporting systems are not new and are of much broader relevance than just the training of physicians. In fact, earlier IOM reports, including many in the Quality Chasm series, contain extensive discussions of these issues and recommendations on how to develop an effective error-reporting and learning system. This report does not repeat those discussions, but rather turns the focus toward residents. Healthcare organizations have been responding to these reports and pressures from the Joint Commission and public bodies. Both the public and healthcare professionals are growing more aware of the importance of identifying errors in understanding how to improve the quality of services and the safety of patients and workers. However, progress in reporting and reducing errors has not been uniform (ACGME, 2008; Hines et al., 2008; Kaldjian et al., 2008).

At the hospital level, to learn from mistakes in patient care involving residents and to prevent similar events in the future requires an error-reporting system with a common set of data standards and a broader definition of what information should be collected that could, perhaps, contribute to the training of doctors. The error-reporting system would have to include training for all residents in what should be reported, how to report incidents, who should report, and how to analyze the reported errors to understand the root causes of the error and the changes needed to prevent future harm to patients (Garbutt et al., 2008). Support and encouragement from executive leadership, methods for reporting errors anonymously, and a feedback loop to the residents, caregivers, and especially the graduate medical faculty are all important elements of the reporting system for promoting its use. Knowledge that the reported information will be used to enhance patient care is critical to motivate doctors and other caregivers to make the effort to report an incident. The perception that there is no follow-up can be a deterrent to reporting (Evans et al., 2006). The use of such reported information to enhance resident training would also be a benefit.

Recommendation 8-2: Graduate medical education-sponsoring institutions should fully involve residents in their safety reporting, learning, and quality improvement systems, and this should become an important part of the residents' educational experience.

Health Information Technology for Clinical Decision Support

Today's residents face a rapidly expanding knowledge base while serving in a learning environment with a growing focus on patient safety measures. As mentioned, fostering relationships between residents and other hospital staff provides important clinical support to residents as they learn to make decisions about patient management. Yet other forms of support can aid them in their clinical decision making as well. Besides seeking help from their peers and supervisors to reduce uncertainty and prevent errors, residents can use an array of information technologies to assist them.

Health information technologies include up-to-date patient-specific data in electronic medical records, clearly documented handovers from other team members, and diagnostic support systems that offer clinicians opportunities to avoid reaching premature closure on diagnoses. Information technology (IT) support systems have been shown to enhance care and reduce errors by alerting patients to drug interactions and providing access to clinical guidelines (Bates et al., 1998; Garg et al., 2005; Petersen et al., 1998). Various studies show that electronic medical records have been observed to help in documentation, thus preventing errors and reducing test ordering by residents (Hier et al., 2005; Keenan et al., 2006; O'Connell et al., 2004; Stair and Howell, 1995).

IT solutions can also enhance communication for supervision in the event that attendings are unable to be on site. Remote access can help attendings monitor the activities of residents as well as patient progress through review of online records. Greater accountability is being required of residency program directors (e.g., monitoring resident hours, privileges for clinical and surgical procedures, workflow management) that require documentation for accreditation purposes and to enhance patient safety (Afrin, 2006).

Despite all the benefits electronic tools can provide, if they are ill suited to an organization's needs or are not used appropriately, unintended or adverse consequences are possible, requiring ongoing maintenance and attention to business processes to prevent such occurrences (IOM, 2006). Not all electronic medical systems are created equally; some are quite advanced while others are more rudimentary, ranging in degrees of content or flexibility of integration with other systems. Those that integrate poorly with other information systems may be more time consuming to use or may create duplicative efforts instead of reducing them (Campbell et al., 2006). Heavy reliance on electronic systems can also decrease general communication skills and the occurrence of face-to-face interactions among clinicians (Ash et al., 2007). In some cases, electronic medical systems can contribute to errors by new users who are learning to navigate these systems and incorrectly fill out information or accidentally press wrong functions, or by program formats that are too cumbersome to enter information in a timely manner (Campbell et al., 2006). For these reasons, training staff in how to use any electronic system is critical to their effectiveness and efficiency (Arora et al., 2007).

Health IT will likely continue to advance and come to be a more widely used tool in hospitals and training facilities. As more residents are exposed to these systems they may be in an ideal position to provide necessary feedback on how to improve their functionality for clinical use. Currently, however, it is beyond the scope of this report to evaluate specific models of electronic medical records or decision support systems for residents. The committee recognizes the potential usefulness of these systems for information transfer, supervision, workload reduction, and enhanced education in the pursuit of patient safety and urges their continued evaluation and adoption. Adoption of these systems can have value to all staff on patient care teams and is not resident specific.

DEVELOPING A TEAM CULTURE

It has been recognized that healthcare structures are complex, "characterized by competing responsibilities and an evolving perception of patient care as a collective responsibility" (Park et al., 2007, p. 111). Residents are increasingly trained and expected to practice in models of integrated care, which rely on the coordination of different services, clinicians, and teams all working together to provide comprehensive care for patients. Facilitating this coordination requires effective communication skills and strategies across and among all units—a fundamental trait of teams and teamwork. A *team* is defined as a distinguishable set of two or more people interacting toward a common goal with specific roles and boundaries on tasks that are interdependent and are completed within a larger organizational context (Kozlowski and Bell, 2003; Salas et al., 1992). The tasks that teams work on tend to require (1) dynamic exchange of team member resources (including information), (2) coordination of activities, (3) adaptability to task demands, and (4) an organizational structure that coordinates members (Salas et al., 1992; Swezey et al., 1994).

Team-based work is an effective strategy not only for combating errors, but also for mitigating the negative impact of high workloads, fatigue, and stress, especially when team members become aware of their own responsibilities in addition to the responsibilities of others (Salas and Cannon-Bowers, 2000b; Salas et al., 2005; Smith-Jentsch et al., 1996). Using a team-based approach for resident work and patient care could suitably address concerns of both continuity and fatigue, reducing potential threats to patient safety.

A study by Singh and colleagues, analyzing malpractice claims in which residents were identified as playing a role in harming patients, concluded that residents "are particularly vulnerable to medical errors owing to teamwork failures" (Singh et al., 2007, p. 2030). From among 240 cases resulting in patient injury, teamwork breakdowns were a factor in 70 percent of them (errors in judgment were a factor in 72 percent and lack of technical competence was a factor in 58 percent) (Singh et al., 2007). It was also found that "lack of supervision and handoff problems were the most prevalent types of teamwork problems [in the malpractice cases], and both were disproportionately more common among errors that involved trainees than those that did not" (respectively, 54 percent vs. 7 percent, $p = .001$, and 20 percent vs. 12 percent, $p = .009$) (Singh et al., 2007, p. 2032). Such data suggest that residents could greatly benefit from a reinforced team structure and training in communication and team practices to prevent patient harm, where supervision is readily available to provide necessary guidance.

Shared Responsibility

For team structures to develop and thrive, it is important to transform the culture of care. By introducing such culture change into residency programs, researchers have noted that "the real challenge of the 80-hour workweek is that it demands a psychological transformation" (Mukherjee, 2004, p. 1824), one that allows residents to tone down expectations of superhuman resistance to long hours and continuous care, and give in to the flexibility of team systems. Residents will continue to strive to be independent practitioners, but given their time constraints and the content of their work, distributing workload among colleagues can help them collectively better manage their time and alleviate demands while on duty. In this way, a team dynamic lends itself to better organization, which has the potential to better sustain continuity of care among multiple health practitioners and, in turn, help improve overall patient care.

There is general agreement among systems experts that a mentality of "shared responsibility" is necessary to successfully adopt interventions for any specifically team-centered goal (Arora et al., 2008). Mutual trust and shared mental models are key components to successfully achieving these goals. Shared mental models refer to an organized knowledge structure among a team for a particular task in which the team is engaged and how team members will interact. This interaction includes anticipating and predicting each other's needs, identifying changes in the team task or teammates, and implicitly adjusting strategies as needed (Salas et al., 2005).

A challenge in adopting this shared mentality is that a variety of team structures exists in hospitals of which residents are a part or with which they need to communicate. Nursing teams, physician teams, resident teams, lab clinicians, pharmacists and other healthcare professionals all exist interdependently with one another and combine into integrated teams to provide comprehensive and continuous care to any given patient. Furthermore, each type of professional (e.g., nurses, doctors) is trained to communicate differently, creating discrepancies in expectations when exchanging information

(Leonard et al., 2004). Targeting residents is a good way to introduce teamwork and shared accountability across these interdependent teams, which can help develop structured communication among all healthcare workers and ultimately reduce gaps or errors in patient care.

Teamwork and Task Performance

The focus of teamwork for residents is individual performance in a team environment; helping each resident perform to his or her fullest capacity, most effectively and efficiently, while creating more learning opportunities. *Teamwork* is defined as a set of interrelated behaviors, cognitions (thoughts), and attitudes (feelings) held by each team member that combine to facilitate adaptive, coordinated performance (Morgan et al., 1986; Salas et al., 2004). Learning and using the five core components of teamwork—specifically, leadership, mutual performance modeling, backup behavior, adaptability, and team orientation (Box 8-2)—can lead to more effective work processes.

BOX 8-2
The Five Core Components of Teamwork

1. **Team leadership:** The ability to direct and coordinate the activities of other team members; assess team performance; assign tasks; develop team knowledge, skills, and abilities; motivate team members; plan and organize; and establish a positive atmosphere.

2. **Mutual performance monitoring:** The ability to develop common understandings of the team environment and apply appropriate task strategies to accurately monitor teammate performance.

3. **Backup behavior:** The ability to anticipate other team members' needs through accurate knowledge of their responsibilities. This includes the ability to shift workload among members to achieve balance during periods of high workload or pressure.

4. **Adaptability:** The ability to adjust strategies based on information gathered from the environment through the use of backup behavior and reallocation of intrateam resources. Altering a course of action or team repertoire in response to changes in conditions (internal or external).

5. **Team Orientation:** The propensity to take others' behavior into account during group interaction and the belief in the importance of the team's goals over individual members' goals.

SOURCE: Salas et al., 2005.

Research by Jung and colleagues has demonstrated that as teams work together applying these components, they can increase their productivity and build shared ideas of how to accomplish a task (Jung et al., 2002). Teamwork depends on each team member's ability and willingness to cooperate toward achieving shared goals. For residents, these goals are providing effective patient care, maximizing learning, and minimizing errors.

An example of using team efforts to achieve these goals is a general medicine residency program that recently developed a team-based teaching program to determine the effects of reducing workload and providing more supervision and teaching upon the quality of resident education and patient care. This was accomplished by creating integrated teams of two attendings (one a primary care physician and the other a hospitalist or subspecialist), two residents, three interns, and two medical students. The team remained together for 2 weeks or more at a time completing daily work and teaching schedules with a cap of 15 patients at any time, who were divided equitably among the interns. Compared to the traditional general medicine resident team, patients treated by the integrated team had a lower mortality rate (1.4 percent vs. 2.4 percent, $p = .053$) and significantly shorter length of stays (LOSs) (4.2 vs. 4.7 days, $p < .01$) (McMahon, 2008). The quality of discharge communications was also higher for the integrated team, and members managed to double their amount of time spent in educational sessions (McMahon, 2008), demonstrating that improved patient care and resident learning can both be facilitated by team structures. Other efforts incorporating interdisciplinary or multidisciplinary team rounds had very similar results of reduced LOS and improved core knowledge and team skills (Curley et al., 1998; O'Mahoney et al., 2007).

Training Residents in Effective Teamwork Strategies

Teamwork skills often need to be learned and numerous reports and publications highlight the importance of team training in realizing goals to enhance patient safety and clinical communication (Barach and Small, 2000; Barach and Weingart, 2004; Jeffcott and Mackenzie, 2008; Leonard et al., 2004). The skills acquired through resident team training can be taught using various techniques (Klein et al., 2009; Rosen et al., 2008; Salas and Cannon-Bowers, 2000a, 2001; Salas et al., 2008; Smith-Jentsch et al., 1998), including simulation strategies that can measure team competency (Zheng et al., 2008), and are valuable because they are applicable to many facets of resident work.

Since there are multiple forms of team structures and methods in which teams can be trained, a straightforward way of introducing team-centered activity and skills into healthcare settings is by training residents as a team around completing specific tasks. Also referred to as "task-tailored training," this type of training can be effective for several processes that take place in hospitals (e.g., surgical procedures, handovers, clinical rounds). For example, a study by Chung et al. (2007) applied a task-tailored team approach to the rounding process (a substantial part of some handover processes) performed by general surgical residents, specifically morning rounds. The strategy focused residents' work during their rounds on three distinct tasks: detecting postoperative complications early or conducting orderly preoperative workup; informing patients of the agenda for the day; and answering patient questions and complaints. In addition to having them focus exclusively on these tasks, the members of the rounding team (nine members: one PGY-5 and eight PGY-1 to PGY-4s) were assigned specific roles. With this direction, residents' work became more purposeful and efficient, and they completed rounds and associated work in 1 hour (Chung and Ahmed, 2007). After a year of implementation, attendings on duty observed substantial increases in resident professionalism and communication, demonstrating the positive effects of team structure and culture, not only on workload but on general attitude as well. Such structure also automated team continuity of care, increasing patient awareness of the resident team and satisfaction with care received (Chung and Ahmed, 2007).

Whichever methods of handover intervention or error reporting are used, residents will have to be trained in the team components of coordination, communication, and cooperation to conduct them most effectively and efficiently. In some facilities, faculty or supervisors may also have to be trained in these matters in order to ensure their organizational adoption and most effective implementation.

CONCLUSION

Redesigning any part of the resident learning process is a challenge. To eliminate preventable adverse events and intercept other errors before they harm the patient, it is important to have in place an environment that is both mindful of errors and nonpunitive, as well as leaders willing to consider redesign of the institutions' systems and processes as necessary to reduce risks. The emphasis on handovers, blame-free error reporting, and teamwork does not mean that individual residents are not expected to develop a sense of loyalty or personal responsibility for individual patient care, but it helps ensure that the best information is available at all times for patient care given that a resident or any caregiver cannot be at the bedside 24 hours a day, 7 days week. It may not be possible to eliminate discontinuity altogether in healthcare settings, but the training system can strive to minimize its effects by enhancing the quality of handovers and error reporting, promoting patient-centered approaches, and improving physician relationships by facilitating communication through team structures. Suggestions for these areas of the system are ones that the committee believes deserve immediate attention and can bear positive results if effectively applied.

REFERENCES

ACGME (Accreditation Council for Graduate Medical Education). 2007. *Common program requirements: General competencies.* http://www.acgme.org/outcome/comp/GeneralCompetenciesStandards21307.pdf (accessed March 17, 2008).

———. 2008. The ACGME duty hour standards: One element for promoting good learning, safe patient care and resident well-being: Data on compliance and effect. Presentation by Ingrid Philibert and Thomas Nasca to the Committee on Optimizing Graduate Medical Trainee (Resident) Hours and Work Schedules to Improve Patient Safety, March 4, 2008, Irvine, CA.

Afessa, B., C. C. Kennedy, K. W. Klarich, T. R. Aksamit, J. C. Kolars, and R. D. Hubmayr. 2005. Introduction of a 14-hour work shift model for housestaff in the medical ICU. *Chest* 128(6):3910-3915.

Afrin, L. B., G. W. Arana, F. J. Medio, A. F. Ybarra, and H. S. Clarke, Jr. 2006. Improving oversight of the graduate medical education enterprise: One institution's strategies and tools. *Academic Medicine* 81(5):419-425.

AHRQ (Agency for Healthcare Research and Quality). 2008. *Hospital survey on patient safety culture: 2008 comparative database report.* Rockville, MD: Agency for Healthcare Research and Quality.

Alvarado, K., R. Lee, E. Christoffersen, N. Fram, S. Boblin, N. Poole, J. Lucas, and S. Forsyth. 2006. Transfer of accountability: Transforming shift handover to enhance patient safety. *Healthcare Quarterly* 9(Supp):75-79.

Arora, V. M., and J. M. Farnan. 2008. Care transitions for hospitalized patients. *Medical Clinics of North America* 92(2):315-324.

Arora, V., and J. Johnson. 2006. A model for building a standardized hand-off protocol. *Joint Commission Journal on Quality & Patient Safety* 32(11):646-655.

Arora, V., J. Johnson, D. Lovinger, H. J. Humphrey, and D. O. Meltzer. 2005. Communication failures in patient sign-out and suggestions for improvement: A critical incident analysis. *Quality and Safety in Health Care* 14(6):401-407.

Arora, V., J. Kao, D. Lovinger, S. C. Seiden, and D. Meltzer. 2007. Medication discrepancies in resident sign-outs and their potential to harm. *Journal of General Internal Medicine* 22(12):1751-1755.

Arora, V. M., J. K. Johnson, D. O. Meltzer, and H. J. Humphrey. 2008. A theoretical framework and competency-based approach to improving handoffs. *Quality and Safety in Health Care* 17(1):11-14.

Ash, J. S., D. F. Sittig, E. G. Poon, K. Guappone, E. Campbell, and R. H. Dykstra. 2007. The extent and importance of unintended consequences related to computerized provider order entry. *Journal of the American Medical Informatics Association* 14(4):415-423.

Barach, P., and S. D. Small. 2000. Reporting and preventing medical mishaps: Lessons from non-medical near miss reporting systems. *BMJ* 320(7237):759-763.

Barach, P., and M. Weingart. 2004. Trauma team performance. In *Trauma: Resuscitation, anesthesia, surgery, and critical care*, edited by W. Wilson, C. Grande, and D. Hoyt. New York: Dekker Inc.

Bates, D. W., L. L. Leape, D. J. Cullen, N. Laird, L. A. Petersen, J. M. Teich, E. Burdick, M. Hickey, S. Kleefield, B. Shea, M. Vander Vliet, and D. L. Seger. 1998. Effect of computerized physician order entry and a team intervention on prevention of serious medication errors. *JAMA* 280(15):1311-1316.

Battles, J. B., and C. E. Shea. 2001. A system of analyzing medical errors to improve GME curricula and programs. *Academic Medicine* 76(2):125-133.

Beach, C., P. Croskerry, and M. Shapiro. 2003. Profiles in patient safety: Emergency care transitions. *Academic Emergency Medicine* 10(4):364-367.

Behara, R., R. L. Wears, and S. J. Perry. 2005. A conceptual framework for studying the safety of transitions in emergency care. In *Advances in patient safety: From research to implementation*, Vol. 2, Concepts and Methodology. Rockville, MD: Agency for Healthcare Research and Quality. Pp. 309-321.

Borowitz, S. M., L. A. Waggoner-Fountain, E. J. Bass, and R. M. Sledd. 2008. Adequacy of information transferred at resident sign-out (in hospital handover of care): A prospective survey. *Quality and Safety in Health Care* 17(1):6-10.

Campbell, E. M., D. F. Sittig, J. S. Ash, K. P. Guappone, and R. H. Dykstra. 2006. Types of unintended consequences related to computerized provider order entry. *Journal of the American Medical Informatics Association* 13(5):547-556.

Care Transitions Program. 2008. *The care transitions program: Improving quality and safety during care hand-offs.* http://caretransitions.org/overview.asp (accessed July 21, 2008).

Carpenter, D. 2008. 2008 innovators. *Healthcare's Most Wired Magazine*, http://www.hhnmostwired.com/hhnmostwired_app/jsp/articledisplay.jsp?dcrpath=HHNMOSTWIRED/Article/data/Summer2008/08SUM_MW_FEA_Innovators&domain= (accessed October 20, 2008).

Carroll, J. S., and M. A. Quijada. 2004. Redirecting traditional professional values to support safety: Changing organisational culture in health care. *Quality and Safety in Health Care* 13(Suppl II):ii16-ii21.

Catchpole, K. R., M. R. De Leval, A. McEwan, N. Pigott, M. J. Elliott, A. McQuillan, C. Macdonald, and A. J. Goldman. 2007. Patient handover from surgery to intensive care: Using Formula 1 pit-stop and aviation models to improve safety and quality. *Pediatric Anesthesia* 17(5):470-478.

Chung, R. S., and N. Ahmed. 2007. How surgical residents spend their training time: The effect of a goal-oriented work style on efficiency and work satisfaction. *Archives of Surgery* 142(3):249-252.

Coleman, E. A., and R. A. Berenson. 2004. Lost in transition: Challenges and opportunities for improving the quality of transitional care. *Annals of Internal Medicine* 141(7):533-536.

Curley, C. M. D. M. S., J. E. M. D. McEachern, and T. P. Speroff. 1998. A firm trial of interdisciplinary rounds on the inpatient medical wards: An intervention designed using continuous quality improvement. *Medical Care* 36(8 Suppl):AS4-AS12.

Evans, S. M., J. G. Berry, B. J. Smith, A. Esterman, P. Selim, J. O'Shaughnessy, and M. DeWit. 2006. Attitudes and barriers to incident reporting: A collaborative hospital study. *Quality and Safety in Health Care* 15(1):39-43.

Fletcher, K. E., S. Saint, and R. S. Mangrulkar. 2005. Balancing continuity of care with residents' limited work hours: Defining the implications. *Academic Medicine* 80(1):39-43.

Forster, A. J., H. J. Murff, J. F. Peterson, T. K. Gandhi, and D. W. Bates. 2003. The incidence and severity of adverse events affecting patients after discharge from the hospital. *Annals of Internal Medicine* 138(3):161-167.

Gandhi, T. K. 2005. Fumbled handoffs: One dropped ball after another. *Annals of Internal Medicine* 142(5):352-358.

Gandhi, T. K., A. Kachalia, E. J. Thomas, A. L. Puopolo, C. Yoon, T. A. Brennan, and D. M. Studdert. 2006. Missed and delayed diagnoses in the ambulatory setting: A study of closed malpractice claims. *Annals of Internal Medicine* 145(7):488-496.

Garbutt, J., A. D. Waterman, J. M. Kapp, W. C. Dunagan, W. Levinson, V. Fraser, and T. H. Gallagher. 2008. Lost opportunities: How physicians communicate about medical errors. *Health Affairs* 27(1):246-255.

Garg, A. X., N. K. Adhikari, H. McDonald, M. P. Rosas-Arellano, P. J. Devereaux, J. Beyene, J. Sam, and R. B. Haynes. 2005. Effects of computerized clinical decision support systems on practitioner performance and patient outcomes: A systematic review. *JAMA* 293(10):1223-1238.

Goldstein, M. J., E. Kim, W. D. Widmann, and M. A. Hardy. 2004. A 360 degrees evaluation of a night-float system for general surgery: A response to mandated work-hours reduction. *Current Surgery* 61(5):445-451.

Halasyamani, L., S. Kripalani, E. Coleman, J. Schnipper, C. van Walraven, J. Nagamine, P. Torcson, T. Bookwalter, T. Budnitz, and D. Manning. 2006. Transition of care for hospitalized elderly patients—Development of a discharge checklist for hospitalists. *Journal of Hospital Medicine (Online)* 1(6):354-360.

Hier, D. B., A. Rothschild, A. LeMaistre, and J. Keeler. 2005. Differing faculty and housestaff acceptance of an electronic health record. *International Journal of Medical Informatics* 74(7-8):657-662.

Hines, S., K. Luna, J. Lofthus, M. Marquart, and D. Stelmokas. 2008. *Becoming a high reliability organization: Operational advice for hospital leaders.* Rockville, MD: Agency for Healthcare Research and Quality.

Horwitz, L. I., H. M. Krumholz, M. L. Green, and S. J. Huot. 2006. Transfers of patient care between house staff on internal medicine wards: A national survey. *Archives of Internal Medicine* 166(11):1173-1177.

Horwitz, L., T. Moin, and M. Green. 2007a. Development and implementation of an oral sign-out skills curriculum. *Journal of General Internal Medicine* 22(10):1470-1474.

Horwitz, L. I., M. Kosiborod, Z. Lin, and H. M. Krumholz. 2007b. Changes in outcomes for internal medicine inpatients after work-hour regulations. *Annals of Internal Medicine* 147(2):97-103.

Horwitz, L. I., T. Moin, H. M. Krumholz, L. Wang, and E. H. Bradley. 2008. Consequences of inadequate sign-out for patient care. *Archives of Internal Medicine* 168(16):1755-1760.

IHI (Institute for Healthcare Improvement). 2007. *Transforming care at the bedside—How-to guide: Creating an ideal transition home for patients with heart failure.* Boston, MA: Institute for Healthcare Improvement.

IOM (Institute of Medicine). 2000. *To err is human: Building a safer health system.* Washington, DC: National Academy Press.

———. 2001. *Crossing the quality chasm: A new health system for the 21st century.* Washington, DC: National Academy Press.

———. 2003. *Keeping patients safe: Transforming the work environment of nurses.* Washington, DC: The National Academies Press.

———. 2004. *Patient safety: Achieving a new standard for care.* Washington, DC: The National Academies Press.

———. 2006. *Preventing medication errors: Quality chasm series.* Washington, DC: The National Academies Press.

Jeffcott, S. A., and C. F. Mackenzie. 2008. Measuring team performance in healthcare: Review of research and implications for patient safety. *Journal of Critical Care* 23(2):188-196.

Joint Commission. 2007. *National Patient Safety Goals: 2007 national patient safety goals hospital version manual chapter, including implementation expectations.* http://www.jointcommission.org/PatientSafety/NationalPatientSafetyGoals/07_hap_cah_npsgs.htm (accessed February 26, 2008).

———. 2008. Behaviors that undermine a culture of safety. *Sentinel Event Alert* (Issue 40, July 9):1. http://www.jointcommission.org/SentinelEvents/SentinelEventAlert/sea_40.htm (accessed July 25, 2008).

Joint Commission International Center for Patient Safety. 2006. Improving hand-off communications: Meeting national patient safety goal 2E. *Patient Safety* 6:9-15.

Jung, D. I., J. J. Sosik, and K. B. Baik. 2002. Investigating work group characteristics and performance over time: A replication and cross-cultural extension. *Group Dynamics* 6(2):153-171.

Kaldjian, L. C., E. W. Jones, B. J. Wu, V. L. Forman-Hoffman, B. H. Levi, and G. E. Rosenthal. 2008. Reporting medical errors to improve patient safety: A survey of physicians in teaching hospitals. *Archives of Internal Medicine* 168(1):40-46.

Kaplan, H. S., and B. Rabin Fastman. 2003. Organization of event reporting data for sense making and system improvement. *Quality and Safety in Health Care* 12(Suppl 2).

Keenan, C. R., H. H. Nguyen, and M. Srinivasan. 2006. Electronic medical records and their impact on resident and medical student education. *Academic Psychiatry* 30(6):522-527.

Klein, C., D. DiazGranados, E. Salas, H. Le, C. S. Burke, R. Lyons, and G. F. Goodwin. 2009. Does team building work? *Small Group Research* 40(2):181-222.

Kozlowski, S. W. J., and B. S. Bell. 2003. Work groups and teams in organizations. In *Handbook of physchology: Industrial and organizational psychology*, Vol. 12, edited by W. Borman, D. Igen, and R. Klimoski. London: Wiley. Pp. 333-375.

Kripalani, S. 2008. *Care transition.* http://webmm.ahrq.gov/printviewperspective.aspx?perspectiveID=52 (accessed April 16, 2008).

Laine, C., L. Goldman, J. R. Soukup, and J. G. Hayes. 1993. The impact of a regulation restricting medical house staff working hours on the quality of patient care. *JAMA* 269(3):374-378.

Landrigan, C. P., J. M. Rothschild, J. W. Cronin, R. Kaushal, E. Burdick, J. T. Katz, C. M. Lilly, P. H. Stone, S. W. Lockley, D. W. Bates, and C. A. Czeisler. 2004. Effect of reducing interns' work hours on serious medical errors in intensive care units. *New England Journal of Medicine* 351(18):1838-1848.

Leonard, M., S. Graham, and D. Bonacum. 2004. The human factor: The critical importance of effective teamwork and communication in providing safe care *Quality and Safety in Health Care* 13(Suppl 1):i85-i90.

McMahon, G. 2008. *Faulkner hospital: Integrated teaching unit—Progress report #3.* Boston, MA: Brigham and Women's Hospital.

Moore, C., J. Wisnivesky, S. Williams, and T. McGinn. 2003. Medical errors related to discontinuity of care from an inpatient to an outpatient setting. *Journal of General Internal Medicine* 18(8):646-651.

Morgan, B. B., A. S. Glickman, E. A. Woodard, A. S. Blaiwes, and E. Salas. 1986. *Measurement of team behavior in a Navy training environment.* Orlando, FL: Naval Training Systems Center, Human Factors Division.

Mukherjee, S. 2004. A precarious exchange. *New England Journal of Medicine* 351(18):1822-1824.

O'Connell, R. T., C. Cho, N. Shah, K. Brown, and R. N. Shiffman. 2004. Take note(s): Differential EHR satisfaction with two implementations under one roof. *Journal of the American Medical Informatics Association* 11(1):43-49.

Okie, S. 2007. An elusive balance—Residents' work hours and the continuity of care. *New England Journal of Medicine* 356(26):2665-2667.

O'Mahony, S., E. Mazur, P. Charney, Y. Wang, and J. Fine. 2007. Use of multidisciplinary rounds to simultaneously improve quality outcomes, enhance resident education, and shorten length of stay. *Journal of General Internal Medicine* 22(8):1073-1079.

Park, J., S. I. Woodrow, R. K. Reznick, J. Beales, and H. M. MacRae. 2007. Patient care is a collective responsibility: Perceptions of professional responsibility in surgery. *Surgery* 142(1):111-118.

Parke, B., and A. Mishkin. 2005. Best practices in shift handover communication: Mars exploration rover surface operations. Paper read at European Space Agency (Special Publication), ESA SP, Nice, France.

Patterson, E. S. 2007. Communication strategies from high-reliability organizations: Translation is hard work. *Annals of Surgery* 245(2):170-172.

———. 2008. Structuring flexibility: The potential good, bad and ugly in standardisation of handovers. *Quality and Safety in Health Care* 17(1):4-5.

Patterson, E. S., E. M. Roth, D. D. Woods, R. Chow, and J. O. Gomes. 2004. Handoff strategies in settings with high consequences for failure: Lessons for health care operations. *International Journal for Quality in Health Care* 16(2):125-132.

Perry, S. J., R. L. Wears, and E. S. Patterson. 2008 (unpublished). *High hanging fruit: Improving transitions in health care*. Washington, DC: Department of Veterans Affairs.

Petersen, L. A., T. A. Brennan, A. C. O'Neil, E. F. Cook, and T. H. Lee. 1994. Does housestaff discontinuity of care increase the risk for preventable adverse events? *Annals of Internal Medicine* 121(11):866.

Petersen, L. A., E. J. Orav, J. M. Teich, A. C. O'Neil, and T. A. Brennan. 1998. Using a computerized sign-out program to improve continuity of inpatient care and prevent adverse events. *Joint Commission Journal on Quality Improvement* 24(2):77-87.

Roberts, K. H., P. Madsen, V. Desai, and D. Van Stralen. 2005. A case of the birth and death of a high reliability healthcare organisation. *Quality and Safety in Health Care* 14(3):216-220.

Roberts, V., and M. M. Perryman. 2007. Creating a culture for health care quality and safety. *Health Care Manager* 26(2):155-158.

Rosen, M. A., E. Salas, T. S. Wu, S. Silvestri, E. H. Lazzara, R. Lyons, S. J. Weaver, and H. B. King. 2008. Promoting teamwork: An event-based approach to simulation-based teamwork training for emergency medicine residents. *Academic Emergency Medicine* 15:1-9.

Sabir, N., S. M. Yentis, and A. Holdcroft. 2006. A national survey of obstetric anaesthetic handovers. *Anaesthesia* 61(4):376-380.

Salas, E., and J. A. Cannon-Bowers. 2000a. The anatomy of team training. In *Training and retraining: A handbook for business, industry, government, and the military*, edited by S. Tobias and J. D. Fletcher. New York: Macmillan Reference. Pp. 312-335.

———. 2000b. Designing training systems systematically. In *The Blackwell handbook of principles of organizational behavior*, edited by E. A. Locke. Malden, MA: Blackwell Publisher Ltd. Pp. 43-59.

———. 2001. The science of training: A decade of progress. *Annual Review of Psychology* 52:471-499.

Salas, E., D. DiazGranados, C. Klein, C. Burke, K. Stagl, G. Goodwin, and S. Haplin. 2008. Does team training lead to improved team outcomes? A meta-analysis. *Human Factors* 50(6):903-933.

Salas, E., T. L. Dickinson, S. A. Converse, and S. I. Tannenbaum. 1992. Towards an understanding of team performance and training. In *Teams: Their training and performance*, edited by R. W. Swezey and E. Salas. Norwood, NJ: Ablex Publishing Corporation.

Salas, E., E. Sims, and C. Klein. 2004. Cooperation and teamwork at work. In *Encyclopedia of applied psychology,* Vol. 1, edited by C. D. Spielberger. San Diego, CA: Academic Press. Pp. 497-505.

Salas, E., D. E. Sims, and C. S. Burke. 2005. Is there a "Big Five" in teamwork? *Small Group Research* 36(5):555-599.

Saultz, J. W. 2003. Defining and measuring interpersonal continuity of care. *Annals of Family Medicine* 1(3):134-143.

Shojania, K. G. 2005. *Interpreting the patient safety literature.* http://webmm.ahrq.gov/printviewperspective.aspx?perspectiveID=5 (accessed April 15, 2008).

Simmons, J., and J. Gonzalez del Rey. 2008. *Family-centered rounds (FCRs): Building systems and skills for success at the bedside of the hospitalized child in a teaching hospital.* Cincinnati, OH: Cincinnati Children's Hospital Medical Center. PowerPoint presentation.

Singer, J. I., and J. Dean. 2006. Emergency physician intershift handovers: An analysis of our transitional care. *Pediatric Emergency Care* 22(10):751-754.

Singer, S. J., D. M. Gaba, J. J. Geppert, A. D. Sinaiko, S. K. Howard, and K. C. Park. 2003. The culture of safety: Results of an organization-wide survey in 15 California hospitals. *Quality and Safety in Health Care* 12(2):112-118.

Singh, H., E. J. Thomas, L. A. Petersen, and D. M. Studdert. 2007. Medical errors involving trainees: A study of closed malpractice claims from 5 insurers. *Archives of Internal Medicine* 167(19):2030-2036.

Smith-Jentsch, K. A., E. Salas, and D. P. Baker. 1996. Training team performance-related assertiveness. *Personnel Psychology* 49(4):909-936.

Smith-Jentsch, K., R. Zeisig, B. Acton, and J. McPherson. 1998. Team dimensional training: A strategy for guided team self-correction. In *Making decisions under stress*, edited by E. Salas and J. A. Cannon-Bowers. Washington, DC: American Psychological Association. Pp. 271-297.

Solet, D. J., J. M. Norvell, G. H. Rutan, and R. M. Frankel. 2005. Lost in translation: Challenges and opportunities in physician-to-physician communication during patient handoffs. *Academic Medicine* 80(12):1094-1099.

Stair, T. O., and J. M. Howell. 1995. Effect on medical education of computerized physician order entry. *Academic Medicine* 70(6):543.

Stevens, D. P., C. S. Sixta, E. Wagner, and J. L. Bowen. 2008. The evidence is at hand for improving care in settings where residents train. *Journal of General Internal Medicine* 23(7):1116-1117.

Sutcliffe, K. M., E. Lewton, and M. M. Rosenthal. 2004. Communication failures: An insidious contributor to medical mishaps. *Academic Medicine* 79(2):186-194.

Swezey, R. W., A. L. Meltzer, and E. Salas. 1994. Some issues involved in motivating teams. In *Motivation: Theory and research*, edited by H. F. O'Neil, Jr., and M. Drillings. Hillsdale, NJ: Lawrence Erlbaum Associates. Pp. 141-169.

TeamSTEPPS. 2007. *Strategies and tools to enhance performance and patient safety: Essentials course*. Rockville, MD: Agency for Healthcare Research and Quality. PowerPoint presentation.

Van Eaton, E. G., K. D. Horvath, W. B. Lober, A. J. Rossini, and C. A. Pellegrini. 2005. A randomized, controlled trial evaluating the impact of a computerized rounding and sign-out system on continuity of care and resident work hours. *Journal of American College of Surgeons* 200(4):538-545.

Vidyarthi, A. 2004. *Fumbled handoff.* http://webmm.ahrq.gov/printview.aspx?caseID=55 (accessed March 18, 2008).

Vidyarthi, A. R., V. Arora, J. L. Schnipper, S. D. Wall, and R. M. Wachter. 2006. Managing discontinuity in academic medical centers: Strategies for a safe and effective resident sign-out. *Journal of Hospital Medicine* 1(4):257-266.

Volpp, K. G., and C. P. Landrigan. 2008. Building physician work hour regulations from first principles and best evidence. *JAMA* 300(10):1197-1199.

Wachter, R. M. 2008. Chapter 8: Transition and handoff errors. In *Understanding patient safety*, edited by J. Shanahan and K. Edmanson. New York: The McGraw-Hill Companies, Inc.

Wachter, R. M., K. G. Shojania, A. J. Markowitz, M. Smith, and S. Saint. 2006. Quality grand rounds: The case for patient safety. *Annals of Internal Medicine* 145(8):629-630.

Warm, E., D. Schauer, T. Diers, B. Mathis, Y. Neirouz, J. Boex, and G. Rouan. 2008. The ambulatory long-block: An Accreditation Council for Graduate Medical Education (ACGME) Educational Innovations Project (EIP). *Journal of General Internal Medicine* 23(7):921-926.

Weick, K. E., and K. M. Sutcliffe, eds. 2001. *Managing the unexpected: Assuring high performance in an age of complexity*. San Francisco: Jossey-Bass.

Zheng, B., P. Denk, D. Martinec, P. Gatta, M. Whiteford, and L. Swanström. 2008. Building an efficient surgical team using a bench model simulation: Construct validity of the legacy inanimate system for endoscopic team training (LISETT). *Surgical Endoscopy* 22(4):930-937.

9

Resources to Implement Improvements for Patient Safety and Resident Training

Improved residency training conditions for patient safety, enhanced resident well-being, and excellent educational outcomes are the committee's goals. Recommendations for duty hour adjustments, enhanced supervision, and workload reduction will best achieve the targeted goals when implemented in concert, and implementation of all the committee's recommendations will require a significant investment in personnel to substitute for the hours that residents are no longer available. To cover some of the excess resident hours with substitute personnel would cost an estimated $1.7 billion dollars, the equivalent of about 9 percent of direct and indirect graduate medical education payments made to teaching facilities from public and private sources. To help institutions implement the changes, the committee recommends that additional funding be provided by all supporters of graduate medical education. Systematic collection of pertinent data would help monitor and evaluate the effects of implementing the recommendations, and research would provide an evidence base for future changes to duty hours or educational strategies.

As the preceding chapters demonstrate, this Institute of Medicine (IOM) committee found the issues of patient safety, resident safety, duty hours, and schedules closely related to broader issues of how graduate medical education is structured, including the work environment, the supervision of residents, and safety practices throughout the training institutions. While recommendations on a particular topic are embedded in the scientific evidence of its corresponding chapter and presented serially in this report, the committee intends the report and all of its recommendations to be considered as a whole. This chapter discusses the possible consequences of implementing a single recommendation (e.g., duty hour limits) in isolation and an approach to phase-in the recommendations in a practical fashion. Thus, the committee presents first a preamble to the recommendations:

Preamble to Recommendations

To promote conditions for safe medical care, improve the education of doctors in training, and increase the safety of residents and the general public, the committee offers the (previous and) following recommendations, which should be implemented with all deliberate speed. While some recommendations should be implemented immediately, changes to duty hours, adjustments in workload, and the funding needed for these changes might require an integrated phase-in. The recommendations will require additional resources—both financial and human. Without the necessary restructuring in resource allocation, attempts to implement the recommendations will fail to have the desired benefits and could even reduce patient safety. The committee believes that the Accreditation Council for Graduate Medical Education and the other organizations charged to implement aspects of the recommendations should begin their work with urgency, and that action on all recommendations should be taken within 24 months.

The committee took a broader perspective than just the duty hours and schedules, looking at related aspects of the work and learning environment of residents. It found little detailed information available on the educational outcomes of training programs since the 2003 limits because many of the first cohort of doctors fully trained under the 2003 limits are just completing their training. Supervision and workload were not addressed when duty hours were set in 2003, and lack of supervision at critical junctures, excess workload, and fatigue can all contribute to error and to reduced learning. Thus, the committee believes that to minimize unintended consequences, recommendations to prevent and mitigate fatigue through adjusting resident duty hours, enhancing supervision, and reducing workload should be implemented in concert. Implementing the duty hour adjustments without the others could prevent achievement of important goals identified by the committee. Recommendations by the committee reflect the best ways to achieve performance and learning goals based on sleep science, learning theory, and the close observations of medical educators.

In designing its recommendations to achieve the targeted goals outlined in the preamble, the committee considered strong evidence from the literature concerning the impact of sleep and fatigue on human performance and the occurrence of error, and based several recommendations on this evidence. Although fatigue creates an unsafe condition in the work environment, there is insufficient evidence to determine the degree to which resident hours of work translates into patient harm. For this reason, the committee did not change the current weekly duty hour limit of 80 hours or the limit of 30 hours for extended duty periods. Rather, it chose to create better opportunities for fatigue prevention and mitigation within the basic duty hour structure and to focus on supervision, handovers, and other systemic changes to enhance learning and safety.

The intent in adjusting the 2003 Accreditation Council for Graduate Medical Education (ACGME) duty hour limits is to:

- Prevent fatigue whenever possible;
- Recognize that some fatigue is inevitable and provide measures to relieve both acute and chronic sleep deprivation and reduce its negative effects;
- Be practical to schedule;
- Be feasible to monitor;
- Enhance the learning experience;
- Preserve the ability of residency programs of various sizes and different specialties to adapt the changes to their circumstances by not mandating a single schedule for all and by allowing limited opportunities for exceptions based on patient need and unusual learning opportunities; and
- Maintain the spirit of residency and the excitement of being a doctor.

The committee was also asked to consider the potential cost impact of its recommendations, and it is the estimated cost and the limited availability of the healthcare workforce that the committee believes are the greatest barriers to further changing resident duty hours. Having an adequate workforce of physician extenders, residents, and physicians alike takes planning to develop incentives and remove disincentives to grow the labor force that the country and individual labor markets require to support reductions in resident hours and serve other healthcare needs. Based on a commissioned cost model, an estimate of the personnel substitution costs associated with several duty hour and workload limits shows that the costs would be substantial, in the ballpark of $1.7 billion in 2008, with variations in that amount depending on who substitutes for residents and how programs choose to schedule residents. Other recommendations of the committee could require additional funds. The committee, while recognizing that funds for health care are in great demand, recommends adequate support by all funders of graduate medical education (GME) and related research so that the recommendations can be fully implemented and have the desired impact.

The committee acknowledges that there are objections from some members of the graduate medical training community to any changes to the

2003 duty hour limits and related aspects of GME, just as there were objections to those limits initially (see Chapters 1 and 2). A crucial objective of graduate medical training is to ensure that the country will produce increasingly competent physicians for independent practice—a long-term patient safety goal. With this in mind, the committee found that there are good reasons to take a deliberate approach to changing resident duty hours.

This chapter reviews current funding for GME and projections for the costs and workforce needs associated with the committee's proposed adjustments to hours and workload. The chapter concludes with a phased implementation strategy and addresses the necessity of further research, data collection, and evaluation that would allow consideration of the appropriateness of resident duty hours in the future.

COST IMPLICATIONS OF CHANGES TO DUTY HOURS

Implementing workload reductions and adjustments to the 2003 duty hour limits will require replacing residents' time with that of other workers and entails substantial costs for society but also potential benefits to patient and resident safety. This section first looks at what the United States invests in graduate medical training and then examines projected estimates of personnel costs to implement the committee's recommendations on hours and workload adjustments.

Funding for Graduate Medical Education

GME is paid for largely through insurance premiums and payroll taxes. A number of parties specifically contribute to GME: the Centers for Medicare and Medicaid Services (CMS), the Department of Veterans Affairs (VA), the Department of Defense (DOD), the Health Resources and Services Administration (HRSA) of the U.S. Public Health Service, states, and private funders. Available data on the level of funding for GME over the past few years is limited. Medicare is the largest single payer of GME ($8.5 billion in 2007), and its funds come in two forms (see Table 9-1). Some of the other funds, particularly from private and state sources, are harder to identify and estimate at the national level. Wynn and colleagues estimate that for all sources of support, direct and indirect expenditures associated with training residents were approximately $18.7 billion in 2003 (Wynn et al., 2006).

Some of this funding comes indirectly through payments for patient care in teaching institutions (e.g., from Medicare indirect medical education [IME] payments, state Medicaid, private payers) to cover the increased costs of care associated with resident training in teaching hospitals. The Medicare IME funds ($5.7 billion in 2007) are provided through higher

TABLE 9-1 Sources of GME Funding

Funding Source	FY 2007 (billion dollars)
CMS	
Medicare	8.50
IME	(5.70)
DGME	(2.80)
VA	1.0
Direct	(0.50)
Indirect	(0.50)
DOD	NA
HRSA	
Children's Hospital GME	0.28
Training in primary care, Medicare, and dentistry	0.05
States—Medicaid	3.20[a]
Private payer (not direct payments, but imputed from higher reimbursement to teaching hospital)	Unknown

NOTE: CMS = Centers for Medicare and Medicaid Services; DGME = Direct graduate medical education payment; GME = Graduate Medical Education; HRSA = Health Resources and Services Administration; IME = Indirect medical education payment; NA = Not available; VA = U.S. Department of Veterans Affairs.
[a]Estimated spending in 2005 (Henderson, 2006).
SOURCES: CBO, 2008; Chang, 2007; HRSA, 2008a,b.

hospital payments. The size of the operating adjustment to a hospital's payment rate is based on teaching intensity and the number of residents per bed, with limits on the rate of increase. It is included in inpatient operating and capital payments under the Medicare inpatient hospital payment system and includes subsidies to hospitals treating patients of the Medical Advantage Program (MedPAC, 2008).

Other funding, such as direct GME (DGME) payments are made to the training institution for support of training such as the residents' stipends, teaching physicians' salaries and benefits, and administrative overhead of GME offices. The DGME payment is based on historic, hospital-specific costs per trainee, with maximum limits on the number of trainees. Some teaching hospitals receive very little, if any, support from the Medicare GME funding stream, even though they train many residents, because they do not serve a large Medicare population (Opas, 2008).

Other federal financial sources of support for GME include the VA and DOD (which both also provide sites for residency training), and HRSA. The VA has approximately 9,500 residency slots in its healthcare facilities (9 percent of U.S. medical resident slots) and, with multiple residents rotating through each slot, participates in the instruction of approximately one-third of U.S. doctors in training (about 34,000 medical residents) per year.[1] The DOD supports approximately 3,000 residency positions in military facilities accredited by the ACGME. HRSA manages the Children's Hospital Graduate Medical Education program, authorized to support the training of stand-alone children's hospitals that do not receive Medicare reimbursements (and therefore do not receive the DGME and IME payments described earlier).

States may voluntarily provide GME funding through their Medicaid programs. Up until at least 2005, all but three states (Illinois, North Dakota, and Texas) did so (Henderson, 2006). However, there has been a debate in recent years about whether the use of federally matched dollars for GME reimbursements is an appropriate use of Medicaid funds. In June 2008, Congress placed a moratorium on a proposed rule until April 1, 2009, to block CMS's prohibition of GME payments from these matched funds (P.L. 110-252, June 30, 2008).

Private payers' contributions are difficult to determine. These payments compensate GME providers for the care received by their beneficiaries rather than paying for GME directly. This is similar in concept to Medicare IME. Private sector sources and the federal government provide the preponderance of funds for GME (Knapp, 2002; Wynn et al., 2006). Finally, economists consider that residents support some of the cost of their education through their own efforts. By providing service in hospitals at a relatively low hourly wage rate, residents, in effect, subsidize the institution for some of the costs of their education. Given the cost of replacing residents with other personnel, they are not a trivial source of support for their own graduate medical training although they pay no tuition.

The committee considered the impact of its recommendations on the financial status of hospitals. Figure 9-1 compares the median margins of total income and patient revenues of hospitals by teaching or non-teaching status as well as whether they are safety net hospitals (Andrews et al., 2007).[2] While there is not a consensus on what constitutes a safety net hospital (Siegel et al., 2004), the study by Andrews and colleagues bases its definition on the proportion of uncompensated care that a hospital provides. For most hospitals, except the category of safety net teaching hospitals, the negative patient revenue margin is smaller than the positive total income margin. Safety net teaching hospitals, however, have a substantial negative patient revenue margin (–9.6 percent), in part because of the uncompensated care they deliver, and a positive total income margin of only 1.2 percent. It is a positive total income margin only because safety net teaching hospitals,

[1]Personal communication, J. P. Bagian, National Center for Patient Safety, July 31, 2008.

[2]"The total income margin is the total income for a hospital (i.e., net patient revenue plus contributions, government appropriations, and other income), divided by the total expenses (i.e., operating costs and other expenses)" (Andrews et al., 2007, p. 13).

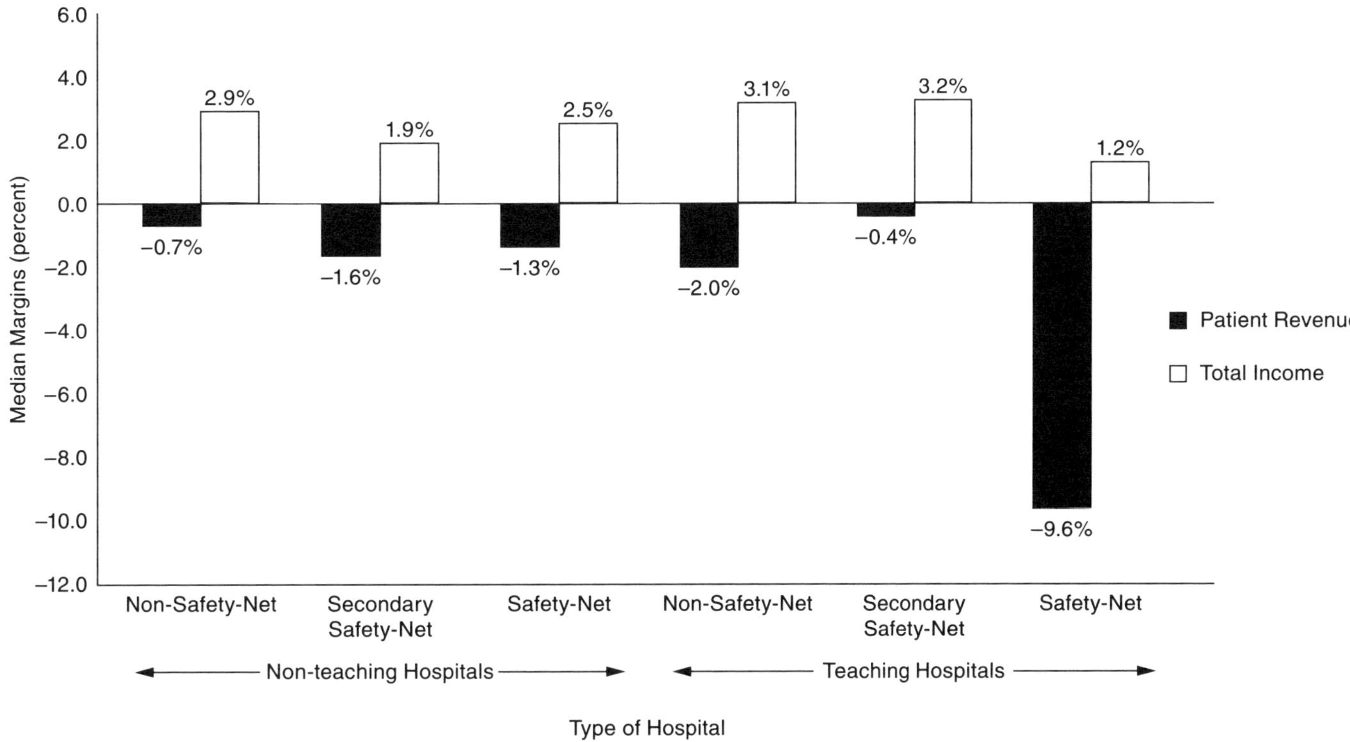

FIGURE 9-1 Median margins of hospitals by teaching status.
SOURCE: Andrews et al., 2007.

on average, successfully obtain sufficient other revenue such as government allocations and other subsidies. Approximately one-third of safety net hospitals, whether teaching or not, have a negative total income margin, and 20 percent of all safety net hospitals are teaching hospitals (Andrews et al., 2007). Teaching hospitals offer a substantial amount of charity care—care provided without expectation of payment. Major teaching hospitals make up only 6 percent of the acute hospitals in the country, yet they deliver 41 percent of all hospital-based charity care, and all teaching facilities provide 71 percent of that charity care. In 2006, the 274 members of the Council of Teaching Hospitals and Health Systems (COTH) provided an estimated $6.3 billion in uncompensated charity care, non-COTH teaching hospitals an estimated $4.5 billion, and non-teaching hospitals $4.5 billion, totaling about $15.3 billion in hospital charity care (COTH, 2008).

The committee is concerned that some safety net teaching hospitals may not have the resources under current funding mechanisms to provide the additional supports necessary to allow residents recommended opportunities to rest, transfer noneducational tasks to others, or offer residents sufficient supervision because their resources already are spread too thinly (Werner et al., 2008).

Cost Model

As previously mentioned, to implement the recommendations of this report, some of the work presently performed by residents will have to be done by others. The committee commissioned a health services researcher, Teryl Nuckols, M.D., MSHS, and a health economist, José Escarce, M.D., Ph.D., both at the David Geffen School of Medicine at the University of California–Los Angeles as well as the RAND Corporation, to construct a model that would provide an estimate of the costs and workforce that would be needed to replace resident work under various duty hour scenarios. This section of the chapter is based on their methods and results. The complete paper is published on the following websites: www.iom.edu/residenthours and www.iom.edu/hcs. A committee member, economist Jayanta Bhattacharya, M.D., Ph.D., performed sensitivity analyses on some of the main assumptions of this model. His discussion and figures are posted with the main paper.

The model, based on existing literature and explicit assumptions, derives estimates from four scenarios, called "components," related to resident hours and workload, which were specified before the committee had formulated its final recommendations. The model provides an indication of the level of expected substitution costs based on 2006 data—approximately $1.6 billion dollars ($1.7 billion when inflated to 2008 dollars)—if all four components are adopted. The costs are reasonably similar whether excess hours of residents' time are replaced by hiring other healthcare providers or additional residents. The simplified components of the model related to the committee's recommendations are:

1. Bringing all residency programs into compliance with the existing 2003 ACGME 80-hour duty limit, since not all programs and residents now comply.
2. Having any extended duty period beyond 21 hours incorporate a 5-hour undisturbed sleep period.
3. Reducing the workload of postgraduate year 1 (PGY-1) residents by 10 percent.
4. Limiting shifts to a maximum of 16 hours for residents beyond PGY-1. The model assumed that after hospitals achieved compliance with the previous three components, they might choose to reduce all shifts to 16 hours. Since they would have to include a 5-hour rest period for shifts lasting longer than 16 hours, eliminating shifts of 16 to 21 hours would require no more substitute hours than would limiting the shifts to 16 hours. Available literature reports that most PGY-1 extended duty shifts exceed 21 hours, so the model assumed hospitals would not limit the shifts of PGY-1s to 16 hours.

Implementing these reforms with the substitution of personnel tailored to the tasks that residents currently perform would require the following increases nationally in full-time equivalents (FTEs): nursing aides, 229; laboratory technicians, 45; licensed vocational nurses, 320; midlevel providers, such as physicians assistants and nurse practitioners, 5,984; and attending physicians, 5,001. Given that there are 1,206 teaching hospitals across the country, less than one FTE would be needed on average per hospital for several of these provider types. If instead more residents were to be added, this would require approximately 8,247 new residents (specialty and subspecialty) in addition to the existing pool of more than 105,000 residents.

Please note that the cost model calculates the economic costs of only certain aspects of the committee's recommendations and does not attempt to predict which elements of the recommendations will be adopted (e.g., keep extended duty periods with protected sleep periods or use only shorter shift schedules). Precise cost estimates of every recommendation were not feasible given time and data constraints and were beyond the committee's statement of task, but the four components do reflect the likely magnitude of costs for a number of the report's major recommendations. Lack of comprehensive nationwide data on the actual hours residents now work, the frequency of their overnight work, and other factors required some assumptions in the model's design based on the existing literature. Sensitivity analyses of selected assumptions allow for a range of estimates to address these assumptions (e.g., different substitution ratios, call frequency during residents' inpatient months, current compliance levels). In addition, the model could not anticipate fully programs' choices of implementation strategies (e.g., if the programs shifted workload from residents with longer hours to those with shorter ones), which could possibly produce lower cost estimates.

Estimating Baseline of Resident Duty Hours, Rate of Violation, and Workload

To estimate the direct annual costs of the proposed reforms, the difference between what residents work at baseline and what they would work under the proposed changes is calculated and called "excess resident work" hours. As discussed in Chapter 2, there has been no recent rigorous or reliable collection of data on the total duty hours of residents in all specialties across the country. The economic model uses the best available data on mean weekly duty hours (66.6 hours) and the hours worked by PGY-1s in excess of 80 hours from the first year of implementation: 29.0 percent of the workweeks were longer than 80 hours, 12.1 percent were 90 or more hours, and 3.9 percent were 100 or more hours (Landrigan et al., 2006). The study did not give details on workweeks that fell below the 80-hour limit. Other more recent studies have found relatively similar rates of duty hours and degree of violations (AMA Division of Market Research and Analysis, 2005; Jagsi et al., 2008). The baseline calculation makes accommodation for the portion of the year that residents spend on inpatient months (e.g., 50.6 percent of PGY-1 residents' months in the Landrigan paper), the number of nights that a resident might be on extended duty, how long residents are staying over the 30-hour limit, and differences between PGY-1s and other years (Nuckols and Escarce, 2008). The values for these assumptions are detailed in the paper describing the model.

Hierarchical Nature of the Model

The costs of the four reform components are estimated in a sequential, hierarchical fashion to prevent counting excess hours twice. Several basic assumptions are built into this hierarchy:

- That achieving compliance with the 80-hour workweek would make it possible to achieve the 30-hour duty period limit at no additional cost because violations of the 30-hour limit are generally what push residents over 80 hours;
- That residency programs would choose to implement a nap only

for residents already working more than 21-hour duty periods at baseline; and
- That reducing average workload by 10 percent would be equivalent to reducing weekly duty hours by the same amount; there are no widely accepted measures or estimates of the workload of residents across specialties (e.g., average daily census, number of admissions or procedures per call day) available.

PGY-1 residents are more likely to violate duty hour limits than more senior residents or fellows because PGY-1s tend to have more inpatient months, more frequent extended duty periods, and more direct patient care responsibilities. Reducing the workload of only first-year residents is factored into the cost model, but the committee recognizes that it may become necessary in some or all of the specialties for workload to be reduced in other years of residency as well. The final component of the hierarchy for cost estimating is a 16-hour shift maximum for residents beyond their intern year.

Substitution Scenarios

As discussed in Chapter 4, many training programs hired replacements to assume "excess resident work" in response to the 2003 limits. This model builds on substitution ratios available in the published literature. Each step, or component, in the hierarchy of hours reduction has its own combination of resident substitutes (Table 9-2). Alternatively, the model estimates having each resident hour replaced by additional residents with no work transferred to other types of personnel. Here the term "resident substitution" means transferring residents' clinical care-related work to other providers or sharing it among a larger population of residents. Mid-level providers (nurse practitioners and physician assistants) have often been considered the prototypical resident substitutes (Stoddard et al., 1994; Whang et al., 2003), but publications following the 2003 ACGME reform and recent testimony from hospital administrators suggest that there are a few basic strategies for reducing resident duty hours or workload:

- *Task-tailored substitutes*: Transferring noneducational patient care tasks to the lowest-level personnel qualified to perform them.
- *Midlevel substitutes*: Transferring work to midlevel providers.
- *Midlevel and attending substitutes*: Transferring work to a mixture of midlevels and attending physicians.
- *Attending substitutes*: Transferring work to attending physicians.
- *Resident substitutes*: Hiring new residents to share the work of existing ones.

TABLE 9-2 Methods: Application of Substitution Strategies to Base Case Scenario

Reform Component	Base Case Scenario Using Substitutes Other Than Additional Residents
1. Achieve compliance with 80-hour workweek	*Specialty residents:* Task-tailored substitutes *Subspecialty residents:* Attending physician substitutes
2. When shifts last 21 to 30 hours, include a 5-hour nap	*All residents:* Attending substitutes
3. Reduce workload of PGY-1 residents by 10%	*PGY-1 specialty residents:* Midlevel substitutes
4. Optional: reduce maximum shift length to 16 hours	*Specialty residents:* 50% midlevel and 50% attendings *Subspecialty residents:* attendings

The committee has concluded that transferring noneducational routine "scut" work to the lowest-level personnel possible ("task-tailored substitution") would enable residents to preserve their educational patient care experiences and reduce their duty hours. This transfer could be accomplished at relatively modest cost compared to hiring other physicians to do residents' work. Substantial reductions in resident duty hours and workload, such as those represented by the cumulative total of the four proposed reform components, would probably not be achievable using midlevel providers alone. Patient care responsibilities would likely require attending supervision for complex or unusual tasks. Further, the work of subspecialty residents is complex, suggesting that only substitution by attending physicians would be appropriate. The nap requirement (Component 2) would likely require the substitution of attending-level physicians so that residents would feel comfortable signing out their patients and would take advantage of naps during extended duty periods. Unwillingness of residents to sign out to other night-float residents has been a barrier to incorporating protected sleep periods into long duty periods (Arora et al., 2006). Similarly, Component 4 would be expected to lead to redesign of overnight call and require greater attending physician presence.

The model assumes that substitution does not affect length of stay, test ordering, or other hospitalization cost determinants. Substitution or resident work by others would occur at a hospital level—meaning that excess work from multiple residents would naturally distribute among substitutes within each hospital. Work would be transferred to substitutes in a 1 hour to 1 hour ratio (i.e., this assumes residents and substitutes would perform tasks at the same speed).

Hourly wage substitutions for the model were based on data from the Bureau of Labor Statistics (BLS). According to BLS, in 2006, mean hourly wages for potential substitutes were as follows: nursing aides, $11.21; medical and clinical laboratory technicians, $16.55; licensed vocational nurses, $18.12; midlevel providers, $37.84; and physicians, $58.76; benefits were 30.1 percent of total compensation (43.1 percent of wages) (U.S. Department of Labor, 2007a,b). Having residents incurs training expenses as well as salary and benefits expenses. The average salary for PGY-1s in the 2007-2008 academic year was $44,747 plus benefits (AAMC, 2007a). The hourly rate of an intern receiving this average stipend and working 66 hours per week throughout the entire year (the mean found in one study after implementation of duty hours) (Landrigan et al., 2006) would be $12.92. However, dividing the estimated total payments for GME by the total number of residents nationally suggests that salary, benefits, and training expenses together cost $187,000 per resident year (this figure includes all public and private sources of direct and indirect payments) (Wynn et al., 2006). Thus, the cost model assumed that the $187,000 costs would apply if additional residents were used to replace reduced hours of current residents. The model includes replacement costs for both "specialty" residents and "subspecialty" fellows. The $187,000 figure was used because no uniform data were available from teaching programs on the incremental cost of training residents.

Costs of Replacing "Excess Resident Duty" Hours

Table 9-3 shows the results of calculations for the four reform scenarios using other personnel as substitutes for the "excess resident work" hours. The cost projections are presented for two groups of teaching hospitals: all hospitals with ACGME-accredited programs (1,206 hospitals) and COTH members (367 hospitals). Three-quarters of residents in academic year 2006-2007 were trained at COTH hospitals (AAMC, 2008c; Nuckols and Escarce, 2008). The total U.S. cost of all four reform components in 2006 dollars is estimated to be $1.6 billion for the 1,206 hospitals with ACGME programs including $1.2 billion for COTH hospitals. In Table 9-3 the total dollars are then broken down into an average cost per hospital and per admission.

As an alternative approach, the model assumes that excess resident hours and work would be distributed among a larger pool of residents than are being trained today. Table 9-4 shows the number and cost of additional residents that would be needed nationally to substitute for the excess resident work hours. At least an additional 8,247 residents would be necessary. Salary and benefits for the 7,639 specialty residents and 608 subspecialty fellows would total more than $500 million when excluding IME payments,

TABLE 9-3 Results: Costs of Hiring Other Providers to Assume Excess Resident Work, Base Case Scenario (2006)

	Hospitals with ACGME Accredited Programs[a]
Baseline Work	Total U.S. Cost
Component 1. Achieve compliance with 80-h workweek	
24,772 PGY-1 specialty residents	$209,742,405
22% of 64,497 specialty residents above PGY-1	$120,139,928
22% of 15,610 subspecialty residents	$45,915,338
Subtotal	$375,797,671
Component 2. When shifts last 21 to 30 h, include 5-h nap	
24,772 PGY-1 specialty residents	$319,707,737
23.1% of 64,497 specialty residents above PGY-1	$192,284,187
23.1% of 15,610 subspecialty residents	$46,537,919
Subtotal	$558,529,843
Component 3. Reduce workload of PGY-1 residents by 10%	
24,772 PGY-1 specialty residents	$391,736,621
Subtotal	$391,736,621
Component 4. Optional: Reduce maximum shift to 16 h	
30.6% of 64,497 specialty residents above PGY-1	$192,950,559
30.6% of 15,610 subspecialty residents	$56,812,524
Subtotal	$249,763,084
Total, components 1-3	$1,326,064,134
Total, components 1-4	$1,575,827,218

NOTE: Totals may reflect rounding.
[a]Including COTH hospitals.

although the total including costs from public and private payers could be $1.5 billion, if the associated patient care costs are included. The cost of training residents is challenging to estimate accurately, as has been explored by other sources (Knapp, 2002; Wynn et al., 2006). Residents' salaries and benefits are significantly lower than the costs involved in training them. The $187,000 estimate for hiring each additional resident, based on the $18.7 billion in payments related to GME (Wynn et al., 2006), may overstate the amount necessary to support training per resident; thus, the $1.5 billion cost for adding residents would be an overestimate. Overall, in 2006-2007, resident stipends and benefits represented 9.0 percent of institutions' operating expenses (AAMC, 2008c).

Expanding the population of residents as an approach to achieving reform has some short-term appeal because residents provide highly skilled labor at a low hourly cost relative to other substitutes, but more precise estimates would be needed to determine the incremental costs of training

		COTH Hospitals		
Cost per Hospital	Cost per Admission	Total U.S. Cost	Cost per Hospital	Cost per Admission
$311,607	$21.37	$281,848,253	$767,979	$33.68
$463,126	$31.76	$418,897,382	$1,141,410	$50.05
$324,823	$22.28	$293,802,466	$800,552	$35.11
$207,100	$14.20	$187,322,313	$510,415	$22.38
$1,099,556	$75.41	$994,548,101	$2,709,940	$118.84
$1,306,656	$89.61	$1,181,870,414	$3,220,355	$141.22

residents on top of their salary and benefits. Such an option would also have long-term implications for the national supply of physicians. In addition, increasing the total number of residents would not necessarily ensure the number and distribution of residents by specialty or geographic area where they are deemed to be needed most.

The cost estimates of the model do not include any assumptions of savings from work and education redesigns. Many institutions may find ways to streamline the work and training of residents to eliminate excess hours without having to hire substitutes for each and every resident hour reduced and without burdening existing residents with increased workloads. Certainly some institutions may experience labor shortages for some of the potential substitutes and many may have serious cost constraints, providing an added incentive to reduce costs through efficiencies and systems redesign. The committee took a conservative approach and did not project or assume any such cost savings.

TABLE 9-4 Results: Reducing Resident Duty Hours by Increasing Number of Residents Nationally

Baseline Work	Number of New Residents Needed	Increase from 2006 (%)	Total Using Annual Salary and Benefits per Resident	Total Using Annual Expenditures by Public Payers Excluding IME[a] per Resident	Total Using Annual Expenditures by All Public and Private Payers per Resident
Component 1. Achieve compliance with 80-h workweek					
Specialty residents	2,121	2.6	$126,108,713	$140,005,876	$396,683,314
Subspecialty residents	187	1.2	$13,144,147	$12,340,648	$34,965,169
Subtotal			$139,252,860	$152,346,524	$431,648,483
Component 2. When shifts last 21 to 30 h, include 5-h nap					
Specialty residents	2,085	2.3	$130,228,151	$137,607,874	$389,888,977
Subspecialty residents	190	1.2	$13,322,373	$12,507,979	$35,439,273
Subtotal			$143,550,524	$150,115,853	$425,328,249
Component 3. Reduce workload of PGY-1 residents by 10%					
Specialty residents	2,477	3.1	$142,992,976	$163,495,200	$463,236,400
Subtotal			$142,992,976	$163,495,200	$463,236,400
Component 4. Reduce maximum shift to 16 h					
Specialty residents	956	1.1	$59,706,617	$63,090,050	$178,755,142
Subspecialty residents	231	1.5	$16,263,676	$15,269,481	$43,263,528
Subtotal			$75,970,293	$78,359,531	$222,018,670
Total for Components 1-4					
Specialty Residents Needed	7,639	8.6	$459,036,458	$504,199,000	$1,428,563,833
Subspecialty Residents Needed	608	3.9	$42,730,197	$40,118,107	$113,667,970
Cost			$501,766,655	$544,317,107	$1,542,241,803

NOTE: Totals may reflect rounding.
[a]Medicare indirect graduate medical education payment.

Sensitivity Analyses

The model of the cost implications of resident duty hour and workload reform requires assumptions about a wide range of parameters. To the extent that such data are available, Nuckols and Escarce derive these parameters from the published literature. However, the committee recognizes that there is uncertainty about these assumptions and that in many cases the published literature contains little relevant information. In other cases, the literature cited by Nuckols and Escarce provides estimates on key parameters such as the frequency of night shifts or extended duty periods or the proportion of a year spent in an inpatient service for only a subset of residents, such as PGY-1s, or from a very limited number of institutions. Hence the committee decided to test the sensitivity of the cost impact estimates by testing some of the assumptions in the model. Jayanta Bhattacharya, an economist and a member of the committee, conducted the sensitivity analyses. The committee provided guidance on the range of assumptions to use for each parameter, with some higher and some lower than the baseline assumption.

Dr. Bhattacharya designed the sensitivity analyses to test one parameter at a time through 11 "thought experiments." The assumptions tested include the following:

- The rate of violations by PGY-1s and more senior residents of the 80-hour week of the 2003 ACGME reform affects the costs of complying with current rules, or Component 1 of the model.
- The frequency of extended duty periods during inpatient rotations for PGY-1s and all other residents, which affects the number of hours that would have to be replaced to accommodate the required sleep period for duty periods lasting from 21 to 30 hours, impacts the costs of Component 2.
- The use of various healthcare professionals as task-tailored substitutes, such as laboratory technicians and licensed vocational nurses; midlevel professionals, such as registered nurses and physician assistants; and attending physicians or other residents affects the cost of covering the excess hours of residents and influences the costs of all four components, based on their assumed substitution patterns.
- In addition to the different hourly costs of various substitutes to cover excess resident hours, the substitution costs could vary depending on the efficiency of the substitutes—whether they produce the same amount of work per hour as the residents or not.
- The rate of compliance with the new reforms of the model also affect the costs of Components 2-4; total costs increase as compliance increases.

Based on the 11 thought experiments used to test these assumptions of the model, it is clear that the cost of implementing each of the four components discussed in the model could vary substantially, depending on the details of the assumptions. The total cost estimates of the potential reforms are surprisingly robust to a wide range of assumptions about current resident work schedules, which providers would perform the work if the potential reforms were adopted, and the relative efficiency of those substitute workers. The $1.6 billion estimated by the model for the cost of

TABLE 9-5 Sensitivity Analyses

No.	Experiment	Sensitivity Range
1	Vary assumption about current compliance with ACGME 80-hour workweek	0-100% as many hours above the 80-hour limit as reported by Landrigan (2006)
2	Vary assumption about noncompliance of residents above PGY-1 with 80-hour workweek	0-100% of PGY-2+ specialty and subspecialty residents worked the hours reported for PGY-1s in Landrigan (2006)
3	Vary assumption about call frequency during inpatient months for PGY-1	Every third night to every seventh night
4	Vary assumption about call frequency during inpatient months for all residents above PGY-1	Every third night to every seventh night
5	Vary assumption about frequency of inpatient rotations among residents above PGY-1	PGY-2+ have 50 to 100% of rotations as PGY-1 in Landrigan (2006)
6	For specialty residents, vary percentage of time transferred to midlevels in Component 1	0-100% transferred to midlevels
7	For all residents, vary percent of time transferred to attendings in Component 4	0-100% transferred to attendings
8	Vary substitution ratio for task-tailored substitutes	Substitution ratio of 0.5 to 3.0 hours for each hour transferred from a resident
9	Vary substitution ratio for midlevel substitutes	Substitution ratio of 0.8 to 2.4 hours for each hour transferred from a resident
10	Vary substitution ratio for attending substitutes	Substitution ratio of 0.5 to 1.0 hour for each hour transferred from a resident
11	Vary assumption about compliance with the changes contemplated in Components 2, 3, and 4	0-100% compliance

implementing all four components is within the range calculated for each parameter in the sensitivity analyses. The cost range for each "experiment" is included in Table 9-5. Assuming at least some compliance with the proposed reforms (as illustrated with components 2, 3, and 4 of the model), the lowest-cost estimate for implementing the potential reforms is \$1.14 billion, while the highest is \$2.52 billion.

The full discussion of the sensitivity analyses, a description of the 11 experiments that Dr. Bhattacharya conducted, graphs of the results, and the

Outcome Range (million dollars)				
Component 1	Component 2	Component 3	Component 4	Total Cost
\$376-\$0	\$559	\$392	\$250	\$1,576-\$1, 200
\$210-\$376	\$559	\$392	\$250	\$1,410-\$1,576
\$376	\$878-\$559	\$392	\$250	\$1,896-\$1,576
\$376	\$797-\$559	\$392	\$500-\$250	\$2,064-\$1,576
\$376	\$439-\$559	\$392	\$125-\$250	\$1,332-\$1,576
\$376-\$381	\$559	\$392	\$250	\$1,576-\$1,581
\$376	\$559	\$392	\$250-\$292	\$1,576-\$1,618
\$362-\$430	\$559	\$392	\$250	\$1,562-\$1,630
\$334-\$665	\$559	\$313-\$940	\$235-\$356	\$1,441-\$2,519
\$305-\$376	\$279-\$559	\$392	\$163-\$250	\$1,138-\$1,576
\$376	\$0-\$559	\$0-\$392	\$0-250	\$376-\$1,576

summary table of cost ranges is on the project's website along with the full paper on the model and can be accessed at www.iom.edu/residenthours or at www.iom.edu/hcs.

Cost of Preventable Adverse Events and Possible Net Costs

One expectation is that reduced hours of work and workload would reduce errors and improve patient safety by reducing preventable adverse events (PAEs). In addition to harming patients, PAEs increase hospitalization costs, outpatient medical care costs, and costs associated with patient disability (Thomas et al., 1999; Zhan and Miller, 2003). With this in mind, the economic analysis was designed to estimate the potential *net* annual costs of the proposed reforms from both direct costs *and* any changes in costs related to PAEs. The framework for this analysis can be described by the equation below:

Net costs = [Cost of resident substitutes] –
[(Baseline costs of PAEs)(Δ in PAE rate)].

If the proposed reforms succeed at reducing PAEs in teaching hospitals, this would likely yield cost offsets both during and after hospitalization. These cost offsets would reduce the gross costs associated with reduced resident hours. However, the possibility exists that the reforms could increase PAEs rather than reduce them. Reducing duty hours has been associated with increased numbers of handovers, which in turn has been associated in one study with a significant increase in the rate of PAEs, for example, although this risk could be mitigated (Petersen et al., 1994, 1998). The committee does not attempt to predict the changes, if any, in PAEs following full implementation of its recommendations. Therefore, this analysis uses a range of possible changes in PAE rates to estimate net costs of implementation of the four components of the model.

The costs of PAEs for teaching hospitals and for society as a whole are likely to differ in their magnitude, according to an analysis of the 2003 duty hours reforms (Nuckols and Escarce, 2005). Teaching hospitals would incur costs resulting from PAEs for additional intensive care unit (ICU), inpatient non-ICU, and physician care. From the societal perspective, considering all costs regardless of who bears them, there would be, in addition to the inpatient event costs, costs for outpatient medical care and the non-medical costs of lost wages and lost household production (Thomas et al., 1999). Thus, the main costs of a PAE occur after the patient leaves the hospital and are borne by society. Since most PAE costs occur after hospital discharge, teaching hospitals are not very likely to experience direct and sizable cost offsets if the proposed duty hour reforms succeed in reducing PAEs. Consequently, the current analysis considers net costs from both the hospital and the societal perspective.

Table 9-6 presents the net costs of reform. Given an absence of literature to suggest the potential effect of the currently proposed reforms, a range of possible changes in PAE rates of –30 percent, –15 percent, 0 percent, +15 percent, and +30 percent is examined. If there is no (0 percent) change in PAEs (column 1), the net cost to the hospital is the same as the gross costs estimated, $1.6 billion (rounded from column 2). Table 9-6 (column 1) shows that a reduction of between 15 and 30 percent in PAEs would be necessary to balance the costs of implementing the four reform components from the hospital perspective. Specifically, the savings from a decline in PAEs of 21.4 percent would totally offset the costs of the four components of the model. From a societal perspective, a reduction of PAEs of less than 15 percent, actually 7.2 percent, would be sufficient to offset the $1.6 billion costs of the four components. Please note that the committee is not *predicting* that all programs will reduce shifts and total hours to those of the model's assumptions. Neither is it *predicting* that harmful errors will drop by 7 percent. However, should this occur, if the other cost assumptions are close to reality then the costs to society of the reforms and the savings from reduced harmful errors, from a societal perspective, would be roughly in balance or cost neutral. Ideally, the current study would be based on the actual rates and costs of PAEs across U.S. teaching hospitals after the 2003 ACGME reform was implemented, but such data are not available at this time.

FUNDING THE COMMITTEE'S RECOMMENDATIONS

The cost model considers hiring additional residents and hiring other types of providers as mutually exclusive options. In reality, a combination of these two substitution strategies is likely to be used as the committee's recommendations are implemented in various ways by programs across the country. The main implication of the economic model is that the proposed reform is costly—$1.7 billion in 2008 dollars. The costs of achieving these reforms relative to the total costs of GME would be approximately 9 percent of current GME payments now borne by all payers ($1.7 billion of the $18.7 billion estimated for 2003 by Wynn [2006]) based on a substitution strategy of either additional residents or other providers. In comparison to Medicare's total outlays of $440.6 billion in 2007 (CBO, 2008), the $1.7 billion of substitution costs are 0.4 percent.

The costs of adapting to resident duty hours in 2003 were borne by teaching institutions under existing funding. Some institutions may have to make relatively few changes to comply with the committee's recommendations; they may have residency programs that do not schedule extended

TABLE 9-6 Results: Net Costs of Proposed Changes, Considering Costs of Resident Substitution and Possible Changes in PAEs (2006)

	Hospitals with ACGME-Accredited Programs			COTH Hospitals		
Possible Change in PAE Rate After Reform	Net Costs to U.S. Hospitals	Cost per Hospital	Cost per Admission	Net Costs to U.S. Hospitals	Cost per Hospital	Cost per Admission
Hospital perspective						
–30%	–$633,547,948	–$525,330	–$36.03	$130,424,709	$355,381	$15.58
–15%	$471,139,635	$390,663	$26.79	$656,147,561	$1,787,868	$78.40
0%	$1,575,827,218	$1,306,656	$89.61	$1,181,870,414	$3,220,355	$141.22
+15%	$2,680,514,801	$2,222,649	$152.43	$1,707,593,266	$4,652,843	$204.04
+30%	$3,785,202,384	$3,138,642	$215.25	$2,233,316,118	$6,085,330	$266.86
Societal perspective						
–30%	–$4,995,355,121	–$4,142,086	–$284.06	–$1,945,367,346	–$5,300,728	–$232.45
–15%	–$1,709,763,951	–$1,417,715	–$97.23	–$381,748,466	–$1,040,187	–$45.62
0%	$1,575,827,218	$1,306,656	$89.61	$1,181,870,414	$3,220,355	$141.22
+15%	$4,861,418,388	$4,031,027	$276.45	$2,745,489,293	$7,480,897	$328.06
+30%	$8,147,009,557	$6,755,398	$463.29	$4,309,108,173	$11,741,439	$514.90

duty periods, that already allow sufficient time off, and that provide excellent supervision, or they may have sufficient resources on hand to fund the necessary changes. Fortunate hospitals might have sufficient operating funds, the ability to create efficiencies and savings, a growing patient base, or the opportunity to raise private funds to support the recommended changes. Not all hospitals, however, would be able to bear the costs of these new proposals.

Determining the financial capacity of teaching institutions to absorb some portion of these costs was beyond the scope of this study. If some hospitals are unable to absorb these costs fully, it could impact other parts of service delivery (e.g., the amount of uncompensated care) or other quality improvements (e.g., adoption of electronic health records). If funds are unavailable to hire substitutes to pick up residents' excess hours, hospitals might increase residents' workload or overburden other staff. This could lead to a decrease in opportunities for learning and indirectly affect patient safety. Without sufficient staff, patients also might have decreased access to hospital services.

The cost projections of the model do not include the additional costs of implementing the committee's recommendations that might stem from duty hour adjustments such as the extra day off per month, safe transportation options, more detailed compliance auditing, and faculty supervision. In New York State, an extensive, detailed duty hour compliance audit costs on average $24,000 per hospital. The cost of oversight would depend on the frequency and nature of any inspections, and there may be local costs of compliance monitoring as well. Also, the model does not estimate the research costs of monitoring the implementation of all the recommendations and evaluating their impact. On the other side of the ledger, the model does not calculate potential savings. Implementation of system redesign strategies could produce savings stemming from efficiencies and streamlining of work and education systems as well as the prevention of PAEs, but these savings are not projected or offset against the costs of the recommendations.

The committee believes that additional funding and personnel should be made available to support workload reduction and compliance with the recommended duty hour limits so that they do not have undesirable effects on patient or resident safety. Residents experienced work compressed into fewer hours after the 2003 duty hour rules and now would have increased pressures. If the recommended duty hour parameters are implemented without additional funds, this would be another unfunded mandate that some training institutions could not afford and workload might be shifted to other staff who are frequently overloaded themselves.

The committee recognizes that this is not an opportune time to be asking for additional funds for the health system. It understands that there are strong political pressures on the Medicare budget and all domestic spending and that MedPAC[3] has recommended reductions in IME funding. It also recognizes that there are many payers that support GME, various places other than health care where the federal budget could be cut, and many patients who want to receive safer hospital care and have better-trained physicians. The committee has responsibly interpreted the available evidence and reached clear conclusions that implementing its recommendations will require additional funds. It urges Congress and other supporters of GME to carefully consider this report and to seek funds to help hospitals implement the duty hour changes.

Recommendation 9-1: All financial stakeholders in graduate medical education, such as the Centers for Medicare and Medicaid Services, Department of Veterans Affairs, Department of Defense, Health Resources and Services Administration, states and local governments, private insurers, and sponsoring institutions, should financially support the changes necessitated by the committee's recommendations to promote patient safety and resident safety and education, with special attention to safety net hospitals.

- **An independent convening body should bring together all the major funders of graduate medical education to examine current financing methodologies and develop a coordinated approach to generate needed resources.**

WORKFORCE IMPLICATIONS

Resident duty hour and workload adjustments, as well as the greater degree of supervision by attending physicians recommended by this committee, will create demand for more residents, midlevel providers, and trained physicians to provide 24-hour coverage in training facilities. This demand for staff will come at a time when other national trends are driving demand for hospital services and personnel to staff these facilities. Technology changes, increasing numbers of patients being hospitalized, and an aging population in the United States are primary contributors to the demand for inpatient services (IOM, 2008; Kozak et al., 2006). Calls for patient safety improvements will also create more demand for hospital-based staff (Shulkin, 2008). For example, some hospitals are recognizing the need to increase staff in hospitals on nights and weekends when there is greater mortality, but currently less comprehensive staffing. In addition, calls for residencies to incorporate more diverse settings than hospitals into

---

[3]Medicare Payment Advisory Commission, 2008 Report to Congress: Medicare Payment Policy, Washington, DC, March 2008.

training programs (e.g., ambulatory care settings) would leave less resident time for inpatient coverage and require more personnel to cover existing inpatient facilities (COGME, 2007). Shifting workload to existing team personnel does not appear to be a solution in many settings because of the workload pressures already experienced in hospitals by all staff (Weissman et al., 2007).

The issue of the proper size of the physician workforce and the adequacy of the supply of particular specialists is a controversial one, lacking consensus (Iglehart, 2008). Many professional organizations, panels of experts, and researchers have identified current and projected workforce shortages for nurses, nurse practitioners, and physicians (AAMC, 2006, 2007b, 2008a,d; American Association of Colleges of Nursing, 2004; ANSR, 2008; Colwill et al., 2008; Larson and Hart, 2007; National League of Nursing, 2005; Salsberg, 2008). There are projections of a physician shortage across the United States by 2020, especially in certain geographic areas and certain specialties, and calls for increasing the size of medical school classes and residency positions to replace an aging physician workforce and serve the greater care needs of an aging population (AAMC, 2007b, 2008a; COGME, 2007; Colwill et al., 2008; IOM, 2008).

On the other hand, a body of evidence indicates that there may be a surplus or at least not a shortage of physicians. Researchers who have analyzed geographic and hospital-specific resource data find that there are vast variations in the size of the physician workforce among geographic regions that are not associated with improved health outcomes and better-quality care (Fisher, 2004; Wennberg et al., 2004). These studies indicate that health care over a period of time beyond just an inpatient stay could be made more efficient, and that effective systems of care provide higher-quality care over the course of a patient's chronic illness. This longitudinal efficiency relates particularly to the lower use of supply-sensitive services such as inpatient hospital days, imaging and diagnostic tests, and physician visits. Studies also show that solely increasing the total supply of physicians is an inefficient way to benefit the specialties and geographic areas that may need more doctors (Goodman, 2004). The reforms necessary to achieve the improvements identified in these studies are beyond the scope of this report and the work of doctors in training.

The pipeline to produce physicians is a long one: a minimum of 4 years of medical school and 3 to 7 or more years of residency. The Association of American Medical Colleges recommends a 30 percent increase in medical school enrollment from 2002 levels by 2015, an increase of 5,000 new positions annually (AAMC, 2008b). The Council on Graduate Medical Education (COGME) has recommended an increase in the number of CMS-funded residency positions by at least 15 percent by 2015, about 3,000 new positions yearly, as well as diversification of training sites based on an

Institute of Medicine model of care delivery (COGME, 2007; IOM, 2003). This 15 percent increase in residency positions would absorb about half of the expanded number of medical school graduates. The remainder would offset the enrollment of international medical graduates who now compose approximately 27 percent of all residents (Brotherton and Etzel, 2007).

The 8,247 additional residents projected by the Nuckols and Escarce economic model, if a larger pool of residents were to cover the excess hours of resident work, is a total incorporating both specialty (7,639) and subspecialty (608) residents (Table 9-5); it is not equivalent to the yearly increase in medical students or residency positions mentioned above. Additional residency positions are one approach to filling the gap in covered hours. Since the Medicare funding cap on residency positions was put in place, only training programs with access to alternative sources of funding (e.g., private resources) have been able to expand their programs to address reduced hours. The committee strongly urges that all possible funding mechanisms be considered, including increasing or eliminating the cap on residency positions. At the same time, the committee recognizes that each institution will have to assess its local labor market, educational capacity, and unique circumstances to determine the most effective way to achieve adherence to the proposed requirements for duty hours, workload, and supervision. Individual institutions may or may not find additional residents to be the preferred approach.

A PHASED IMPLEMENTATION OF DUTY HOURS, ITS EVALUATION, AND FURTHER RESEARCH

Phase-in of Recommendations

To promote safe medical care, improve the education of doctors in training, and increase the safety of residents and the general public, the committee offers its recommendations, which should be implemented with all deliberate speed. The committee believes action is needed urgently because U.S. hospitals still have a too-high error rate and too many patients are harmed during their stay. The committee realizes that its recommendations will not prevent all patient harm and that residents are not responsible for the whole problem. However, resident duty hours and schedules is a risk that can be ameliorated. Fatigue contributes to unsafe conditions and can increase the risk of errors. Fatigue among residents is something that can be reduced through a judicious use of periods for rest and sleep between duty periods and by limits on extended long duty periods. Other benefits are likely to result from the recommendations, including fewer automobile accidents caused by tired residents, a better environment for learning and working that enhances the acquisition of needed competencies, and a greater participation by residents in the hospital's culture of safety.

While some recommendations could and should be implemented immediately, changed duty hours, workload, and funding issues might require an integrated phase-in. The recommendations will require additional resources—both financial and human. Without the necessary restructuring in resource allocation, attempts to implement certain recommendations will fail to have the desired benefit and could even produce conditions that are less favorable to patient safety. The committee believes that the ACGME and the other organizations charged to implement aspects of the recommendations should undertake their work with urgency and that all institutions with residency programs take action to begin implementation of all recommendations within 24 months.

The committee proposes several recommendations that should be considered as a package and implemented in concert. The recommended parameters concerning the reduction in duty hours and fatigue mitigation are particularly intertwined. Although the 80-hour week and the 30-hour limit on extended duty are unchanged from the current rules, the recommended length and scheduling of rest periods during the month are crucial to making those duty hour limits supportable. The recommended periods for rest during extended duty and periods away from the hospital to allow for rest and sleep recovery are intended to establish safer working conditions, to protect residents from excessive fatigue, and to protect patients from fatigue-induced errors. The recommended duty hour parameters are also closely linked to the recommendations concerning workload, supervision, and funding. Supervision enhancements and workload reductions, however, could be put in place before duty hours are changed. Some institutions will probably be able to implement the recommended changes independently, but others would need outside funds to help support the hiring of additional staff to assume the excess duties (workload and hours) of residents.

Not only would the benefits of these duty hour parameters be less likely to materialize if they are implemented piecemeal and in a disjointed fashion, but also unintended and potentially harmful consequences could result without the accompanying committee recommendations on workload and funding. For example, if duty hours are reduced again but workload remains at current levels, residents will be under greater stress as they rush to complete work, which increases the likelihood of making errors and could further reduce the amount of time they have for educational activities; this could negatively impact the safety of both current and future patients. The potential benefits to society of well-trained physicians and fewer people suffering from PAEs argue for all funders of GME to contribute appropriately to support these GME reforms.

Some recommendations should be implemented immediately, such as limits on moonlighting, enhanced supervision, improved handovers, error reporting, and the provision of transportation to residents finishing a long duty period and those too tired to drive safely. Also, programs should quickly enhance their formal education efforts on sleep to include the latest scientific research on fatigue, sleep deprivation, and methods to mitigate fatigue.

Research and Evaluation Plans

The lack of systematic data collection before and after the 2003 rules hampered the committee's ability to determine their impact fully and to assess how much of the complaints about duty hour reform represent rhetoric and resistance to change rather than valid criticisms. The literature that exists too often comes from single-institution studies with insufficient statistical power to determine effects on patient outcomes and is often specific to one specialty, making findings difficult to generalize.

Collecting baseline information now on the current situation concerning residents' duty hours and workload would permit evaluations of the impact of this report's recommendations once they are implemented. When designing evaluation studies, it will be important to include a sufficient number of programs so that the studies can produce a national picture across all programs as well as evaluations targeted to individual specialties.

Ongoing data collection at the national level will be useful for monitoring the full impact of the committee's recommendations. Because major policy changes cannot be tested in the laboratory and it is impossible to predict all the effects of those changes on the healthcare system, there could be some unintended and unanticipated reactions to the recommendations. For example, some specialties might find that programs reduce hours and workloads without accompanying redesign of education, causing fewer opportunities for residents to achieve procedural or medical competency or requiring a longer residency. Or the less continuous care provided by residents could increase their detachment and reduce their commitment to their patients if work is not adequately restructured to permit enhanced team coverage. While the committee certainly does not want to *increase* workload or hours for residents, some programs might try to meet the recommended parameters of rest and work periods by increasing the frequency of overnight duty periods and reducing the current amount of time off duty. When Residency Review Committees (RRCs) implement specialty-specific workload caps, it would be useful for them to monitor the impact of the caps on both residents and their learning as well as on the costs, coverage, and access to services at training institutions.

Similarly, when designing and evaluating innovative projects to test creative ways to meet the intent of the committee's recommendations while allowing for alternative approaches that might better suit a type of program or specialty, researchers must take care to include a sufficient number of sites to generate statistically powerful findings and a design that will support comparisons of the impact of the recommendations as implemented. While many experts have told the committee that one rule does not fit all and that every specialty has different characteristics and needs, it is difficult to have a realistic understanding of their differences and similarities without reliable data.

Some of the ideas for the types of data that would be useful have been detailed in the preceding chapters. Below are some key research topics that the committee recommends for future consideration. These and other research issues have been discussed throughout the report. It is important for all the stakeholders in GME, contributors of ideas as well as funds, to be included in a discussion of an evaluation and research agenda, and involved in the necessary priority setting. It will be a challenge to researchers to tease apart the various and overlapping contributing factors to patient safety to determine the extent of their individual impacts, but the results will have widespread benefits and would be of use well beyond the scope of the current study.

- *The relationship between improved processes for handovers and shorter and longer duty hours.* This report has discussed many of the multiple factors in the resident work and learning environment that contribute to error and potential patient harm. Understanding the handover process and the risks and benefits it poses for patients is an important aspect of the issue that is ripe for systematic research, primarily to learn ways to improve on current methods, but also to assess whether the risks to patients from handovers are greater than the risks of being cared for by a fatigued resident who has more familiarity with the patient.
- *Resident fatigue and patient harm, as well as residents' own safety within various scheduling and fatigue mitigation approaches.* On the assumption that there will be considerable variation in the way individual programs choose to meet the parameters of the duty hour recommendation, further examination of scheduling effects on PAEs and resident safety for specialties and program sizes is important, along with analysis of the buffering effect of enhanced supervision and teamwork. Research could investigate whether duty hours can be further reduced from current recommendations.
- *Resident workload and its impact on patient safety.* Resident workload has been understudied for its effect on short-term and long-term patient safety. The workload of residents also needs to be assessed for its educational value.
- *Measuring and achieving competence by specialty within reduced*

duty hours and workload. Although competency-based education is a broader topic than could be covered by this committee and relates to the restructuring of GME more generally, it is of concern to this study as a method for assessing the impact of the reduction of duty hours and workload on resident attainment. More efficient methods of teaching and conveying information and procedural skills, as well as the assessment of residents' knowledge and skills would contribute to the positive impact of the committee's recommendations. Achieving competence also means that residents need sufficient exposure to an appropriate range of patient experiences including, depending on specialty, outpatient exposure.

- *Opportunities for and limitations to substitution of other providers for residents.* Better workforce data are needed to assess the impact of implementing the recommendations on both the existing workforce and future workforce needs nationally and in specific geographic areas. Researchers should also assess whether and the extent to which additional residents are needed.
- *Impact of the IOM recommendations after implementation.* Clearly the impact of implementation on patient safety is the overriding concern; however, other impacts, such as resident safety, are also important. The variations in impacts based on different approaches to implementation by different programs and specialties are also key to the fine-tuning of the rules in the future.

The committee believes that it is essential to build an information base to evaluate what happens going forward. With mechanisms to monitor and evaluate the ongoing implementation of the committee's recommendations, as well as innovative experiments, it should be possible to adjust the rules periodically as needed. For example, grounds for ACGME's granting exceptions might have to be tightened or expanded; or further measures might be necessary after workloads are reduced through the elimination of noneducational activities, if the work remaining is consistently of too high an intensity. Additionally, such information, as previously mentioned, could also help to avoid major problems or unintended consequences, such as an aspect of the duty hour parameters producing insurmountable challenges to certain specialties more than others that cannot be handled on an exceptions basis; or residents not sleeping during the prescribed 5-hour rest period who may choose to catch up on paperwork instead of mitigating their fatigue; or some small residency programs having to close because of insufficient staff to cover the excess hours of residents.

Recommendation 9-2: To gather the data necessary to monitor implementation of these recommendations and to prepare for future adjust-

ments as needed to achieve the desired objectives, ACGME should convene a meeting of stakeholders and potential funders to set priorities for research and evaluation projects. The Centers for Medicare and Medicaid Services, Agency for Healthcare Research and Quality, National Institutes of Health, Department of Defense, Department of Veterans Affairs, and other funders should support this work as a high priority.

Because so many individuals and organizations have strong economic and professional interests in GME, and resident duty hours in particular, it will be a challenge to come up with an agenda for research projects. It will be even more challenging to design research projects that can produce sound scientific evidence of use to policy makers. Given the likelihood that research and evaluation funds will be quite limited, it is especially important that they be spent wisely to focus on priority issues and to do so in a methodologically sound manner that is acceptable to the key organizations that will have to be involved in implementation of future policy changes. The results of these research, monitoring, and evaluation projects should, in the future, indicate areas for further refinement of the rules and alternatives that could better enhance the goals of patient safety, resident safety, and training.

REFERENCES

AAMC (Association of American Medical Colleges). 2006. *AAMC statement on the physician workforce, 2006.* http://www.aamc.org/workforce/workforceposition.pdf (accessed June 4, 2008).

———. 2007a. *AAMC survey of housestaff stipends, benefits and funding: Autumn 2007 report.* Washington, DC: Association of American Medical Colleges.

———. 2007b. *Recent studies and reports on physician shortages in the U.S., 2007.* http://www.aamc.org/workforce/recentworkforcestudies2007.pdf (accessed June 4, 2008).

———. 2008a. *Help wanted: More U.S. doctors projections indicate America will face shortage of M.D.s by 2020.* http://www.aamc.org/workforce/helpwanted.pdf (accessed June 4, 2008).

———. 2008b. *Medical school enrollment plans: Analysis of the 2007 AAMC survey.* Washington, DC: Association of American Medical Colleges. https:services.aamc.org/publications/showfilecfm?file=version111.pdf&prd_id=229&prv_id=279&pdf_id=111 (accessed June 4, 2008).

———. 2008c. *Teaching hospitals: About AAMC-COTH membership.* http://www.aamc.org/teachinghospitals.htm (accessed August 12, 2008).

———. 2008d. *U.S. medical school enrollment projected to rise 21 percent by 2012: Both new and existing schools will fuel growth.* http://www.aamc.org/newsroom/pressrel/2008/080501.htm (accessed May 2, 2008).

AMA (American Medical Association) Division of Market Research and Analysis. 2005. *Medical students and residents work-hour survey May.* Chicago, IL: MemberConnect.

American Association of Colleges of Nursing. 2004. *Nursing faculty shortage fact sheet.* http://www.aacn.nche.edu/Media/backgrounders/facultyshortage.htm (accessed June 3, 2008.

Andrews, R. M., D. E. Stull, I. Fraser, B. Friedman, and R. L. Houchens. 2007. *Serving the uninsured: Safety-net hospitals, 2003.* Rockville, MD: Agency for Healthcare Quality and Research.

ANSR (Americans for Nursing Shortage Relief). 2008. *Testimony regarding FY 2009 appropriations for Title VIII—Nursing workforce development programs.* https://www.ncsbn.org/FY09_Senate_Testimony.pdf (accessed August 26, 2008).

Arora, V., C. Dunphy, V. Y. Chang, F. Ahmad, H. J. Humphrey, and D. Meltzer. 2006. The effects of on-duty napping on intern sleep time and fatigue. *Annals of Internal Medicine* 144(11):792-798.

Brotherton, S. E., and S. I. Etzel. 2007. Graduate medical education, 2006-2007. *JAMA* 298(9):1081-1096.

CBO (Congressional Budget Office). 2008. *The budget and economic outlook: An update.* Washington, DC: Congressional Budget Office. Supplemental data of mandatory outlays. http://www.cbo.gov/budget/factsheets/2008b/medicare.pdf (accessed September 23, 2008).

Chang, B. K. 2007. Presentation to the Committee on Optimizing Graduate Medical Trainee (Resident) Hours and Work Schedules to Improve Patient Safety, December 3, 2007, Washington, DC.

COGME (Council on Graduate Medical Education). 2007. *19th report: Enhancing flexibility in graduate medical education.* Rockville, MD: Bureau of Health Professions, U.S. Department of Health and Human Services.

Colwill, J. M., J. M. Cultice, and R. L. Kruse. 2008. Will generalist physician supply meet demands of an increasing and aging population? *Health Affairs* 27(3):w232-w241.

COTH (Council of Teaching Hospitals and Health Systems). 2008. *AAMC teaching hospitals and health systems.* Washington, DC: Association of American Medical Colleges. Pp. 14-38.

Fisher, E. S., D. E. Wennberg, T. A. Stukel, and D. J. Gottlieb. 2004. Variations in the longitudinal efficiency of academic medical centers. *Health Affairs* (Supplement Web Exclusives): VAR19-VAR32.

Goodman, D. C. 2004. Twenty-year trends in regional variations in the U.S. physician workforce. *Health Affairs* (Supplement Web Exclusives):VAR90-VAR97.

Henderson, T. M. 2006. *Medicaid direct and indirect graduate medical education payments: A 50-state survey.* Washington, DC: Association of American Medical Colleges.

HRSA (Health Resources and Services Administration). 2008a. *Other health professions programs: Children's hospitals graduate medical education payment program.* http://www.hrsa.gov/about/budgetjustification09/childrenshospital.htm (accessed October 16, 2008).

———. 2008b. *Training in primary care medicine and dentistry.* http://www.hrsa.gov/about/budgetjustifcation09/hptraining.htm (accessed October 16, 2008).

Iglehart, J. K. 2008. Medicare, graduate medical education, and new policy directions. *New England Journal of Medicine* 359(6):643-650.

IOM (Institute of Medicine). 2003. *Health professions education: A bridge to quality*, edited by A. C. Greiner, E. Knebel, and the Committee on the Health Professions Education Summit. Washington, DC: The National Academies Press.

———. 2008. *Retooling for an aging America: Building the health care workforce.* Washington, DC: The National Academies Press.

Jagsi, R., D. F. Weinstein, J. Shapiro, B. T. Kitch, D. Dorer, and J. S. Weissman. 2008. The Accreditation Council for Graduate Medical Education's limits on residents' work hours and patient safety: A study of resident experiences and perceptions before and after hours reductions. *Archives of Internal Medicine* 168(5):493-500.

Knapp, R. M. 2002. Complexity and uncertainty in financing graduate medical education. *Academic Medicine* 77(11):1076-1083.

Kozak, L. J., C. J. DeFrances, and M. J. Hall. 2006. *National hospital discharge survey: 2004 annual summary with detailed diagnosis and procedure data.* Hyattsville, MD: National Center for Health Statistics.

Landrigan, C. P., L. K. Barger, B. E. Cade, N. T. Ayas, and C. A. Czeisler. 2006. Interns' compliance with Accreditation Council for Graduate Medical Education work-hour limits. *JAMA* 296(9):1063-1070.

Larson, E. H., and L. G. Hart. 2007. Growth and change in the physician assistant workforce in the United States, 1967-2000. *Journal of Allied Health* 36(3):121-130.

MedPAC (Medicare Payment Advisory Commission). 2008. *Hospital acute inpatient services payment system.* http://www.medpac.gov/publications/other_reports/Sept06_MedPAC_Payment_basics_hospital.pdf (accessed October 14, 2008).

National League of Nursing. 2005. *Nurse faculty shortage fact sheet.* http://www.nln.org/governmentaffairs/pdf/NurseFacultyShortage.pdf (accessed June 3, 2008).

Nuckols, T. K., and J. J. Escarce. 2005. Residency work-hours reform. A cost analysis including preventable adverse events. *Journal of General Internal Medicine* 20(10):873-878.

———. 2008 (unpublished). *Cost analysis for the Institute of Medicine project "Optimizing Graduate Medical Trainee (Resident) Hours and Work Schedules to Improve Patient Safety."* RAND Corporation. www.iom.edu/residenthours or www.iom.edu/hcs.

Opas, L. 2008. IOM workshop: Cost impact of duty hour change. Presentation to the Committee on Optimizing Graduate Medical Trainee (Resident) Hours and Work Schedules to Improve Patient Safety, March 4, 2008, Irvine, CA.

Petersen, L. A., T. A. Brennan, A. C. O'Neil, E. F. Cook, and T. H. Lee. 1994. Does housestaff discontinuity of care increase the risk for preventable adverse events? *Annals of Internal Medicine* 121(11):866.

Petersen, L. A., E. J. Orav, J. M. Teich, A. C. O'Neil, and T. A. Brennan. 1998. Using a computerized sign-out program to improve continuity of inpatient care and prevent adverse events. *Joint Commission Journal on Quality Improvement* 24(2):77-87.

Salsberg, E. 2008. Addressing healthcare workforce issues for the future. Statement before the Committee on Health, Education, Labor, and Pensions (HELP), United States senate, February 12, 2008.

Shulkin, D. J. 2008. Like night and day—Shedding light on off-hours care. *New England Journal of Medicine* 358(20):2091-2093.

Siegel, B., M. Regenstein, and P. Shin. 2004. Health reform and the safety net: Big opportunities; major risks. *Journal of Law, Medicine & Ethics* 32(3):426-432.

Stoddard, J. J., D. A. Kindig, and D. Libby. 1994. Graduate medical education reform. Service provision transition costs [erratum appears in *JAMA* 272(16):1256]. *JAMA* 272(1):53-58.

Thomas, E. J., D. M. Studdert, J. P. Newhouse, B. I. W. Zbar, K. M. Howard, E. J. Williams, and T. A. Brennan. 1999. Costs of medical injuries in Utah and Colorado. *Inquiry* 36(3):255-264.

U.S. Department of Labor: Bureau of Labor Statistics. 2007a. *Employer cost for employee compensation, annual data for 2006.* http://data.bls.gov (accessed August 12, 2008).

———. 2007b. *National compensation survey: Occupational wages in the United States.* http://www.bls.gov/ncs/ocs/sp/ncbl0910.pdf (accessed August 12, 2008).

Weissman, J. S., J. M. Rothschild, E. Bendavid, P. Sprivulis, E. F. Cook, R. S. Evans, Y. Kaganova, M. Bender, J. David-Kasdan, P. Haug, J. Lloyd, L. G. Selbovitz, H. J. Murff, and D. W. Bates. 2007. Hospital workload and adverse events. *Medical Care* 45(5): 448-455.

Wennberg, J. E., E. S. Fisher, T. A. Stukel, and S. M. Sharp. 2004. Use of Medicare claims data to monitor provider-specific performance among patients with severe chronic illness. *Health Affairs* (Supplement Web Exclusives):VAR5-VAR18.

Werner, R. M., L. E. Goldman, and R. A. Dudley. 2008. Comparison of change in quality of care between safety-net and non-safety-net hospitals. *JAMA* 299(18):2180-2187.

Whang, E. E., M. M. Mello, S. W. Ashley, and M. J. Zinner. 2003. Implementing resident work hour limitations: Lessons from the New York State experience. *Annals of Surgery* 237(4):449-455.

Wynn, B., C. Guarino, L. Morse, and M. Cho. 2006 (unpublished). *Alternative ways of financing graduate medical education*. RAND Corporation, Santa Monica, CA.

Zhan, C., and M. R. Miller. 2003. Excess length of stay, charges, and mortality attributable to medical injuries during hospitalization. *JAMA* 290(14):1868-1874.

Appendix A

Statement of Task

The Institute of Medicine will form a consensus committee to (1) synthesize current evidence on medical resident schedules and healthcare safety and (2) develop strategies to enable optimization of work schedules to improve safety in the healthcare work environment. The strategies recommended will take into account the learning and experience that residents must achieve during their training. The recommendations will be structured to optimize both the quality of care and the educational objectives.

The committee is asked to deliver its report in 12 months, and thus will focus on two priority tasks—each with component tasks as well as related issues to be considered as relevant to the main task but not necessarily studied in depth. Although the issues to be studied are broad ones, to permit comprehensive coverage of the priority issues in the specified timeframe, the scope is limited to medical residents (versus all physicians or all healthcare workers) and their work schedules (versus all work processes or the entire work environment). The committee is asked to consider the impact of recommended actions on costs; however, a detailed cost analysis is outside the scope of the study.

Task #1: Review and Synthesize Evidence on Optimal Resident Work Schedules, including:

- Evidence on the relationship between resident work schedules, resident performance, and the quality of care delivered by residents—specifically patient safety. Consider also evidence on the safety of the residents, the education and training experience of the residents, the quality of the interactions from both the resident and patient perspective, and other aspects of safety and quality of care such as care hand-offs and transitions.

- As relevant, consider evidence on the relationship between sleep, fatigue, work schedules, and performance for other health care professionals as well as generally.

- Evidence on the strategies, practices, interventions, and tools that have been employed in the United States, Australia, Canada, Europe, New Zealand, and elsewhere to optimize the work schedules for residents to assure the safety and quality of patient care. Identify barriers to change and strategies for overcoming them. Examine how related issues are handled such as staffing, financial costs, and other resources. Consider also other approaches to the nature of resident work and the role of the resident (such as assigning tasks traditionally assigned to medical residents to other healthcare professionals) and resident training (such as use of simulations).
  - As relevant, consider approaches to similar issues in other healthcare work environments and other industries as well as more general issues such as teamwork and organizational culture.

Task #2: Develop Strategies for Implementing Optimal Resident Work Schedules

- Make recommendations for how the strategies, practices, interventions, and tools identified in Task #1 can be implemented to optimize resident schedules to improve the safety of the healthcare work environment and the quality of care.
- Recommend actions for stakeholders including residents, hospitals, professional societies, accrediting bodies, administrators and funders of residency training programs, federal and state agencies, and policy makers at all levels. Identify actions that can be taken in the short and long term. The recommendations should specify who should take what actions to create a care environment that is safe for patients, residents, and other health workers. Recommendations should also address anticipated barriers to change such as the culture of medical education and health care institutions.
  - Consider and describe the consequences of these recommended actions for the cost of medical training and of health care. As discussed above, costs are to be considered in general terms—the task is not to develop explicit cost estimates for recommended changes.

Appendix B

Comparison of Select Scheduling Possibilities Under Committee Recommendations and Under 2003 ACGME Duty Hour Rules

These examples are offered as possible monthly schedules for a single resident under the committee's proposed duty hour parameters and under current Accreditation Council for Graduate Medical Education (ACGME) rules.

Tables B-1a and B-1b provide a comparison of schedules with extended duty periods. The benefits of the committee's recommendations are protected on-duty sleep time per week (e.g., 7.5 hours on average), regularity of days off, and less possible variability in the week-to-week duty hour totals (60 to 80) than under ACGME rules (60 to 90). Under the committee's recommendations, extended duty periods may not be more often than every third night, while under ACGME rules they might be scheduled closer together because averaging is allowed. Under current rules, residents may go for more than a week without a day off. The committee recommends 5 days off per month rather than 4, with at least one day off each week without averaging.

B-1a: Possible Extended Duty (q4) Monthly Schedule for a Single Resident Under Committee Proposal. Extended duty period every fourth night *no averaging* (e.g., in hospital from 7 a.m. to 1 p.m. next day); 30-hour maximum consists of 16 hours for admitting patients, 5-hour sleep period, and 9 hours for completion of work, educational activities, and transfer of responsibilities.

Mon	Tues	Wed	Thurs	Fri	Sat	Sun	Duty Hours
30 (16 + 5 sleep + 9)		10	10	30 (16 + 5 sleep + 9)		Off duty	80 (including 10 hours for sleep periods)
10	30 (16 + 5 sleep + 9)		10	10	Off duty	Off duty	60 (including 5 hours for sleep period)
30 (16 + 5 sleep + 9)		10	10	30 (16 + 5 sleep + 9)		Off duty	80 (including 10 hours for sleep periods)
10	10	Off duty	30 (16 + 5 sleep + 9)		10	10	70 (including 5 hours for sleep period)
							TOTAL HOURS: 290 AVERAGE DUTY HOURS PER WEEK: 72.5 AVERAGE PROTECTED ON-DUTY SLEEP PERIOD: 7.5 hours/week DAYS OFF: 5 full days per month

B-1b: Possible Extended Duty (q4) Monthly Schedule for a Single Resident Under Current ACGME Rules. Extended duty period every fourth night *on average* (e.g., in hospital from 7 a.m. to 1 p.m. next day); 30-hour maximum consists of 24 hours for admitting patients + 6 hours for completion of work, educational activities, and transfer of responsibilities. *Day off per week is averaged.*

Mon	Tues	Wed	Thurs	Fri	Sat	Sun	Duty Hours
30 (24 + 6)		10	10	30 (24 + 6)		10	90
10	30 (24 + 6)		10	10	Off duty	Off duty	60
30 (24 + 6)		10	30 (24 + 6)		10	Off duty	80
10	Off duty	10	30 (24 + 6)		10	10	70
							TOTAL HOURS: 300 AVERAGE DUTY HOURS PER WEEK: 75.0 DAYS OFF: 4 full days per month

Tables B-2a and B-2b provide a comparison of schedules with extended duty periods every fifth night. The benefits of the committee's recommendations are protected sleep time per week (e.g., 6.25 hours on average) and regularity of days off. In this example with extended call every fifth night, the variability in work hours from week to week is the same under the committee's recommendations and under the current rules.

B-2a: Possible Extended Duty (q5) Monthly Schedule for a Single Resident Under Committee Proposal. Extended duty period every fifth night (e.g., in hospital from 7 a.m. to 1 p.m. next day); 30-hour maximum consists of 16 hours of work, 5-hour sleep period, and 9 hours for completion of work and transfer of responsibilities.

Mon	Tues	Wed	Thurs	Fri	Sat	Sun	Duty Hours
Off duty	30 (16 + 5 sleep + 9)		10	10	10	30 (16 + 5 sleep + 9)	90 (including 10 hours for sleep periods)
Continuation of 30-hour shift	10	10	10	30 (16 + 5 sleep + 9)		Off duty	60 (including 5 hours for sleep period)
10	10	30 (16 + 5 sleep + 9)		10	Off duty	Off duty	60 (including 5 hours for sleep period)
30 (16 + 5 sleep + 9)		10	10	10	10	Off duty	70 (including 5 hours for sleep period)
							TOTAL HOURS: 280 AVERAGE DUTY HOURS PER WEEK: 70.0 AVERAGE PROTECTED ON-DUTY SLEEP PERIOD: 6.25 hours/week DAYS OFF: 5 full days per month

B-2b: Possible Extended Duty (q5) Monthly Schedule for a Single Resident Under Current ACGME Rules. Extended duty period every fifth night (e.g., in hospital from 7 a.m. to 1 p.m. next day); 30-hour maximum consists of 24 hours for admitting patients and 6 hours for completion of work, educational activities, and transfer of responsibilities; *day off per week is averaged.*

Mon	Tues	Wed	Thurs	Fri	Sat	Sun	Duty Hours
Off duty	30 (24 + 6)		10	10	10	30 (24 + 6)	90
Continuation of 30-hour shift	10	10	10	30 (24 + 6)		Off duty	60
10	10	30 (16 + 5 + 9)		10	Off duty	Off duty	60
30 (24 + 6)		10	10	10	10	10	80
							TOTAL HOURS: 290 AVERAGE DUTY HOURS PER WEEK: 72.5 DAYS OFF: 4 full days per month

Tables B-3a and B-3b provide a comparison of schedules using a daytime shift length of 10 hours. The benefits of the committee's recommendations are regularity of days off and, thus, less variability in week-to-week duty hour totals (50 to 60) than under ACGME rules (50 to 70). Under current ACGME rules, a resident could go for more than 2 weeks without a day off.

B-3a: Possible 10-Hour Daytime Schedule for a Single Resident Under Committee Proposal

Mon	Tues	Wed	Thurs	Fri	Sat	Sun	
10	10	10	Off duty	10	10	10	60
Off duty	10	10	10	10	10	10	60
10	10	10	10	10	Off duty	Off duty	50
10	10	10	10	10	10	Off duty	60
							TOTAL HOURS: 230 AVERAGE DUTY HOURS PER WEEK: 57.5 DAYS OFF: 5 full days per month

B-3b: Possible 10-Hour Daytime Schedule for a Single Resident Under Current ACGME Rules

Mon	Tues	Wed	Thurs	Fri	Sat	Sun	
10	10	10	Off duty	10	10	10	60
10	10	10	10	10	10	10	70
10	10	10	10	10	10	Off duty	60
10	10	10	10	10	Off duty	Off duty	50
							TOTAL HOURS: 240 AVERAGE DUTY HOURS PER WEEK: 60.0 DAYS OFF: 4 full days per month

Tables B-4a and B-4b provide a comparison of schedules using a nighttime shift length of 12 hours. The committee's proposal offers more recovery sleep time during night shift rotations by providing 48 hours off after 4 nights on so that there are 8 days off duty rather than 4. The ACGME currently has no limits on the number of consecutive night shifts.

B-4a: Possible 12-Hour Nighttime Schedule for a Single Resident Under Committee Proposal. After every fourth night on duty, resident receives 2 days off.

Mon	Tues	Wed	Thurs	Fri	Sat	Sun	
12	12	12	12	Off duty	Off duty	12	60
12	12	12	Off duty	Off duty	12	12	60
12	12	Off duty	Off duty	12	12	12	60
12	Off duty	Off duty	12	12	12	12	60
							TOTAL HOURS: 240 AVERAGE DUTY HOURS PER WEEK: 60.0 DAYS OFF: 8 full days per month

B-4b: Possible 12-Hour Nighttime Schedule for a Single Resident Under Current ACGME Rules. No limits on the number of consecutive night shifts.

Mon	Tues	Wed	Thurs	Fri	Sat	Sun	
12	12	12	12	Off duty	12	12	72
12	12	12	Off duty	12	12	12	72
12	12	Off duty	12	12	12	12	72
12	Off duty	12	12	12	12	12	72
							TOTAL HOURS: 288 AVERAGE DUTY HOURS PER WEEK: 72.0 DAYS OFF: 4 full days per month

Appendix C

International Experiences Limiting Resident Duty Hours

Altering the resident experience through reduced duty hours is an ongoing process in many countries. Some have implemented regulations to adjust resident schedules, while others have adopted more guideline-based methods. These diverse approaches accommodate very different healthcare systems, medical education programs, and cultures than in the United States. For most of the countries examined by the committee, the primary reason presented for modification of total resident hours is worker safety. Each country has experienced difficulties implementing its intended reductions due to workforce shortages and some have faced strains on educational training, including reduced clinical contact and procedural experience. The committee reviewed the strategies used by these countries to overcome the challenges encountered from reducing resident duty hours and identified educational redesign, scheduling flexibility, and a period of phase-in to adjust to changes as relevant lessons for the U.S. graduate medical education system.

The desire to identify optimal resident duty hours is not unique to the United States. Duty hour regulation has been in place for more than 20 years in some countries (NZRDA, 2007). The statement of task for this Institute of Medicine (IOM) report specified an examination of the experiences of Europe (primarily the United Kingdom), New Zealand, Australia, and Canada to gain insights on additional strategies, practices, interventions, and tools employed by these countries in their efforts to adapt to changes in resident schedules. International efforts to modify the duty hours and work environment of medical residents provide useful perspectives on alternative systems. It is difficult to draw direct solutions from these experiences given that medical education, medical training, and the overall infrastructure of health systems vary substantially between other countries and the United States. However, lessons learned from abroad might help residency programs in the United States anticipate potential challenges of implementing additional changes to resident duty hours and new scheduling practices, if they were to occur.

Therefore, this review first provides a snapshot of current duty hour regulations and available patient safety data in the aforementioned countries. Duty hour regulations are then discussed in more detail, along with comments on compliance with and enforcement of those regulations. Consequences of reduced duty hours on the resident workforce labor supply, and the impacts of hour reductions on resident education and training are also examined, with a final look at efforts to overcome these challenges through program redesign.

GENERAL OVERVIEW OF COUNTRY EXPERIENCES

The rationale for regulations to reduce duty hours in New Zealand and Europe and for duty hour reform in Australia has been primarily from the perspective of worker safety, more so than patient safety, as it has been in the United States or Canada. Yet a combination of these concerns has caused each of the countries examined to make various changes in its medical resident training, and each has used different approaches to implement them. It is important to note that many of those changes are taking place within an extended training system compared to U.S. training programs (Foundation Programme, 2007; Jarvis, 2002). The average duration of medical school in most European countries is 6 years, similar to New Zealand and Australia, with some schools including a year or two of internship training (World Health Organization, 2007). The duration of medical school in Canada and the United States is normally 4 years. Upon being awarded a medical degree, graduates in the United States and Canada spend at least 3 years of residency training for general practice and additional years of training thereafter (usually 3 or more) depending on the specialty pursued (Medical Council of Canada, 2008; National Recruitment Office for General Practice Training, 2008; NHS, 2008a; Royal Australasian College of Physicians, 2007). One example that highlights how different training in other countries can be is the New Zealand experience. After 6 years of medical school, residents there typically spend another 4 years in basic training for general practice before qualifying to undertake specialty training (e.g., surgery, pediatrics, pathology). New Zealand residents who then train in those specialties (known there as "registrars") "can continue providing services to patients for more than 10 years while they complete their training" (NZRDA, 2008a, p. 1). In contrast, most specialty residency programs in the United States typically last 3-5 years, similar to programs in Canada or the United Kingdom.

The maximum weekly duty hours permitted by various recommendations, regulations, and legislation also vary considerably across the countries examined: from 37 hours per week to an unlimited number of hours per week that residents may work (Table C-1). Much of the available information on duty hours comes from national websites of medical professional organizations, resident organizations, or government agencies.

TABLE C-1 Resident Duty Hour Regulations in Various Countries, 2008

Country	How Hours Are Regulated	Maximum Hours Averaged per Week	Maximum Consecutive Hours per Shift	Minimum Hours of Rest Between Shifts
Europe[a]	European Comission and collective agreements	48	13	11
France[b]	Government	52.5	10	—
United Kingdom[c]	Ministerial agreements	56-64	14-24	8-12
Denmark[d]	Legislation and collective agreements	37	13-16	11
New Zealand[e]	Collective agreements	72 (not averaged)	16	8
Australia[f]	Collective agreements	Unrestricted	NA	NA
Canada[g]	Provincial collective agreements	Unrestricted nationally	Varies by province	Varies by province
Manitoba	—	89	24 + 2	NA
British Columbia	—	NA	24	8
Ontario	—	60 (ICU, ED)	24 + 1 (Anes, OB/GYN, ICU, CCU)	NA

NOTE: Anes = anesthesiology; CCU = cardiac care unit; ED = emergency department; ICU = intensive care unit; NA = Not applicable; OB/GYN = obstetrics-gynecology.
[a]European Trade Union Confederation, 2006.
[b]*Code du Travail*, 2006; Woodrow et al., 2006.
[c]NHS Employers, 2008b.
[d]Danish Medical Association, 2008; Ministry of Science Technology and Innovation, 2008.
[e]NZRDA, 2007.
[f]Australian Medical Association, 2005.
[g]PAIRO-CAHO, 2005; PAR-BC, 2008; PARIM, 2008.

The committee performed a qualitative analysis of the evidence found for the countries investigated in this report that related to residents' duty hours and effects on resident training, education outcomes, and patient outcomes.[1] The committee found limited documentation of the impact of reduced resident duty hours on patient outcomes, but a substantial literature on resource management and medical training outcomes linked to duty hour reductions. For example, despite differences in the duty hour regulations of each country, a common implementation problem that all have faced is labor supply shortages. Providing residents with sufficient exposure to learning opportunities within the guidelines has also been difficult, but some countries are working on ways to address the issues. Discussion of these challenges follows the country descriptions of current regulations and rates of compliance.

Patient Safety Events in the United States and Abroad

Regarding the international evidence on patient safety, there have been a few large epidemiologic studies describing adverse events (AEs) and preventable adverse events (PAEs) in several countries. Each study has sought to characterize the nature and causes of these events, although none measured the contribution of resident fatigue or duration of work hours. Table C-2 represents the results of major studies conducted in six different countries.

Among these studies, only three countries identified admissions from teaching hospitals in their sample and stratified their data by hospital type. AE rates of teaching hospitals compared to non-teaching hospitals were available for Canada, Australia, and the United States (those of the United States are only for the states of Utah and Colorado), and only the Canadian study provided PAE rates as well (Baker et al., 2004; Thomas et al., 2000b). The AE rates of major teaching hospitals in Australia did not differ from those in non-teaching or private hospitals (10.8 percent vs. 10.7 percent) (Thomas et al., 2000b). However, the difference in AE rates of major teaching hospitals in the United States compared to non-teaching hospitals was greater (4.0 percent vs. 2.5 percent) (Thomas et al., 2000b), as it was for Canadian teaching hospitals (10.3 percent vs. 5.2-6.0 percent) (Baker et al., 2004). Baker and colleagues were the only authors that commented on the difference in rates between hospital types, highlighting the following explanations: (1) differences in acuity of patient populations

[1]IOM staff searched Medline and Embase databases using a combination of the following terms: junior doctors, doctors in training, residency, resident work, workload, patient outcomes, adverse medical events, mortality, morbidity, medical education, medical training, Europe, United Kingdom, France, Denmark, Germany, Australia, New Zealand, and Canada. Websites of national medical and resident organizations were also searched.

TABLE C-2 International Comparison of Adverse Events (AEs) and Preventable Adverse Events (PAEs)

Country (year data collected)	Number of Hospitals	Hospital Medical Records	AE Definition	AE Rate from Medical Records (%)	Portion of AE Rate That Was Preventable (%)
United Kingdom[a] (1999-2000)	2	1,014	Unintended injury caused by medical management rather than by disease process	10.8	47.0
Denmark[b] (1998)	17	1,097	Unintended injury caused by medical management that resulted in disability, death, or prolonged hospital stay	9.0	40.4
New Zealand[c] (1998)	13	6,579	Unintended injury or complication that resulted in disability, death, or prolonged hospital stay and was caused by healthcare management rather than by the underlying disease	12.9	37.1
Australia[d,g] (1992)	28	14,179	Same as New Zealand Adverse Events Study	10.6	NA
Canada[e] (2000)	20	3,745	Same as New Zealand Adverse Events Study	7.5	36.9
United States					
New York[f] (1984)	51	30,195	Unintended injury caused by medical management that resulted in disability	3.7	27.6
Utah and Colorado[d,g] (1992)	28	14,700	Injury caused by medical management that resulted in prolonged hospital stay or disability at discharge	3.2	NA

[a]Vincent et al., 2001.
[b]Schiøler et al., 2001.
[c]Davis et al., 2003.
[d]Thomas et al., 2000b.
[e]Baker et al., 2004.
[f]Brennan et al., 1991.
[g]The data in these two rows come from the Thomas et al. (2000b) study referenced above, which attempted to harmonize the variance in study methods between two earlier and separate 1992 studies: one on AEs in Australia (Wilson et al., 1995) and the other on AEs in Utah and Colorado in the United States (Thomas et al., 2000a). The AE and PAE rates reported in those earlier studies are AE 16.6%, PAE 51% (Wilson et al., 1995); AE 2.9% (Utah, Colorado), PAE 32.6% (Utah), 27.4% (Colorado) (Thomas et al., 2000a).

between settings are difficult to capture with precise accuracy, (2) teaching hospitals may receive patients at different points in their care that place them at higher risk for AEs than at other hospitals, and (3) care delivered by multiple health professionals in teaching hospitals may increase risk of AEs due to miscommunication or lack of coordination of care (Baker et al., 2004). These same conclusions could be drawn to explain the differences in AE rates among U.S. hospital types, although insufficient data exist to do so. Insufficient data also prevented the committee from being able to conclude whether a correlation exists between duty hours and AEs.

The rates of AEs and PAEs found in each of the jurisdictions studied represent a distillation of local practices, documentation, and culture. They also demonstrate methodological differences between studies, temporal changes in care, and the changing nature of defining AEs. In trying to determine a relationship between duty hours and AE rates, inspection of Tables C-1 and C-2 reveals that patients in countries with lower duty hours were not necessarily at less risk of AEs or PAEs than patients in countries with longer duty hours. The AE rates for the United States, represented by Utah and Colorado, are lower than those of all the other countries by more than 50 percent. More recent data on amenable mortality rates for international populations under the age of 75 (Nolte and Mckee, 2008) also show no apparent association between preventable events and duty hours across countries. Although these studies do not establish whether a correlation exists between AEs and duty hours, the results appear inconsistent with general expectations that reduced duty hours improve patient safety. Furthermore, they underscore the variability of health systems and patient safety across countries and the complexity of achieving improved patient outcomes.

CURRENT RESIDENT DUTY HOUR REGULATIONS

Europe

The main objective of the European Working Time Directive (EWTD), issued by the European Council, "is to promote health and safety at work, given the clear evidence that people who work long hours run higher risks of illness and accidents" (European Trade Union Confederation, 2006). The EWTD was first established in 1993 to place limits on all workers' hours throughout Europe. That directive included physicians but excluded "doctors in training." In 2000, a new directive passed to include the "junior doctor" constituency after European resident groups lobbied for the change (Woodrow et al., 2006). The amended article established that medical residents are subject to all laws of the EWTD, accompanied by a requirement that by 2009 all health systems in the European Union limit resident work to a maximum of 48 hours averaged per week (European Trade Union Confederation, 2006). Regulated transition periods allow countries to reduce duty hour limits gradually until they reach the 2009 goal (NHS, 2007). For example, the United Kingdom is currently in a transitional phase consisting of 56-hour workweek schedules (NHS, 2007), while Denmark already meets the 48-hour workweek goal (Ministry of Science Technology and Innovation, 2008). Some countries, whose limits were much higher in the past and are working toward achieving the directive targets for 2009, have found it more difficult to restructure resident schedules to fit the new requirements while maintaining service to patients and educational opportunities for residents. Some reasons for disparities in transitions include workforce supply issues (e.g., physician-to-patient ratios and resident-to-physician ratios of each country), the general health status of a country's population and subsequent effects on physician workload, and the different organizational structures of entire health systems.

The EWTD that will apply to European residents by 2009 also includes the following (Council Directive No. 93/104/EC, 1993):

- A minimum daily rest period of 11 consecutive hours,
- A minimum rest period of 1 day (24 hours) per week,
- A maximum of 8 hours of night work on average per 24 hours, and
- A right to 4 weeks of paid annual leave.

Prior to the establishment of the EWTD, each European country had different duty hour regulations for its medical residents, ranging from a 65-hour-per-week maximum in Ireland, to 56 hours averaged over 24 weeks in Germany, to no hour restrictions at all in Denmark and France (Australian Medical Association, 1998).

New Zealand

New Zealand has been enforcing duty hour regulations for many years in attempting to address the issue of overworked medical residents. Since 1985, the maximum permissible duty for New Zealand residents has been 72 hours a week, with a set limit of 16 consecutive hours a day. They are also not to work more than 12 consecutive days without a 48-hour break (NZRDA, 2007). However, the regulations allow residents to be scheduled to work seven consecutive night shifts of 10 hours or more roughly once a month, which has been identified as being counter productive to resident well-being and performance (Dula et al., 2001; Powell, 2004).

Australia

Australia reacted very differently to the issue of duty hour restrictions for its medical residents. In 1996, the Australian Medical Association adopted federal policy for safer working environments for medical residents and, in 1999, further adopted a National Code of Practice for them to follow (Scallan, 2003). Its purpose is to provide "practical guidance on how to eliminate or minimise risks arising from the hazards associated with shiftwork and extended working hours," which include guidelines on performance-based scheduling, incident reporting, and education on fatigue mitigation techniques (Australian Medical Association, 2005, p. 4). This code is not a legal regulation nor is it mandated by any organization within or outside the medical community. The code is simply a set of guidelines that hospitals and doctors are strongly encouraged to follow according to their best judgment and acknowledge the responsibilities of employers and employees under Australia's Occupational Health and Safety legislation (Australian Medical Association, 2005). The committee that composed the code based it on available empirical evidence on sleep and fatigue and their impact on work performance. This evidence showed that extended hours of work had negative effects on medical training for three main reasons: "lack of time for formal and independent study, lack of motivation due to fatigue, and work patterns that failed to provide necessary supervision and feedback for effective learning" (Scallan, 2003, p. 910).

Canada

In Canada, individual provinces establish duty hour regulations through collective agreements negotiated by resident associations with hospitals, resulting in variations in these regulations across the country. Currently, the only province that has an explicit hour limit is Manitoba, which enforces a maximum of 89 hours per week averaged over 4 weeks (Fok et al., 2007; PARIM, 2008). In the provinces of Alberta, Quebec, and the Maritimes, collective agreements limit resident duty to 12 hours per routine day (not averaged over a week). Ontario and Saskatchewan have no overall hour limit, but the Professional Association of Internes of Ontario does have limitations of 60 hours per week for residents' performing in-hospital shift work, such as in intensive care units (ICUs) and emergency departments (EDs) (Fok et al., 2007). Shifts in EDs cannot exceed 12 hours, and in ICUs they cannot exceed 24 hours (plus handover); in other hospital areas, after 24 hours on call residents cannot admit new patients and have to be at home by noon (PAIRO-CAHO, 2005).

ENFORCEMENT OF DUTY HOUR REGULATIONS

To ensure compliance with the regulations outlined in the previous section, each country has developed its own method for enforcing them; none seem to constitute an ACGME equivalent.

Europe

In the United Kingdom, National Health Service (NHS) Employers enforce compliance with hour regulations. They are under contractual obligation to monitor compliance by collecting and reporting duty hour patterns of their residents on a semiannual basis, which the NHS Executive and the British Medical Association review jointly. Recent compliance reports show that most health authorities perform at near-perfect compliance (NHS Employers, 2008a). However, a 2004 independent study on resident compliance in a large urban U.K. hospital showed residents working beyond the duty hour regulations (whose weekly maximum was 56 hours at the time): the average number of hours worked by physicians was 63.6 hours per week and by surgeons 70.2 hours per week (Jagsi and Surender, 2004). Such mixed results imply that monitoring efforts may not be completely rigorous or accurate, perhaps a common problem shared with the United States.

Data from a 2002 survey gathered by the Royal College of Physicians indicated that residents and physicians in most European countries worked total hours or continuous duty periods in excess of those established in the EWTD (many worked 30 or more continuous hours) despite claims that these countries were compliant with the directive (Royal College of Physicians, 2002). Such excesses are consistent with the previously stated findings in the United Kingdom. The committee did not find more current compliance rates for other European countries.

The committee also had difficulty uncovering a clear picture of enforcing bodies for EWTD regulations in the remaining European countries (Mayor et al., 2004). However, an article reporting, in part, on France, whose national government is the regulatory force for duty hour limits, mentioned that it has not assigned an organization the responsibility for enforcing the established limits, nor does it currently monitor compliance rates itself (Woodrow et al., 2006). In Germany, since residents are fully licensed physicians and regular employees of the hospitals, their duty hours are enforced by the same state and local institutions that are responsible for enforcing the duty hours of all employees (trade supervisory board). These few examples indicate that monitoring and enforcement methods likely vary significantly across European countries.

New Zealand

The New Zealand Resident Doctors' Association (NZRDA) negotiates resident hours for its members through collective employer agreements and regulates the provisions therein (NZRDA, 2008b). In 2006, the average duty hours of New Zealand doctors was 45.8, and the average worked by doctors aged 24 years or younger was 57.7 hours per week (Medical

Council of New Zealand, 2006). Both numbers fall well below the national maximum of 72 hours. A nationwide survey of residents (63 percent response rate: 1,366 responses) conducted by Gander and colleagues in 2003 showed that 57 percent of residents worked between 50 and 70 hours a week and that 13 percent worked more than 70 hours, substantially more than the average hours reported in 2006 to the Medical Council (Gander et al., 2007).

Australia

Australia's approach to resident work scheduling is unique and may reflect a culture with expectations that resident hours should be similar to those of other workers in the population. Resident duty hour guidelines are advisory, as opposed to being binding rules, and there is no designated enforcement body. The guidelines of the code imply that working more than 50 hours per week puts a resident at "significant risk" of fatigue and associated negative consequences, while working 70 hours or more is considered to put residents at "higher risk" (Australian Medical Association, 2005; Scallan, 2003). The Australian Medical Association takes it upon itself to conduct national surveys of physician duty hours as a way to gauge current practice. Results of the 2001 national survey showed that 70 percent of Australian medical residents worked an average of more than 50 hours a week (Scallan, 2003). The independent study by Gander et al. (2007) had similar results. In May 2006, the Australian Medical Association conducted another national survey representing all doctors (not just junior doctors), and results indicated that 62 percent of hospital doctors are in the "significant" and "higher risk" categories for doctor safety based on hours of work, with 85 percent of surgical doctors falling in those categories. Although the later survey noted some relative improvements (e.g., the longest continuous period of work was 39 hours, down from 63 hours in 2001), it is inaccurate to compare the results of the two surveys since the earlier one surveyed only residents, whereas the latter surveyed all physicians. Also, there was no mention in either survey of any fatigue-mitigating methods used by doctors during extended shifts on duty, such as periods for sleep. The authors of the later report concluded that 39 hours is still too many to work consecutively and that hospitals need to continue taking steps to mitigate any safety risks to residents and patients (Australian Medical Association, 2006).

Canada

The Royal College of Physicians and Surgeons of Canada (RCPSC) and the College of Family Physicians of Canada (CFPC) are the regulatory bodies for duty hours in Canada. "However, the RCPSC has neither developed nor endorsed specific policies regarding duty hours. As a result, there has been no national drive to implement duty hour restrictions, but most provinces have nonetheless adopted them independently of each other within a relatively short time frame" (Woodrow et al., 2006, p. 1047). Just as duty hour regulations vary according to province, so does enforcement of those regulations.

Data on actual hours worked in Canada are very limited, but a 2007 survey of British Columbia first-year residents is helpful in describing some general patterns that are occurring at least in that part of Canada. The survey reported that these residents work an average of 65.4 hours a week (including on-call work) and sleep an average of 41.9 hours a week (Fok et al., 2007). However, when looked at separately, surgical residents reported working many more hours than their non-surgical colleagues. On average, surgical residents worked 80.4 hours per week compared to 57 hours per week for non-surgical residents. Of the surgical residents, orthopedic surgeons and general surgeons work the most hours, 102 hours and 88.9 hours a week, respectively (Fok et al., 2007).

Overall

Although governing bodies try to enforce compliance with duty hour limits, there is some indication that the number of hours reported officially by residents through their institutions are not necessarily accurate. Generally, the residents in separate surveys seem to report working more than the guidelines suggest. Therefore, the committee cannot conclude that these countries uniformly enforce their stated hour limits or that the reported compliance rates are accurate. The following section discusses some reasons why countries may encounter difficulties achieving compliance with their own regulations along with other challenges that medical training has faced because of reducing duty hours.

CONSEQUENCES AND BARRIERS

Reducing resident duty hours abroad has led to several changes in health systems that have pushed countries to create new ways to provide continuous services to patients and maintain educational opportunities for residents. The most significant organizational change among hospitals to achieve this has been the trend to replace traditional 24-hour (or longer) shifts with shorter shifts of 12 or 16 hours in accordance with local requirements or preferences (BBC, 2004). Concerns about shift-based schedules regarding continuity of care and resident training have been articulated (Carr, 2003; Jagsi and Surender, 2004; Powell, 2004; Royal College of Surgeons of England, 2008), as in the United States, and countries are taking steps to address related changes (e.g., improving handover protocol). However, another persistent problem caused by this change in work system is that of labor supply shortage.

Labor Supply Shortages

Employing shift systems, with explicit or implicit reductions in total duty hours, requires more doctors to provide 24-hour coverage. The implementation of more shift-based schedules to comply with the duty hour regulations and recommendations in Europe, New Zealand, and Australia has thus created or exacerbated workforce shortages and strained financial resources. This has been identified as one of the most difficult challenges that countries face as a result of restricting work hours (Child and Old, 2004; Powell, 2004; Sheldon, 2004).

New Zealand hospitals faced a sudden shortage of residents after the new regulations passed in 1985; they were unable to ensure resident staffing to care for patients at all times. Hospitals decided to employ more residents because increasing resident staffing was the most immediate solution available to provide sufficient coverage of care while being asked to comply quickly with reduced hour restrictions (Child and Old, 2004). However, as the demand for more residents began to outstrip the supply of medical graduates, it became necessary to employ more senior doctors and nurses to cover gaps in working hours (Powell, 2004). Residents in New Zealand are currently in short supply, not only because of increased demand from hour restrictions but also because of increased service demands on the healthcare system by a rapidly growing elderly population (a trend that countries worldwide are facing).

In Europe, recruiting more residents to fill the needed positions requires additional funding to support their salaries, but hiring higher-level health professionals (with accompanying higher salaries) to fill the same gaps significantly increases the costs associated with supplying enough workers. In an attempt to comply with the mandated changes of the EWTD, "many countries maintain that a significant increase in physicians and residents will be needed to adequately cover service requirements. Figures ranging from 6,000 to 27,000 have been cited at a cost of up to EUR 1.75 billion" (Sheldon, 2004; Woodrow et al., 2006, p. 1048). The United Kingdom alone estimated the need for an additional 15,000 doctors to staff the NHS in order to meet the requirements of the Working Time Directive (Child and Old, 2004).

Australia's health workforce is also under pressure as doctors work fewer hours and show preferences for work-life balance, such as having more time to spend with family (Productivity Commission, 2005). "The medical workforce shortage [in Australia] is both absolute and relative: we do not train enough doctors, and once trained, doctors are not working in areas of greatest need. Chronic vacancies exist across the health care system" (Skinner, 2006, p. 35). In February 2004, the state of New South Wales reported more than 900 vacancies at the resident and registrar levels (Greater Metropolitan Clinical Taskforce Metropolitan Hospitals Locum Issues Group, 2005). Exacerbating the problem is the decision of many junior doctors to opt out of the public hospital workforce to work locum positions instead (part-time fill-in work), which pay nearly three times more than full-time salaried public positions. This, in turn, significantly increases hospital budgets and reduces the number of doctors training in needed specialties (Chisholm, 2008) since locum certification does not require specialty training and many are general practitioner positions. However, a recent recruitment assessment in Australia indicates that limiting shifts to 12 hours is a positive and effective solution to recruitment and retention challenges among critical care nurses that may hold promising solutions for residents as well (Dwyer et al., 2007).

Impact on Training and Education

Limiting duty time has the potential to reduce the educational opportunities residents have (Carr, 2003). There is speculation from various U.K. sources that "shift working has had a detrimental effect on learning for doctors in training" and that new rotations "have led to an erosion of the team system and a loss of mentorship in training" (Carr, 2003, p. 623). Residents also perceive adverse effects on continuity of care and on resident-patient relationships and resident-supervisor relationships (Ardagh, 2003; Jagsi and Surender, 2004). Supervision seems to be equally important for resident learning abroad as it is here (Scallan, 2003), and its importance receives attention in Chapter 4 of this report.

Surgeons particularly express significant concern about the potential effects of hour limits on resident performance because of reduced operative experience (Royal College of Surgeons of England, 2007b, 2008; Thorne et al., 2006) or reduced follow-through with postoperative patient care (Mestres et al., 2006). A large British survey in 2002 showed that a third of orthopedic surgical residents were not taught in the operating theater or in clinic (British Orthopaedic Association, 2002), and a senior resident in the British training system commented that "many senior house officers arrive at posts halfway through their rotations without any real competence in operative skills as basic as suturing and tying knots" (Chikwe et al., 2004, pp. 418-419). Another large survey of cardiothoracic residents throughout the United Kingdom reported a unanimously perceived negative impact on their training since EWTD implementation (West et al., 2007). Less hands-on training and an increase in shift-based duty has led to reports of residents' dissatisfaction with their degree of learning or the teaching style provided (Rawnsley et al., 2004; West et al., 2007). The committee listened to testimony by Dr. Bernard Ribeiro, president of the Royal College of Surgeons of England, that reiterated many of these same sentiments and presented additional data on reduction in the number of operations performed across surgical specialties since the 2004 EWTD, showing some residents performing as much as 25 percent fewer procedures (Royal College of Surgeons of England, 2008). These findings imply that decreasing residents' time to perform surgical procedures may decrease their general level of competence in these tasks due to the demands and intricacies of the work (Chikwe et al., 2004). The committee found no other objective measures of such outcomes, and as noted below, evaluation of these impacts on training has yet to be published.

To compensate for reductions in training time for U.K. surgical residents, redesign efforts have focused on providing more time concentrated on procedural activities than other activities. For example, the study by Lim and colleagues (2006) suggests that reorganizing institutional structure can maintain the competency levels of cardiac surgical trainees in these activities despite the reduction in work hours. Their institution adopted a team-based model to achieve the desired results, ensuring adequate time with trainers to maximize learning (Lim and Tsui, 2006). However, others in the field have noted the difficulty that less robust programs may have in implementing such changes and the likelihood that many programs would not be able to do so, given their lack of resources (Mestres et al., 2006; West, 2007). In response to these issues, the Royal College of Surgeons has developed its own set of rotation guidelines in compliance with the EWTD for surgeons in training to follow, hoping to maximize patient safety and resident surgeon learning (Royal College of Surgeons of England, 2007a). This issue has yet to be resolved in the United Kingdom, and no surgeons have yet emerged from the shortened training system that would allow the NHS to evaluate its overall impact and costs.

Observations of resident training in Australia suggest that much resident time is spent on administrative tasks and providing service demands for the health system (Gleason et al., 2007). A consequence of this has been variable quality in resident education because of insufficient time to attend some of their didactic sessions (Gleason et al., 2007). Australian medical education leaders have observed that "the existing systems for delivery of education and training are inefficient, under-resourced and under pressure, and they will not be sustainable into the future" (McGrath et al., 2006, p. 348). This may not necessarily be a result of reduced hours, but reducing hours could further affect such limitations.

The committee concludes that reports from other countries indicate that reducing total duty hours places strain on the healthcare workforce and creates challenges to maintaining the quality of resident learning environments. Several countries have experienced financial and manpower strains trying to staff hospitals sufficiently to provide adequate 24-hour care, and many have reported insufficient time to teach residents or to have them practice procedures. Noneducational resident workload has also been found to be an impediment to the resident learning experience.

APPLICABLE LESSONS FOR RESIDENCY PROGRAMS IN THE UNITED STATES

Retaining Flexibility in Work Schedules

A key lesson from these countries is the need for flexibility in how programs arrange schedules within the confines of established regulations. In the United Kingdom, since the introduction of the Working Time Regulations, the NHS has moved toward achieving the goal of safer work conditions through reduced hours for medical residents (NHS, 2007). While considering how best to implement the newest directive with further reduced hours, the NHS has taken into account an array of rotation schedules with various combinations of night shifts, weekend shifts, day shifts, and rest hours that its hospitals could adopt (Royal College of Physicians, 2006). In each instance, individual hospitals need to determine the most effective combination of workforce (e.g., numbers of residents and other staff on duty in a given shift) and required hours to achieve safe patient care and quality resident education. Similar to the medical education system in the United Kingdom, different types of residency programs in the United States have national requirements to fulfill regarding educational content within the duty hour restrictions. Therefore, the different organizational methods across programs require flexibility in the way work schedules are arranged.

The committee concludes that, given the differences among healthcare institutions, different specialties, and the needs they must satisfy, maintaining flexibility in the way work hours are scheduled is necessary. This echoes the findings in Chapter 3 of the experiences of programs in the United States as they adjusted to the 2003 duty hour limits. It appears that other countries have experienced problems with adherence to established duty hour limits, as have programs in the United States. (See Chapter 2 on adherence to duty hours and its enforcement in the United States.)

Time for Phase-in

The international experience indicates that a well-planned period of transition would be necessary as part of any major changes in total work hour limits. In Europe, 4 years after residents became subject to the EWTD laws, as programs were still transitioning to current requirements, multiple countries in the European Union (EU) were protesting the mandated schedule claiming that they lacked the resources, in either workforce, finances, or both, to successfully comply with the regulations by the given deadline (Sheldon, 2004).

Based on the international experience, a phase-in of committee-recommended changes would be beneficial. Changes specifically related to resident duty hours and schedules might require some time for planning to accommodate constraints in workforce and other resources. Chapter 7 contains the committee's recommendations for adjusting duty hours, and Chapter 9 includes macro-level estimates of the potential costs, as well as the type and number of clinical personnel that would be needed to replace existing residents under further adjusted duty hours.

Redesign of Resident Education and Training Systems

The introduction of duty hour restrictions for residents in other countries has created an impetus for changing the fundamentals of medical education and training programs. Complying with new regulations caused hospital providers to alter their work practices, staffing, and delivery methods in order to ensure a degree of quality care in light of these significant changes.

For example, in Australia the National Code clearly states that it was created as "one part of a broader education and awareness program to change the current individual and organisational beliefs and culture that support working hours and patterns that would be considered unacceptable in most other industry sectors" (Australian Medical Association, 2005, p. 4). Culture change is at its core. To achieve this goal, the Australian Medical Association developed a risk assessment strategy to evaluate the extent of hazard caused by residents' working extended hours (working more than 50 hours per week is considered "significant risk," working 70 hours or more is "higher risk") and recommends principles that should be at the foundation of work schedule design to minimize risk to patient and resident safety. Aside from these assessments, high service demands and inadequate funding of education and skills training have also been identified as risks to resident training (Gleason et al., 2007). As institutions try to increase the number of residents to satisfy service demands, the authors of the study note that merely increasing the numbers of residents will not alleviate their work intensity, nor will it solve training deficiencies unless adequate resources are provided. They conclude that a strong emphasis on training, supported by sufficient resources, is necessary to deal with these changes and reference the U.K. efforts to redesign its educational program (examined below). Similarly, although restrictions on hours vary by province in Canada, many medical training programs throughout the country have been focused on improving patient outcomes through improved resident training strategies for several years. Strategies have consisted of increasing the educational value and success of residency programs rather than using them to produce residents as a cheaper form of healthcare workforce (Landau, 2007).

Competency-Based Training

Perhaps the foremost issue posed by reduced duty hours for residents is how to ensure competence during and at the end of training. To address this issue and create long-term improvements, changes need to occur at the educational level. This has much to do with the fact that the duty hour restrictions provide less time to train residents. As a result, both the United Kingdom and New Zealand have been significantly reshaping their residency programs toward competency-based or performance-based training, which determines a trainee's level of expertise by their ability to demonstrate specified required skills at a given point in training, as opposed to a more time-based educational model, which focuses on completing physician training in a certain number of years. Their programs now center on providing an experience with more educational value for residents and have been fundamentally restructured to ensure the desired results (NHS, 2008b; Workforce Taskforce, 2007).

One such example from the United Kingdom is the new Foundation Programme, which constitutes the first 2 years of residency training upon graduating medical school. Intended to "bridge the gap between undergraduate school and specialist/medical training" (Foundation Programme, 2007, p. 5), the Foundation Programme focuses on patient safety and improving the quality of care by having residents demonstrate competence in communication and consultation skills, patient safety, and teamwork, in addition to more typical clinical skills (Major shake-up in medical training, 2005). This training replaces the previous introductory 2-year program and is a departure from that program, which focused nearly exclusively on demonstrating competence in clinical skills (Foundation Programme, 2007). No assessments have been made to date as to whether the quality of resident work has been affected by the new program, although one study has reported concerns on behalf of both the trainees and the educators about the generic nature of some of the skills trainees were expected to acquire

(O'Brien et al., 2006). Still in its nascent years, the Foundation Programme has opportunities to modify its curricula.

Monitoring Quality Measures and New Models of Care

The NHS has also developed several pilot programs that not only alter resident rotations and schedules, but also incorporate new quality improvement practices such as regularly assessing patient and staff satisfaction and reporting patient safety measures, as well as measuring the compliance of residents with the time limits. In some instances the pilot programs show that they achieved reductions in patient wait times for treatment or surgery, reductions in the time between prescribing and administering certain medications, shorter lengths of stay in hospital wards, and improved discharging processes, even with reduced resident hours (NHS, 2004).

One pilot created the Hospital at Night program, a new model of care that reduces the risk of adverse events by improving the quality of care taking place during nighttime hours in hospitals, while also aiming to enhance resident learning during daytime hours (Hospital at Night and The NHS Information Centre, 2008; Institute for Healthcare Improvement, 2007; NHS, 2008c). With the completion of more pilots, the NHS anticipates providing additional solutions to the EWTD that are transferable across residency programs and improve resident learning and the quality of patient care across the entire health system.

Based on the above review of educational changes, it is evident that some countries have begun to redesign their educational system, focusing changes in both curriculum and competency-based aspects, and that these changes have been in response to decreased duty hours and some negative impacts on resident training.

CONCLUSION

When considering the experiences of the countries examined in this chapter with duty hour regulations, it is important to remember that most of these countries have lower duty hour limits than the United States (e.g., Europe, 48; New Zealand, 72). Even Australia, where no national maximum limit exists, tries to have residents work no more than 70 hours per week. With these lower limits, key stakeholders in those countries report considerable challenges as they adjust to the reduced availability of residents to provide 24-hour care. The concerns about workforce shortages and financial constraints because of these changes are particularly noteworthy, as are the frequently voiced concerns that the educational experience of residency is deteriorating. However, some countries have produced innovative designs and promoted culture change.

Yet based on the evidence examined from other countries, it is not possible to determine an ideal number of hours that residents should be required to work. There are limited or no data to show that decreasing duty hours improves patient safety, and many redesign efforts have yet to be fully developed or evaluated for effect on resident learning or patient safety. Furthermore, the international data on the quality of patient safety indicate that duty hour regulations may not be the greatest factor affecting quality of care and that other facets of delivery systems or educational programs likely deserve more attention as areas for improvement.

Despite the limited evidence from other countries, the committee was able to draw lessons on retaining scheduling flexibility, allowing phase-in time, and redesigning educational programs for residency training. However, the committee concluded that no single system is directly applicable to that of the United States given the different construct and culture of our healthcare system. Although the committee lacks systematic multinational evidence on resident education and patient outcomes as a result of implementing duty hour regulations, from the evidence gathered it appears that any changes in duty hour limits would necessitate modifications to redesign and enhance the medical training system, to have an available workforce to substitute for hours that residents are not available to staff, and to create substantial financial resources. A similar call to redesign medical education and training programs exists in the United States (Meyers et al., 2007). It may be important to maintain some degree of flexibility as residency training programs determine which innovative educational approaches and new scheduling designs to adopt for their healthcare delivery and resident learning needs.

REFERENCES

Ardagh, M. 2003. Beyond Ashburton: Junior hospital doctor employment in New Zealand. *The New Zealand Medical Journal* 116(1168):1-2.

Australian Medical Association. 1998. *Overseas experience in regulating hours of work of doctors in training*. Kingston ACT, Australia: Australian Medical Association.

———. 2005. *National code of practice—Hours of work, shiftwork and rostering for hospital doctors*. Kingston ACT, Australia: Australian Medical Association Ltd.

———. 2006. *Safe hours = safe patients: AMA safe hours audit 2006*. Kingston, Australia: Australian Medical Association.

Baker, G. R., P. G. Norton, V. Flintoft, R. Blais, A. Brown, J. Cox, E. Etchells, W. A. Ghali, P. Hebert, S. R. Majumdar, M. O'Beirne, L. Palacios-Derflingher, R. J. Reid, S. Sheps, and R. Tamblyn. 2004. The Canadian Adverse Events Study: The incidence of adverse events among hospital patients in Canada. *CMAJ: Canadian Medical Association Journal* 170(11):1678-1686.

BBC (British Broadcasting Corporation). 2004. Q&A: Junior doctors' hours. *BBC News*. http://news.bbc.co.uk/1/hi/health/3941073.stm (accessed January 23, 2008).

Brennan, T. A., L. L. Leape, N. M. Laird, L. Hebert, A. R. Localio, A. G. Lawthers, J. P. Newhouse, P. C. Weiler, and H. H. Hiatt. 1991. Incidence of adverse events and negligence in hospitalized patients. Results of the Harvard Medical Practice Study I. *New England Journal of Medicine* 324(6):370-376.

British Orthopaedic Association. 2002. *Education and training for SHOs: A snap-shot of the moment and recommendations for the future.* London: British Orthopaedic Association.

Carr, S. 2003. Education of senior house officers: Current challenges. *Postgraduate Medical Journal* 79(937):622-626.

Chikwe, J., A. C. de Souza, and J. R. Pepper. 2004. No time to train the surgeons. *BMJ* 328(7437):418-419.

Child, S., and A. Old. 2004. Resident medical officer working conditions in New Zealand: Results of a recent survey. *New Zealand Medical Journal* 117(1204).

Chisholm, D. 2008. *Junior doctors opting out.* http://www.stuff.co.nz/sundaystartimes/4464669a20455.html (accessed May 27, 2008).

Code du Travail (Labor law in France). 2006. Paris, France: Ministère du travail, des relations sociale, et de la famille et de la solidarité.

Council Directive No. 93/104/EC. 1993. *European working time directive.* Council of the European Union, Brussels.

Danish Medical Association. 2008. *Duties—in brief.* http://www.laeger.dk/portal/page/portal/LAEGERDK/LAEGER_DK/ENGLISH/DUTIES_IN_BRIEF (accessed June 11, 2008).

Davis, P., R. Lay-Yee, R. Briant, W. Ali, A. Scott, and S. Schug. 2003. Adverse events in New Zealand public hospitals II: Occurrence and impact. *New Zealand Medical Journal* 116(1183).

Dula, D. J., N. L. Dula, C. Hamrick, and G. C. Wood. 2001. The effect of working serial night shifts on the cognitive functioning of emergency physicians. *Annals of Emergency Medicine* 38(2):152-155.

Dwyer, T., L. Jamieson, L. Moxham, D. Austen, and K. Smith. 2007. Evaluation of the 12-hour shift trial in a regional intensive care unit. *Journal of Nursing Management* 15(7):711-720.

European Trade Union Confederation. 2006. *Working time directive.* http://www.etuc.org/a/504 (accessed November 2, 2007).

Fok, M. C., A. Townson, B. Hughes, and W. C. Miller. 2007. Work hours, sleep deprivation, and fatigue: A British Columbia snapshot. *British Columbia Medical Journal* 49(7):387-392.

Foundation Programme. 2007. Curriculum, Academy of Medical Royal Colleges, London.

Gander, P., H. Purnell, A. Garden, and A. Woodward. 2007. Work patterns and fatigue-related risk among junior doctors. *Occupational and Environmental Medicine* 64(11):733-738.

Gleason, A. J., J. O. Daly, and R. E. Blackham. 2007. Prevocational medical training and the Australian curriculum framework for junior doctors: A junior doctor perspective. *Medical Journal of Australia* 186(3):114-116.

Greater Metropolitan Clinical Taskforce (GMCT) Metropolitan Hospitals Locum Issues Group. 2005. *Meeting the medical locum challenge—Options for action within the NSW public hospital system.* Sydney, Australia: GMCT.

Hospital at Night and the NHS Information Centre. 2008. *The case for hospital at night—The search for evidence.* Manchester, England: NHS.

Institute for Healthcare Improvement. 2007. *The hospital at night project: Reducing risks at our most vulnerable time of day.* Boston, MA: Institute for Healthcare Improvement.

Jagsi, R., and R. Surender. 2004. Regulation of junior doctors' work hours: An analysis of British and American doctors' experiences and attitudes. *Social Science & Medicine* 58(11):2181-2191.

Jarvis, J. 2002. Response from an RMO. *New Zealand Medical Journal* 155(1166).

Landau, L. I. 2007. Quality junior doctor training, improved workforce outcomes and patient safety. *Australian Health Review* 31(Suppl 1):S106-S108.

Lim, E., and S. Tsui, on behalf of the Registrars and Consultant Cardiac Surgeons of Papworth Hospital. 2006. Impact of the European Working Time Directive on exposure to operative cardiac surgical training. *European Journal of Cardiothoracic Surgery* 30(4):574-577.

Major shake-up in medical training, UK. 2005 *Medical News Today*. http://www.medicalnewstoday.com/articles/29074.php (accessed November 12, 2007).

Mayor, S., J. Burgermeister, K. Kosner, T. Villanueva, A. Tuffs, B. Spurgeon, F. Turone, M. Houston, and T. Sheldon. 2004. Over the limit? *Student BMJ* 12(September).

McGrath, B. P., I. S. Graham, B. J. Crotty, and B. C. Jolly. 2006. Lack of integration of medical education in Australia: The need for change. *Medical Journal of Australia* 184(7):346-348.

Medical Council of Canada. 2008. *Medical training/licensure system in Canada.* http://www.img-canada.ca/en/licensure_overview/licensure.html#undergrad (accessed February 19, 2008).

Medical Council of New Zealand. 2006. *Annual report 2006: Medical workforce survey.* http://www.mcnz.org.nz/portals/0/publications/2006.pdf (accessed May 29, 2008).

Mestres, C.-A., J.-M. Revuelta, and A. C. Yankah. 2006. The European working time directive: Quo vadis? A well-planned and organized assassination of surgery. *European Journal of Cardiothoracic Surgery* 30(4):571-573.

Meyers, F. J., S. E. Weinberger, J. P. Fitzgibbons, J. Glassroth, F. D. Duffy, and C. P. Clayton. 2007. Redesigning residency training in internal medicine: The consensus report of the Alliance for Academic Internal Medicine Education Redesign Task Force. *Academic Medicine* 82(12):1211-1219.

Ministry of Science Technology and Innovation. 2008. *Working hours.* http://www.workindenmark.dk/Working_hours (accessed June 10, 2008).

National Recruitment Office for General Practice Training. 2008. *GP careers: The training program.* http://www.gprecruitment.org.uk/gpcareers/programme.htm (accessed September 18, 2008).

NHS (U.K. National Health Service). 2004. *Working time directive: Pilots programme report.* http://www.healthcareworkforce.nhs.uk/pilotprojects.html (accessed November 14, 2007).

———. 2007. *European working time directive (EWTD).* Coventry, UK: National Library for Health.

———. 2008a. *Modernising medical careers: Training in 2008.* http://www.mmc.nhs.uk/default.aspx?page=318 (accessed September 18, 2008).

———. 2008b. *The past—The principles of MMC.* http://www.mmc.nhs.uk/default.aspx?page=310 (accessed June 2, 2008).

NHS Employers. 2008a. *Junior doctors' terms and conditions of service and associated documents.* http://www.nhsemployers.org/pay-conditions/pay-conditions-467.cfm (accessed May 30, 2008).

———. 2008b. *National Health Service hospital medical and dental staff and doctors in public health medicine and the community health service (England and Wales): Terms and conditions of service.* NHS.

———. 2008c. *Hospital at night.* http://www.healthcareworkforce.nhs.uk/hospitalatnight.html (accessed June 2, 2008).

Nolte, E., and C. M. McKee. 2008. Measuring the health of nations: Updating an earlier analysis. *Health Affairs* 27(1):58-71.

NZRDA (New Zealand Resident Doctors' Association). 2007. *Collective agreement.* Auckland, New Zealand: New Zealand Resident Doctors' Association.

———. 2008a. Resident doctors' association fact sheet. Auckland, New Zealand: New Zealand Resident Doctors' Association.

———. 2008b. *What is NZRDA: About us.* http://www.nzrda.org.nz/Site/About_Us/What_is_NZRDA.ashx (accessed June 5, 2008).

O'Brien, M., J. Brown, I. Ryland, N. Shaw, T. Chapman, R. Gillies, and D. Graham. 2006. Exploring the views of second-year Foundation Programme doctors and their educational supervisors during a deanery-wide pilot Foundation Programme. *Postgraduate Medical Journal* 82(974):813-816.

PAIRO-CAHO. 2005. 2005-2008 PAIRO-CAHO agreement, edited by the Professional Association of Internes and Residents of Ontario (PAIRO) and the Council of Academic Hospitals of Ontario (CAHO). Ontario, Canada.

PAR-BC (Professional Association of Residents of British Columbia). 2008. *Collective agreement & constitution: Scheduling.* http://www.par-bc.org/collectiveagreement.php/20 (accessed May 28, 2008).

PARIM (Professional Association of Residents and Interns of Manitoba). 2008. *Collective agreement: Article 14—duty hours.* http://www.parim.org/50 (accessed February 22, 2008).

Powell, D. 2004. Key issues facing Resident Medical Officers (RMOs). *New Zealand Medical Journal* 117(1204).

Productivity Commission. 2005. *Australia's health workforce: Research report.* Canberra, Australia: Commonwealth of Australia.

Rawnsley, A., K. Hurst, and M. Robinson. 2004. Clinical and education implications of shift work. *Medical Teacher* 26(1):71-73.

Royal Australasian College of Physicians. 2007. *Requirements for physician training: New Zealand 2008 handbook.* Wellington, NZ: The Royal Australasian College of Physicians.

Royal College of Physicians. 2002. *European working time directive: European dimensions.* http://www.rcplondon.ac.uk/news/statements/doc_ewtd_european.asp (accessed June 10, 2008).

———. 2006. *Designing safer rotas for junior doctors in the 48-hour week.* Prepared for on behalf of a multidisciplinary working group by N. Horrocks and R. Pounder. London, England: Royal College of Physicians of London.

Royal College of Surgeons of England. 2007a. *Rota planning: Guidance from the working time directive working party.* London, England: Professional Standards and Regulation, Royal College of Surgeons of England.

———. 2007b. *Safe shift working for surgeons in training: Revised policy statement from the working time directive working party.* London, England: Professional Standards and Regulation, Royal College of Surgeons of England.

———. 2008. Lessons from surgery. Presentation by Bernard Ribeiro to Committee on Optimizing Graduate Medical Trainee (Resident) Hours and Work Schedules to Improve Patient Safety, March 4, 2008, Irvine, CA.

Scallan, S. 2003. Education and the working patterns of junior doctors in the UK: A review of the literature. *Medical Education* 37(10):907-912.

Schiøler, T., H. Lipczak, B. L. Pedersen, T. S. Mogensen, K. B. Bech, A. Stockmarr, A. R. Svenning, and A. Frølich. 2001. Incidence of adverse events in hospitalized patients. The Danish Adverse Event Study (DAES). *Ugeskrift for Laeger* 163(39):5378.

Sheldon, T. 2004. Pressure mounts over European working time directive. *BMJ (Clinical Research Ed.)* 328(7445):911.

Skinner, C. A. 2006. Re-inventing medical work and training: A view from generation x. *Medical Journal of Australia* 185(1):35-36.

Thomas, E. J., D. M. Studdert, H. R. Burstin, E. J. Orav, T. Zeena, E. J. Williams, K. M. Howard, P. C. Weiler, and T. A. Brennan. 2000a. Incidence and types of adverse events and negligent care in Utah and Colorado. *Medical Care* 38(3):261-271.

Thomas, E. J., D. M. Studdert, W. B. Runciman, R. K. Webb, E. J. Sexton, R. M. Wilson, R. W. Gibberd, B. T. Harrison, and T. A. Brennan. 2000b. A comparison of iatrogenic injury studies in Australia and the USA I: Context, methods, casemix, population, patient and hospital characteristics. *International Journal for Quality in Health Care* 12(5):371-378.

Thorne, L., S. Burn, S. Shaw, B. Arvin, and R. Bradford. 2006. Neurosurgical trainees operative experience before and after introduction of the new deal for junior doctors. *British Journal of Neurosurgery* 20(1):31-35.

Vincent, C., G. Neale, and M. Woloshynowych. 2001. Adverse events in British hospitals: Preliminary retrospective record review. *BMJ* 322(7285):517-519.

West, D. 2007. European working time directive implementation and cardiothoracic training: Larger centers may optimise training [comment]. *European Journal of Cardio-thoracic Surgery* 31(5):958.

West, D., M. Codispoti, and T. Graham. 2007. The European Working Time Directive and training in cardiothoracic surgery in the United Kingdom: A report of the specialty advisory board in cardiothoracic surgery of the Royal College of Surgeons of Edinburgh. *The Surgeon* 5(2):81-85.

Wilson, R. M., W. B. Runciman, R. W. Gibberd, B. T. Harrison, L. Newby, and J. D. Hamilton. 1995. The quality in Australian health care study. *Medical Journal of Australia* 163(9):458-471.

Woodrow, S. I., C. Segouin, J. Armbruster, S. J. Hamstra, and B. Hodges. 2006. Duty hours reforms in the United States, France, and Canada: Is it time to refocus our attention on education? *Academic Medicine* 81(12):1045-1051.

Workforce Taskforce. 2007. *Reshaping medical education and training to meet the challenges of the 21st century: A report to the Ministers of Health and for tertiary education from the Workforce Taskforce.* Wellington, New Zealand: Ministry of Health.

World Health Organization. 2007. *World directory of medical schools: Final updates through December 2007.* http://www.who.int/hrh/wdms/en/ (accessed September 19, 2008).

Appendix D

Glossary, Acronyms, and Abbreviations

GLOSSARY

Actigraphy: A relatively non-invasive method of monitoring human rest and activity cycles. A small actigraph unit (often in a form similar to a wrist-watch), is worn by someone to continually measure motor activity. The data are later transmitted to a computer where it can be analyzed.

Ad libitum: The means of performing an activity, such as sleep, at one's own discretion.

Adaptability: In teamwork, an ability to adjust strategies based on information gathered from the environment through the use of backup behavior and reallocation of intrateam resources. Altering a course of action or team repertoire in response to changing conditions (internal or external).[1]

Adverse Event: An unintended physical injury resulting from or contributed to by medical care rather than the underlying condition of the patient, that requires additional monitoring, treatment, or hospitalization or results in death. Not all adverse events are caused by errors.[2]

Attending physician: A physician who has completed his or her medical residency and is fully licensed to practice medicine independently. Attending physicians serve as supervisors to residents as they complete their training and bear responsibility for the clinical work of residents that are assigned to their team.

Backup behavior: Ability to anticipate other team members' needs through accurate knowledge of their responsibilities. This includes the ability to shift workload among members to achieve balance during high periods of workload or pressure.[1]

Burnout: A state of exhaustion or extreme fatigue accompanied by three common symptoms: (1) emotional exhaustion—depleted energy from overwhelming work demands, (2) depersonalization—personal detachment from one's job or surroundings, and (3) lack of personal accomplishment due to self-perceptions of inefficiency.

Competencies: Specific knowledge, skills, behaviors, and attitudes and the appropriate educational experiences required of residents to complete graduate medical education (GME) programs.[3]

Cross-coverage: The availability of other residents to care for admitted patients when the resident who has had primary responsibility for these patients' care is not on duty.

Didactic: A kind of systematic instruction by means of planned learning experiences, such as conferences or grand rounds.[3]

Duty hours: Applied to medical residents, this means all time spent in clinical and academic activities related to the program, that is: patient care (both inpatient and outpatient), administrative duties relative to patient care, provision for transfer of patient care, time spent in-house during call activities, and scheduled activities, such as conferences. Duty hours do not include reading and preparation time spent away from the duty site.[3]

Extended duty period: Also known as "long call," refers to the 30-hour (24 + 6) maximum continuous duty period allowed under the 2003 Accreditation Council for Graduate Medical Education (ACGME) limits.

Float (day or night): A shift of residents that are not assigned to a single service but "float" across services or teams to help with admissions and follow-up.

Handover: The transfer of physician responsibility for a patient's care along with the transfer of patient information from one healthcare provider to another. Also, commonly referred to as a "handoff," "transfer," or "sign-out."

Health care safety net: Those providers that organize and deliver a significant level of health care and other related services to uninsured, Medicaid, and other vulnerable patients.[4]

Long call: Also known as "extended duty period," refers to the 30-hour (24 + 6) maximum continuous duty period allowed under the 2003 ACGME limits.

Midlevel provider: A term used to categorize clinicians such as nurse practitioners (NPs) and physician assistants (PAs). Midlevel providers are often referred to as physician extenders.

Mutual performance monitoring: The ability to develop common understandings of the team environment and apply appropriate task strategies to accurately monitor teammate performance.[1]

On call: Duty hours beyond the normal workday when residents are required to be immediately available in the assigned institution. Also referred to as in-house call.[3]

Percutaneous injuries: Injuries that penetrate the skin (e.g., needlesticks, cuts).

Physician extender: A term used to categorize clinicians such as nurse practitioners (NPs) and physician assistants (PAs). Physician extenders are often referred to as midlevel providers.

Preventable Adverse Event: An adverse event attributable to error.[2]

Resident: Any physician enrolled in a GME program. A resident has received a medical degree and practices medicine under the supervision of fully licensed physicians, in an accredited graduate medical education hospital or clinic, as a physician in training. These trainees are often referred to by their training year, PGY-1 being a first year resident (also know as postgraduate year one, or an *intern*), PGY-2 a second year resident, and so forth through PGY-6.

Shift: A scheduled period of work, whether during the day, evening, or night.

Task-tailored substitute: The lowest-qualified-level personnel to whom non-educational patient care responsibilities can be transferred.

Team: A distinguishable set of two or more people with specific roles and boundaries interacting toward a common goal on tasks that are interdependent and are completed within a larger organizational context.[5]

Team leadership: Ability to direct and coordinate the activities of other team members; assess team performance; assign tasks; develop team knowledge, skills, and abilities; motivate team members; plan and organize; and establish a positive working atmosphere.[1]

Team orientation: Propensity to take other's behavior into account during group interaction and belief in the importance of the team's goals over individual members' goals.[1]

Teamwork: Set of interrelated behaviors, cognitions (thoughts), and attitudes (feelings) held by each team member that combine to facilitate adaptive, coordinated performance.[6]

NOTES

[1]Salas, E., D. E. Sims, and C. S. Burke. 2005. Is there a "big five" in teamwork? *Small Group Research* 36(5):555-599.

[2]IOM (Institute of Medicine). 2000. *To err is human: Building a safer health system.* Washington, DC: National Academy Press.

[3]ACGME. 2008. Glossary of terms. http://www.acgme.org/acWebsite/about/ab_ACGMEglossary.pdf (accessed November 7, 2008).

[4]IOM. 2000. *America's health care safety net: Intact but endangered.* Washington, DC: National Academy Press.

[5]Kozlowski, S. W. J., and B. S. Bell. 2003. Work groups and teams in organizations. In *Handbook of psychology: Industrial and organizational psychology.* Vol. 12, edited by W. Borman, D. Igen, and R. Klimoski. London: Wiley. Pp. 333-375.

[6]Salas, E., E. Sims, and C. Klein. 2004. Cooperation and teamwork at work. In *Encyclopedia of applied psychology.* Vol. 1, edited by C. D. Spielberger. San Diego, CA: Academic Press. Pp. 497-505.

ACRONYMS AND ABBREVIATIONS

ACRONYMS

AAMC	Association of American Medical Colleges
ABIM	American Board of Internal Medicine
ABMS	American Board of Medical Specialties
ABNS	American Board of Neurological Surgery
ABSITE	American Board of Surgery In-Training Examination
ACGME	Accreditation Council for Graduate Medical Education
ACS	American College of Surgeons
AE	adverse event
AHRQ	Agency for Healthcare Research and Quality
ALOS	average length of stay
AMA	American Medical Association
AMI	acute myocardial infarction
ANSR	Americans for Nursing Shortage Relief
APDIM	Association of Program Directors in Internal Medicine
AY	academic year
BLS	Bureau of Labor Statistics
CFPC	College of Family Physicians of Canada
CHGME	Children's Hospital Graduate Medical Education (program)
CIR	Committee of Interns and Residents
CMI	case mix index
CMS	Centers for Medicare and Medicaid Services
COGME	Council on Graduate Medical Education
COTH	Council of Teaching Hospitals and Health Systems
DGME	direct graduate medical education (payment)
DO	doctor of osteopathic medicine
DOD	U.S. Department of Defense
DSH	disproportionate share hospital (payment)
ED	emergency department
EM	emergency medicine
EOG	electro-oculography
EU	European Union
EWTD	European Working Time Directive

FAA	Federal Aviation Administration
FMCSA	Federal Motor Carrier Safety Administration
FRA	Federal Railroad Administration
FTE	full-time equivalent
GME	graduate medical education
HCUP	Healthcare Cost and Utilization Project
HCUP NIS	Healthcare Cost and Utilization Project Nationwide Inpatient Sample
HHS	U.S. Department of Health and Human Services
HIPAA	Health Insurance Portability and Accountability Act
HOS	hours of service
HRO	high-reliability organization
HRSA	Health Resources and Services Administration
ICU	intensive care unit
IHI	Institute for Healthcare Improvement
IM	internal medicine
IME	indirect medical education (payment)
IOM	Institute of Medicine
IPPS	Inpatient Prospective Payment System (Medicare)
IPRO	the designated quality improvement organization for New York, Medicaid and Medicare review, and other tasks (originally the Island Peer Review Organization)
IV	intravenous
JC	Joint Commission (formerly JCAHO)
LOS	length of stay
MBI	Maslach Burnout Inventory
MD	doctor of medicine
MEDPAR	Medicare Provider Analysis and Review File
NHS	National Health Service (United Kingdom)
NICU	neonatal intensive care unit
NIS	Nationwide Inpatient Sample (HCUP)
NZRDA	New Zealand Resident Doctors Association
OB/GYN	obstetrics-gynecology
OJT	on-the-job training
OSHA	Occupational Safety and Health Administration

PAE	preventable adverse event
PGY	postgraduate year
PGY-1, PGY-2	postgraduate year 1, postgraduate year 2 of medical residency
PICU	pediatric intensive care unit
PIF	program information form
RCA	root-cause analysis
RCPSC	Royal College of Physicians and Surgeons of Canada
RRC	Resident Review Committee
SEIU	Service Employees International Union
TEMIS	Trauma and Emergency Medicine Information System
U.K.	United Kingdom
U.S.	United States
USMLE	U.S. Medical Licensing Exam
VA	U.S. Department of Veterans Affairs

ABBREVIATIONS

CI	confidence interval
g	grams
mL	milliliters
mph	miles per hour
ms	milliseconds
OR	odds ratio

Appendix E

Committee Member Biographies

Michael M. E. Johns, M.D. (*Chair*), is chancellor at Emory University. Until recently, he was the executive vice president for health affairs of Emory University, chief executive officer (CEO) of the Robert W. Woodruff Health Sciences Center, chairman of the Board of Emory Healthcare, and professor in the Departments of Otolaryngology (School of Medicine) and Health Policy (Rollins School of Public Health), Emory University. He was in charge of Emory's affiliations with Grady Memorial Hospital and the Emory Healthcare Hospital Affiliation Program with 60 hospitals in Georgia and surrounding states. Dr. Johns received his bachelor's degree and graduate studies in biology at Wayne State University in Detroit and an M.D. from the University of Michigan Medical School. He joined the Medical Service Corps of the U.S. Army, serving at Walter Reed Army Medical Center. He joined the Department of Otolaryngology and Maxillofacial Surgery at the University of Virginia Medical Center before being recruited to Johns Hopkins as professor and chair of otolaryngology-head and neck surgery. He served 6 years as dean of the Johns Hopkins School of Medicine and vice president for medical affairs at Johns Hopkins University. Dr. Johns is internationally recognized for his work as a cancer surgeon of head and neck tumors and for his studies of treatment outcomes.

James Bagian, M.D., was chosen as the first director of the Department of Veterans Affairs (VA) National Center for Patient Safety (NCPS), which was established in 1999. He is also the chief patient safety officer for the VA. NCPS develops, leads, and oversees activities and programs concerned with improving patient safety throughout the VA healthcare system. A National

Astronautics and Space Administration (NASA) astronaut for 15 years, Dr. Bagian was a crew member on two Space Shuttle missions, Discovery, March 1989, and Columbia, June 1991. Following the 1986 Challenger space shuttle explosion, he supervised the capsule's recovery from the ocean floor. He served as an investigator for the Challenger mishap and, in 2003, as the medical consultant-chief flight surgeon for the Columbia Accident Investigation Board. Dr. Bagian holds a B.S. degree in mechanical engineering from Drexel University and a doctorate in medicine from Thomas Jefferson University and is board certified in preventive medicine. Dr. Bagian is on the faculties of the Uniformed Services University of the Health Sciences and the University of Texas Medical Branch. He is a member of the National Academy of Engineering and the Institute of Medicine.

Jayanta Bhattacharya, M.D., Ph.D., is an assistant professor of medicine and a Center for Health Policy-Center for Primary Care Outcomes Research core faculty member. His research focuses on the constraints that vulnerable populations face in making decisions that affect their health status, as well as the effects of government policies and programs designed to benefit vulnerable populations. He has published empirical economics and health services research on medical residents and the impact of their work hours, the elderly, adolescents, HIV/AIDS, and managed care. Most recently, he has researched the regulation of the viatical-settlements market (a secondary life insurance market that often targets HIV patients) and summer-winter differences in nutritional outcomes for low-income American families. He is also working on a project examining the labor market conditions that help determine why some U.S. employers do not provide health insurance. He worked for 3 years as an economist at the RAND Corporation in Santa Monica, California, where he also taught health economics as a visiting assistant professor at the University of California-Los Angeles. He received a Ph.D. in economics and an M.D. from Stanford University.

Maureen Bisognano, M.S., is the executive vice president and chief operating officer (COO) of the Institute for Healthcare Improvement (IHI), a position she has held since 1995. Prior to joining IHI, she was senior vice president of the Juran Institute, where she consulted with senior leaders worldwide on strategy and improvement in healthcare settings, and was CEO of the Massachusetts Respiratory Hospital. She has served on the boards of the Massachusetts Hospital Association, the Lean Enterprise Institute, the National Center for Healthcare Leadership, and the American Society for Quality, among others. She currently serves on the board of the Luther Midelfort Clinic and since 2005 has been a member of the Commonwealth Fund's Commission on a High Performance Health System. She has taught at the Harvard School of Public Health since 1997 and in 2007 was appointed an instructor of medicine at Harvard Medical School and a research associate in the Division of Social Medicine and Health Inequalities at the Brigham and Women's Hospital. Ms. Bisognano began her career in health care as a staff nurse at Quincy City Hospital, eventually becoming chief operating officer there. She holds a B.S. from the State University of New York and an M.S. from Boston University.

Pascale Carayon, Ph.D., is Procter & Gamble Bascom Professor in Total Quality in the Department of Industrial and Systems Engineering and the director of the Center for Quality and Productivity Improvement at the University of Wisconsin-Madison where she leads the Systems Engineering Initiative for Patient Safety. Her research on human factors engineering and patient safety has been funded by the Agency for Healthcare Research and Quality, the National Institutes of Health (NIH), and the Robert Wood Johnson Foundation. She is the editor of the recently published *Handbook of Human Factors and Ergonomics in Health Care and Patient Safety*. She is the North American editor for *Applied Ergonomics* and a member of the editorial boards of the *Journal of Patient Safety*, *Behaviour and Information Technology*, and *Work and Stress*. In 2006, she was elected the secretary general of the International Ergonomics Association and a fellow of the Human Factors and Ergonomics Society. Dr. Carayon received her engineer diploma from the Ecole Centrale de Paris, France, in 1984 and her Ph.D. in industrial engineering from the University of Wisconsin-Madison in 1988. Dr. Carayon is internationally recognized for her research in human factors and systems engineering, in particular in the area of healthcare quality and patient safety.

Jordan J. Cohen, M.D., is currently professor of medicine and public health at George Washington University and president emeritus of the Association of American Medical Colleges (AAMC). During his 12 years as the president of the association (1994-2006), Dr. Cohen launched new initiatives in each of the association's mission areas of education, research, and patient care; expanded and modernized the AAMC's services for medical students, applicants, residents, and constituents; strengthened the association's communications, advocacy, and data-gathering efforts, and established many initiatives for improving medical education and clinical care. Prior to becoming president of the AAMC, Dr. Cohen spent 40 years in academic medicine, as dean of the medical school and professor of medicine at the State University of New York at Stony Brook, president of the medical staff at University Hospital, professor and associate chairman of medicine at the University of Chicago-Pritzker School of Medicine, and physician-in-chief and chairman of the Department of Medicine at the Michael Reese Hospital and Medical Center. He also held medical faculty positions at Harvard,

Brown, and Tufts universities and was president of the medical staff at the New England Medical Center Hospital in Boston. He is a graduate of Yale University and Harvard Medical School and completed his postgraduate training in internal medicine on the Harvard service at the Boston City Hospital.

David F. Dinges, Ph.D., is professor and chief of the Division of Sleep and Chronobiology in the Department of Psychiatry at the University of Pennsylvania School of Medicine. His research focuses on physiological, neurobehavioral, and cognitive effects of sleep loss and circadian biology and their relationship to health and safety. He has scientifically developed and validated behavioral, technological, and pharmacological interventions for these effects. During the past 30 years his research has been supported by NIH, NASA, the Department of Defense, Department of Transportation, and Department of Homeland Security. He has advised federal and private regulatory policies regarding duty hours and fatigue management. He currently leads the Neurobehavioral and Psychosocial Factors Team for the NASA-funded National Space Biomedical Research Institute. He is currently a member of the NIH-NINR (National Institute for Nursing Research) Council. He has been president of the Sleep Research Society and the World Federation of Sleep Research and Sleep Medicine Societies, and has served on the Board of Directors of the American Academy of Sleep Medicine and the National Sleep Foundation. He is currently editor-in-chief of the scientific journal *SLEEP*. He has received numerous awards, including the 2004 Decade of Behavior Research Award from the American Psychological Association and the 2007 NASA Distinguished Public Service Medal.

Javier A. Gonzalez del Rey, M.D., M.Ed., is currently professor of pediatrics, associate director Division of Emergency Medicine, and director of Pediatric Residency Training Programs at Cincinnati Children's Hospital Medical Center (CCHMC), University of Cincinnati College of Medicine. Dr. Gonzalez del Rey's major areas of interests include pediatric residency, pediatric emergency medicine education, and international pediatric training. He has won numerous teaching awards including the Cincinnati Children's Hospital Medical Center Faculty Teaching Award, the University of Cincinnati Department of Emergency Medicine Golden Apple Award, and most recently, the Parker J. Palmer Courage to Teach Award from the Accreditation Council of Graduate Medical Education (ACGME). He received his university and medical school education at the National University Pedro Henriquez Urena in the Dominican Republic, completed his pediatric residency at the University of Connecticut Primary Care Program, and did his fellowship training in general academic pediatrics and pediatric emergency medicine at the Cincinnati Children's Hospital Medical Center. He is currently certified in pediatrics and pediatric emergency medicine (PEM). He has completed a master's of medical education. He is currently a member of the National PEM Fellows Conference, the chair of the American Academy of Pediatrics PREP-EM course, and the organizer of many international educational exchange programs.

Peter J. Kolesar, Ph.D., is professor emeritus at Columbia University and research director of Columbia's Deming Center for Quality, Productivity and Competitiveness. He holds degrees in physics and mathematics from Queens College (City University of New York) and a Ph.D. in industrial engineering and operations research from Columbia University. He has been on the faculties of the Imperial College of Science & Technology (London), the Université de Montréal, and the City University of New York and on the technical staffs of the RAND Corporation and Bell Labs. Professor Kolesar held joint appointments with Columbia's Graduate School of Business and School of Engineering and Applied Science, teaching courses in optimization, statistics, quality, and production management. Dr. Kolesar's research and teaching have also focused on the effective implementation of process improvement methodology, including extensive applications in many manufacturing industries and a wide variety of services. Dr. Kolesar has twice been an examiner for the Malcolm Baldrige U.S. National Quality Award and has been a member of the Council of the Operations Research Society of America. He is a fellow of the American Association for the Advancement of Science and of the Institute for Operations Research and Management Science and was a member of the boards of the Juran Institute and the Montana Fly Company.

Brian W. Lindberg, M.B.A., has served as the executive director of the Consumer Coalition for Quality Health Care since 1993. The coalition advocates for programs and policies that address the critical need for a healthcare system that provides meaningful choices and information, consumer participation, grievance and appeals rights, consumer advocacy, and independent quality oversight and improvement. Mr. Lindberg served on the Planning Committee for the National Quality Forum (appointed by Vice President Gore), and currently serves as the chair of its Consumer Council. He has also served on its Board of Directors. He represents consumer viewpoints on various panels, including the consumer advisory panels of the Joint Commission and the National Committee for Quality Assurance. Mr. Lindberg also provides public policy consultation for the Gerontological Society of America (GSA), the National Academy of Elder Law Attorneys (NAELA), the National Association of State Long-Term Care Ombudsman Programs (NASOP), Experience Wave, and other organizations. Mr. Lindberg worked in Congress for 10 years on the House Select Committee on Aging and the Senate Special Committee on Aging. He holds a bachelor of social work degree from Temple University and a master's degree in management of human services from Brandeis University, and has studied social and health-care policy at the University of Stockholm's International Graduate School.

Kenneth M. Ludmerer, M.D., is professor of medicine and a professor of history at Washington University in St. Louis where he has won awards for his outstanding bedside teaching and practice of internal medicine. Dr. Ludmerer received an A.B. from Harvard College and an M.A. and M.D. from the Johns Hopkins School of Medicine. After medical school he did a residency in internal medicine at Barnes Hospital in St. Louis and graduate work in history at Harvard. Other positions held by Dr. Ludmerer include American College of Physicians teaching and research scholar 1980-1983; Henry J. Kaiser Family Foundation faculty scholar in general internal medicine 1981-1986; Kaiser Family Foundation research grants 1986-1992; Macy Foundation research grant 1989-1994; and Spencer Foundation research grant 1992-1995. Dr. Ludmerer is also present or past member of many editorial boards including *History of Education Quarterly*, *Pharos*, *Annals of Internal Medicine*, *Academic Medicine*, and *American Journal of Medicine*. He is best known for his work in medical education and health-care policy, authoring books about the creation and evolution of American medical education (*Learning to Heal* and *Time to Heal*). He received the Abraham Flexner Award for Distinguished Service to Medical Education from the Association of American Medical Colleges for this work.

Daniel Munoz, M.D., is a fellow in the Division of Cardiology at the Johns Hopkins University School of Medicine. He completed his residency training in internal medicine at the Johns Hopkins Hospital, where he will return as chief resident in medicine in 2009-2010. He obtained his medical degree from the Johns Hopkins School of Medicine (class of 2005) and also has a master's in public administration from Harvard University's John F. Kennedy School of Government where he concentrated on health economics and public policy. He has a bachelor of arts in economics from Princeton University where he graduated with honors. He spent the summers of 1999 and 2001 working in the U.S. Senate in the Health Policy Office of Senator Edward M. Kennedy in Washington, DC. He is a regular columnist for *Hopkins Medicine Magazine* and a frequent contributor to the *Baltimore Sun*.

Christopher S. Parshuram, M.D., graduated from Otago University of New Zealand, with prizes in medicine and pharmacology. After a residency in pediatrics at the Royal Children's Hospital in Melbourne, Australia, he moved to Canada where he completed specialist fellowship training in pediatric critical care medicine and clinical pharmacology in Toronto and Edmonton. He completed his Ph.D. in clinical epidemiology in 2005, on the subject of patient safety. Dr. Parshuram was appointed as a staff physician in the Department of Critical Care Medicine in the Hospital for Sick Children in 2003, and is a scientist in child health evaluation sciences in the Research Institute. In addition to formal training in systems of healthcare delivery, Dr. Parshuram has expertise in cardiac arrest prevention, reducing errors that are associated with medications, and preventing fatigue in healthcare workers. He has received peer-reviewed research funding from the Heart and Stroke Foundation of Canada, the Society of Critical Care Medicine, and the Canadian Institutes of Health Research. He is a career scientist of the Ministry of Health and Long Term Care, the director of the Centre for Safety Research, and an assistant professor at the University of Toronto in the Faculty of Medicine.

Ann E. Rogers, Ph.D., R.N., is an associate professor at the University of Pennsylvania School of Nursing and holds a joint appointment at the Center for Sleep and Respiratory Neurobiology, University of Pennsylvania School of Medicine. She holds a bachelor's degree in nursing from the University of Iowa College of Nursing, a master's degree from the University of Missouri-Columbia, and a doctorate from Northwestern University. She is one of six nurses in the United States who have been recognized (or have earned a certificate) as a diplomate of the American Board of Sleep Medicine. Dr. Rogers is the principal investigator of a seminal study on the effects of staff nurse fatigue on patient safety. In addition, Dr. Rogers wrote a paper entitled "Work Hour Regulations in Safety-Sensitive Industries" commissioned by the Institute of Medicine (IOM). This paper was included in the IOM Committee on the Work Environment for Nurses and Patient Safety report *Keeping Patients Safe: Transforming the Work Environment of Nurses* released on November 4, 2004. She is a fellow of both the American Academy of Nursing and the American Academy of Sleep Medicine.

Denise M. Rousseau, Ph.D., earned her graduate degrees in psychology at the University of California, Berkeley. She currently chairs master's programs in healthcare management and medical management and is the faculty director of the Institute for Social Entrepreneurship and Innovation at the Carnegie Mellon University's H. John Heinz II School of Public Policy and Management and the Tepper School of Business. She also directs a project on evidence-based organizational practices and conducts research and consults in a variety of settings. Before joining Carnegie Mellon, she served on the faculties of Northwestern University's Kellogg School of Management, the University of Michigan's Department of Psychology and Institute for Social Research, and the Naval Postgraduate School at Monterey. She has also been a visiting professor at universities in the United Kingdom, Singapore, Thailand, and China.

Eduardo Salas, Ph.D., is university trustee chair and Pegasus Professor of Psychology at the University of Central Florida. He also holds an appointment at the Institute for Simulation & Training. Previously, he was a senior research psychologist and head of the Training Technology Development Branch of the Naval Air Systems Command (NAVAIR)-Orlando for 15 years. During this period, Dr. Salas served as a principal investigator for numerous R&D programs focusing on teamwork, team training, advanced training technology, decision making under stress, learning methodologies, and performance assessment. His expertise includes helping organizations foster teamwork, design and implement team training strategies, facilitate training effectiveness, manage decision making under stress, develop performance measurement tools, and design learning and simulation-based environments. He is currently working on designing tools and techniques to minimize human errors in aviation and medical environments. He has consulted for a variety of manufacturing, pharmaceutical laboratories, industrial, and government organizations. Dr. Salas is a fellow of the American Psychological Association (Division's 14, 19, and 21) and the Human Factors and Ergonomics Society. He was editor (2000-2004) of the *Human Factors* journal and is currently associate editor of the *Journal of Applied Psychology*. He received his Ph.D. degree in industrial and organizational psychology from Old Dominion University.

Bruce Siegel, M.D., Ph.D., is a research professor and director of the Center for Health Care Quality in the Department of Health Policy at the George Washington University School of Public Health and Health Services. There he oversees the Aligning Forces for Quality Initiative of the Robert Wood Johnson Foundation. Much of his work has sought to understand and improve the quality of health care received by Americans, with a focus on its most vulnerable populations. His work has included developing innovation in reducing crowding and improving hospital patient flow, eliminating ethnic and racial disparities in care, and supporting the safety net. Dr. Siegel has previously held the positions of New Jersey commissioner of health, president of the New York City Health and Hospitals Corporation, and president of Tampa General Healthcare. In addition, he served as a director of the ACGME, as a senior fellow at New School University, and as an adviser to the Institute of Medicine, the World Bank, and other health-related organizations. Dr. Siegel received his A.B. degree from Princeton University, M.D. from Cornell University Medical College, and M.P.H. from the Johns Hopkins School of Hygiene and Public Health. He is board certified in preventive medicine.

Appendix F

Public Meeting Agendas

AGENDA

Committee on Optimizing Graduate Medical Trainee (Resident) Schedules to Improve Patient Safety
December 3-4, 2007

National Academy of Sciences
2100 C Street, NW, Lecture Room, Washington, DC

Monday, December 3

CLOSED SESSION (Committee Members and IOM Staff Only)

8:00 AM **Committee Deliberations**

10:45 **Break**

END CLOSED SESSION—START OPEN SESSION

11:00 **Welcome**
Michael Johns, M.D., Committee Chair

11:10 **Remarks from Study Sponsor and Committee Questions**
Carolyn Clancy, M.D., Director, AHRQ

11:50	**Lunch** Discussion of Sponsor's Presentation
12:50	**Break**
1:00 PM	**Workshop Begins** Introduction: Michael Johns, Committee Chair
1:05-1:40	**Panel 1: The Accreditation Council for Graduate Medical Education's Duty Hours Requirements** **Paul Friedmann, M.D.,** former ACGME official **Ingrid Philibert,** Sr. VP, Department of Field Activities, ACGME [did not participate because of telephone malfunction]
1:40-2:30	**Panel 2: Impact of Duty Hours Requirements on Education** **Steven Weinberger, M.D.,** Sr. VP for Medical Education, American College of Physicians **Tom Whalen, M.D.,** Regent, American College of Surgeons [not present because flight was cancelled] **Michael Ehlert, M.D.,** President, American Medical Students Association **Sunny Ramchandani, M.D.,** past Chair of Residents and Fellows Section, AMA
2:30-3:15	**Panel 3: Work Hours, Patient Safety, and Enforcement** **L. Toni Lewis, M.D.,** Executive VP, Committee of Interns and Residents **Peter Lurie, M.D.,** Public Citizen **Ethan Fried, M.D.,** Vice Chair for Education, Residency Director-Internal Medicine, St. Luke's-Roosevelt Hospital Center
3:15-3:30	**Break**
3:30-4:25	**Panel 4: Sleep and Outcomes Research** **Charles Czeisler, M.D.,** Director of Sleep Medicine, Harvard Medical School and Brigham and Women's Hospital

	Christopher Landrigan, M.D., Director, Sleep and Patient Safety, Brigham and Women's Hospital; Research and Fellowship Director, Children's Hospital Boston, Assistant Professor of Pediatric Medicine, Harvard Medical School
	Kevin Volpp, M.D., Ph.D., Assistant Professor of Medicine and of Health Care Systems, Wharton School, University of Pennsylvania and faculty, Center for Health Equity Research and Promotion at Philadelphia VA Medical Center
4:25-5:00	**Panel 5: The Federal Role Funding Graduate Medical Education**
	Miechal Lefkowitz, Technical Advisor, Division of Acute Care, Cent for Medicare and Medicaid Services
	Barbara K. Chang, M.D., Director of Medical and Dental Education, Office of Academic Affiliations, Department of Veterans Affairs
5:00	**Open mike for audience**
5:30	**Recess**

END OPEN SESSION—START CLOSED SESSION

6:45	**Dinner Meeting, Committee deliberations**
8:45	**Adjournment**

Tuesday, December 4, Board Room (CLOSED SESSION)

8:00 AM	**Committee deliberations**
3:00 PM	**Adjournment**

AGENDA

Committee on Optimizing Graduate Medical Trainee (Resident) Schedules to Improve Patient Safety
March 4-5, 2008

100 Academy Drive, Beckman Center, Huntington Room, Irvine, CA

Tuesday, March 4

CLOSED SESSION (Committee Members and IOM Staff Only)

8:00 AM-12:20 PM

END CLOSED SESSION—START OPEN SESSION

12:20 PM	**Workshop Begins** Introduction, Michael Johns, Committee Chair
12:25-1:25	**Panel 1: Lessons from Residency Directors** **Jimmy Hara, M.D., F.A.A.F.P.,** Family Medicine Residency Director, Former DIO, Kaiser Permanente Center for Medical Education, Los Angeles, CA **Pamela L. Dyne, M.D.,** Residency Program Director, UCLA/Olive View-UCLA Emergency Medicine Residency Program; Co-Director, UCLA/Olive View-UCLA Combined Emergency Medicine and Internal Medicine Residency Program; Associate Professor of Medicine/Emergency Medicine, David Geffen School of Medicine at UCLA **Vineet Arora, M.D., M.A.P.P.,** Associate Program Director, Internal Medicine Residency Program and Assistant Dean, Pritzker School of Medicine, University of Chicago
1:25-2:30	**Panel 2: Findings and Strategies from Research Literature and Other Industries** **Mark R. Rosekind, Ph.D.,** President and Chief Scientist, Alertness Solutions **Arpana Vidyarthi, M.D.,** Director of Quality, Division of Hospital Medicine; Director, Patient Safety and Quality Programs, GME, UCSF School of Medicine **Ingrid Philibert, M.H.A., M.B.A.,** Senior Vice President, Field Activities, ACGME **Thomas Nasca, M.D., M.A.C.P.,** Chief Executive Officer, ACGME
2:30-2:45	**Break**

2:45-4:10 **Panel 3: Lessons from Surgery**

Gerald B. Healy, M.D., F.A.C.S., President, American College of Surgeons; Otolaryngologist-in-Chief, Children's Hospital Boston; Healy Chair in Otolaryngology, Harvard Medical School, Boston, Massachusetts

Bernard F. Ribeiro, C.B.E., President, Royal College of Surgeons of England, London

H. Hunt Batjer M.D., F.A.C.S., Michael Marchese Professor and Chair, Department of Neurological Surgery, Northwestern University Feinberg School of Medicine; Chairman, American Board of Neurological Surgery

Christian de Virgilio, M.D., Vice Chair, Education; Director, General Surgery Residency Program Harbor-UCLA Medical Center; Co-Chair, College of Applied Anatomy; Professor of Surgery, UCLA School of Medicine

4:10-5:15 **Panel 4: Cost Impact of Duty Hour Change**

Mark Noah, M.D., F.A.C.P., Director of Residency Training Program, Department of Medicine; Medical Director, Graduate and Continuing Medical Education, Cedars-Sinai Medical Center

Richard J. Liekweg, M.B.A., Chief Executive Officer, UC-San Diego Medical Center

Charles Daschbach, M.D., M.P.H., Director of Academic Affairs and DIO, St. Joseph's Hospital and Medical Center, Phoenix, AZ

Lawrence M. Opas, M.D., Associate Dean for GME, DIO, and Chief, Department of Pediatrics USC/Los Angeles County and USC Medical Center

5:15-5:55 **Panel 5: Ethical and Consumer Perspectives**

Elizabeth M. Imholz, J.D., Special Projects Director, Consumer Union of the United States

Felicia Cohn, Ph.D., Director of Medical Ethics, University of California-Irvine School of Medicine

5:55 **Open mike for audience**

END OPEN SESSION—START CLOSED SESSION

Wednesday, March 5

CLOSED SESSION (Committee Members and IOM Staff Only)

8:00 AM-3:00 PM

AGENDA

Committee on Optimizing Graduate Medical Trainee (Resident) Schedules to Improve Patient Safety
May 8-9, 2008

National Academy of Sciences
500 Fifth Street, NW, Keck Building, Room 201, Washington, DC

Thursday, May 8

CLOSED SESSION (Committee Members and IOM Staff Only)

7:30 AM-1:30 PM

END CLOSED SESSION—START OPEN SESSION

1:30 **Public Panel Begins**
Introduction, Michael Johns, Committee Chair

Panel on Education in Training Programs, Accreditation for Patient Safety, and Educational Outcomes
Debra Weinstein, M.D., representing AAMC; Vice President for GME at Partners Healthcare
Paul Schyve, M.D., Senior Vice-President, Joint Commission
Kevin Weiss, M.D., M.P.H., President and CEO of American Board of Medical Specialties

END OPEN SESSION—START CLOSED SESSION

3:00 PM-6:30 PM

Friday, May 9

CLOSED SESSION (Committee Members and IOM Staff Only)

7:30 AM-8:30 AM

END CLOSED SESSION—START OPEN SESSION

8:30 **Challenges in Resident Workload**
Lisa Bellini, M.D., Associate Dean for Graduate Medical Education; Vice Chair for Education and Inpatient Services, Department of Medicine, University of Pennsylvania Health System

END OPEN SESSION—START CLOSED SESSION

9:15 AM–3:00 PM

Index

A

B

C

D

E

F

G

H

I

J

L

M

P

Q

R

W

Y